Public-Key Cryptography and Computational Number Theory

Public-Key Cryptography and Computational Number Theory

Proceedings of the International Conference
organized by the
Stefan Banach International Mathematical Center
Warsaw, Poland, September 11 – 15, 2000

Editors

Kazimierz Alster
Jerzy Urbanowicz
Hugh C. Williams

Walter de Gruyter · Berlin · New York 2001

Editors

Kazimierz Alster
Institute of Mathematics
Polish Academy of Sciences
ul. Sniadeckich 8
00-950 Warsaw
Poland

Jerzy Urbanowicz
Institute of Mathematics
Polish Academy of Sciences
ul. Sniadeckich 8
00-950 Warsaw
Poland

Hugh C. Williams
Department of Mathematics
and Statistics
University of Calgary
2500 University Drive
Calgary, Alberta
Canada T2N 1N4

Mathematics Subject Classification 2000:
11–06; 11A25, 11G20, 11T71, 11Y16, 14G50, 94A60

Key words:
Public-Key Cryptography, Computational Number Theory, efficiency and security of cryptosystems, DL-cryptosystems and signatures, elliptic (hyperelliptic) curves, Jacobians, algebraic groups, IQ cryptography, error correcting codes, NTRU, XTR, Fermat numbers, Wieferich primes, Wilson primes, gcd algorithms, Euler's ϕ-function

∞ Printed on acid-free paper which falls within the guidelines of the ANSI to ensure permanence and durability.

Library of Congress – Cataloging-in-Publication Data

Public-key cryptography and computational number theory : September 11–15, 2000, Warsaw, Poland / editors, Kazimierz Alster, Jerzy Urbanowicz, Hugh C. Williams.
p. cm.
ISBN 3-11-017046-9
1. Coding theory–Congresses. 2. Public-key cryptography–Congresses. I. Alster, Kazimierz. II. Urbanowicz, Jerzy, 1951– . III. Williams, Hugh C. IV. International Conference on Public-Key Cryptography and Computational Number Theory (2000 : Warsaw, Poland)
QA268 .P83 2001
003′.54–dc21

2001047563

Die Deutsche Bibliothek – CIP-Einheitsaufnahme

Public-key cryptography and computational number theory : proceedings of the international conference ; Warsaw, Poland, September 11–15, 2000 / organized by the Stefan Banach International Mathematical Center. Ed. K. Alster .... – Berlin ; New York : de Gruyter, 2001
ISBN 3-11-017046-9

Printed in Germany.
Cover design: Thomas Bonnie, Hamburg.
Typeset using the authors' TeX files.
Printing and binding: Hubert & Co. GmbH & Co. KG, Göttingen.

To the memory of
Enigma's breakers
Marian Rejewski,
Jerzy Różycki and
Henryk Zygalski

The Editors

In the gardens of the castle Les Fouzes in southern France in 1941. Right to left: Marian Rejewski, Jerzy Różycki and Henryk Zygalski.

Preface

The International Conference on Public-Key Cryptography and Computational Number Theory was organized by the Stefan Banach International Mathematical Center and held in Warsaw, Poland, on September 11–15, 2000. Eighty one mathematicians from twelve countries attended the conference.

These proceedings of that conference contain twenty selected and refereed research and survey papers delivered by conference speakers and by a few invited mathematicians who were unable to come to the meeting. Fifteen articles on public-key cryptography are concerned with efficiency and security of a variety of systems. Considerable attention is devoted to ordinary discrete logarithm (DL) cryptosystems, DL-cryptosystems based on elliptic curves, the Jacobians of hyperelliptic curves, and algebraic groups and class groups of imaginary quadratic orders. Other new ideas in cryptography are also presented, for example on connections between cryptography and error correcting codes, and on new cryptosystems (NTRU and XTR). Five papers on computational number theory are concerned with computations related to elliptic curves, Fermat numbers, Wieferich and Wilson primes, extended binary gcd algorithms and Euler's ϕ-function. We thank all the authors and referees for all their contributions to the Proceedings.

Bogdan Bojarski, the Director of the Institute of Mathematics of the Polish Academy of Sciences and Gerhard Frey of the Institute of Experimental Mathematics of the University of Essen opened the conference. We gratefully acknowledge their guidance and help.

Many people helped with the organization of the conference and in editing of the Proceedings. We thank Neal Koblitz and Harald Niederreiter, who played an important role in the initial planning of the conference, all the members of the Program Committee for their help, and Aleksander Wittlin for the idea of organizing the conference. We wish to thank Jan K. Kowalski for offering valuable suggestions for improving the presentation of material. We gratefully acknowledge the support of our sponsors: National Bank of Poland, Power Telecommunication Tel-Energo, Telekomunikacja Polska and Vestigator Consulting.

The editors would like to dedicate the volume to the memory of the Polish mathematicians Marian Rejewski (1905–1980), Jerzy Różycki (1909–1942) and Henryk Zygalski (1907–1978), who broke the military version of the famous Enigma in December, 1932 and January, 1933. In early January, 1933 they were able to read the first intercepted messages. In July, 1939 they conveyed a reconstructed version of the Enigma device, together with documentation, to the British and French governments, an action which greatly accelerated the Allied cryptanalytic effort and shortened the war. This key contribution of these three mathematicians was classified for many years, and as a consequence they are still not receiving the recognition for their discoveries that is due.

Our special thanks go to Andrew Odlyzko for writing a foreword for this volume on the progress made in cryptography from the time of Enigma until the present and to Arkadiusz Orłowski for preparing the photograph of Enigma's breakers for publication and to the authors for their consent to dedicate the volume.

June, 2001

Kazimierz Alster
Jerzy Urbanowicz
Hugh C. Williams

Mathematics, cryptology, and technology

Andrew Odlyzko

The start of the 21st century is a golden age for applications of mathematics in cryptology. The beginnings of this age can be traced to the work of Rejewski, Różycki, and Zygalski on breaking Enigma. Their work was a breakthrough in several ways. It made a tremendous practical contribution to the conduct of World War II. At the same time, it represented a major increase in the sophistication of the mathematical tools that were used. Ever since, mathematics has been playing an increasingly important role in cryptology. This has been the outcome of the intricate relationships of mathematics, cryptology, and technology, relationships that have been developing for a long time.

While codes and ciphers go back thousands of years, systematic study of them dates back only to the Renaissance. Such study was stimulated by the rapid growth of written communications and the associated postal systems, as well as by the political fragmentation in Europe. In the 19th century, the electric telegraph provided an additional spur to the development of cryptology. The biggest impetus, though, appears to have come with the appearance of radio communication at the beginning of the 20th century. This technological development led to growth of military, diplomatic, and commercial traffic that was open to non-intrusive interception by friend or foe alike. The need to protect such traffic, from interception was obvious, and led to the search for improved codes and ciphers. These, in turn, stimulated the development of cryptanalytic methods, which then led to development of better cryptosystems, in an endless cycle. What systems were built has always depended on what was known about their security, and also on the technology that was available.

Between the two world wars, the need for encrypting and decrypting ever greater volumes of information reliably and securely, combined with the available electromechanical technology, led many cryptosystem designers towards rotor system. Yet, as Rejewski, Różycki, and Zygalski showed, the operations of rotor machines created enough regularities to enable effective cryptanalysis through mathematical techniques. This was yet another instance of what Eugene Wigner has called the "unreasonable effectiveness of mathematics," in which techniques developed for abstract purposes turn out to be surprisingly well suited for real applications.

The sophistication of mathematical techniques in cryptography continued increasing after World War II, when attention shifted to cryptosystems based on shift register sequences. A quantum jump occurred in the 1970s, with the invention of

Public-Key Cryptography and Computational Number Theory
K. Alster, J. Urbanowicz and H. C. Williams (Eds.)

public key cryptography. This invention was itself stimulated by technological developments, primarily the growth in information processing and transmission. This growth was leading to explosive increases in the volume of electronic transactions, increases that show no signs of tapering off even today, a quarter century later. The large and heterogeneous populations of users that were foreseen in developing civilian settings were leading to problems, such as key management and digital signatures, that previously had not been as severe in smaller and more tightly controlled military and diplomatic communications. At the same time, developments in technology were offering unprecedented possibilities for implementing complicated algorithms. Mathematics again turned out to provide the tools that were used to meet the challenge.

The public key schemes that were invented in the 1970s used primarily tools from classical number theory. Yet as time went on, the range of applicable mathematics grew. Technology continued improving, but in uneven ways. For example, while general computing power of a personal computer grew explosively, there was also a proliferation of small, especially wireless devices, which continued to have stringent power and bandwidth limitations. This put renewed emphasis on finding cryptosystems that were thrifty with both computation and transmission. At the same time, there was growth in theoretical knowledge, which led to breaking of numerous systems, and required increases in key sizes of even well-trusted schemes such as RSA.

The outcome of the developments in technology and science is that today we are witnessing explosive growth in applications of sophisticated mathematics in cryptology. This volume is a collection of both surveys and original research papers that illustrate well the interactions of public key cryptography and computational number theory. Some of the systems discussed here are based on algebra, others on lattices, yet others on combinatorial concepts. There are also some number theoretic results that have not been applied to cryptography yet, but may be in the future. The diversity of techniques and results in this volume does show that mathematics, even mathematics that was developed for its own sake, is helping solve important problems of our modern society. At the same time, mathematics is drawing valuable inspiration from the practical problems that cryptology poses.

Table of contents

A survey on IQ cryptography

Johannes Buchmann and Safuat Hamdy

Abstract. This paper gives a survey on cryptographic primitives based on class groups of imaginary quadratic orders (IQ cryptography, IQC). We present IQC versions of several well known cryptographic primitives, and we explain, why these primitives are secure if one assumes the hardness of the underlying problems. We give advice on the selection of the cryptographic parameters and show the impact of this advice on the efficiency of some IQ cryptosystems.

1. Introduction

The term IQ cryptography (IQC) refers to cryptography based on class groups of imaginary quadratic orders. IQC has been invented in 1988 [6]. Therefore, IQC is of about the same age as ECC (elliptic curve cryptography) [23, 18], yet IQC did not get the same attention. What was most lacking was a comprehensive guide for using IQC. That is, there is still no formal document that describes how to select the cryptographic parameters and which cryptographic schemes to use. Advice on the selection of the cryptographic parameters for IQC has been given in [14]; a summary of some results of that paper has been included in this work. The aim of this paper is to present IQC versions of some well known cryptographic schemes, to describe the algorithms that can be used for the underlying arithmetic, and to discuss the performance of IQC schemes using those algorithms. Thus, this paper is another step towards a standardization of IQC.

Although class groups are ordinary finite abelian groups, some cryptographic schemes based on discrete logarithms cannot be used with class groups in a straightforward way. The reason is that the order of class groups (or odd divisors thereof) cannot be computed efficiently. However, some discrete logarithm based cryptographic schemes, for example any signature scheme of ElGamal type, require the knowledge of the group order. In order to use these schemes, it is necessary to modify them. In this paper we present such modifications for DSA. It also turns out that class groups are well suited for the Guillou–Quisquater signature scheme.

Finally, the performance of IQ cryptosystems has never been compared to the performance of established cryptosystems. Since the performance depends on the size of the cryptographic parameter, which in turn depends on the desired security level, the selection of the cryptographic parameter had to be investigated first. This

Public-Key Cryptography and Computational Number Theory
K. Alster, J. Urbanowicz and H. C. Williams (Eds.)

has been done in [14]. In this paper first realistic benchmarks for IQ cryptosystems are presented, where the cryptographic parameter has been chosen of such a size that solving the discrete logarithm problem in the class group is about as hard as solving the integer factoring problem for integers of certain size. The result is that IQ cryptosystems appear to be practical. Since comparably little research has been spent on efficient IQ arithmetic, it is reasonable to expect significant improvements in this area in the future.

This paper is organized as follows: Section 2 recalls some relevant facts and notations from the theory of imaginary quadratic number fields. In Section 3 we discuss the computational problems on which IQ cryptosystems are based, and we discuss some properties of class groups that are relevant to cryptographic applications. In Section 4 we present IQ versions of some well known cryptographic schemes. Finally, in Section 5 we present benchmarks for an IQC variant of DSA and compare them with benchmarks of traditional cryptosystems.

2. Basic notation

We shall briefly recall some notations that we shall use in the sequel (see [16] or [4] for full details). Let Δ be a negative integer such that $\Delta \equiv 0, 1 \pmod 4$. Then the ring $\mathcal{O}_\Delta = \mathbb{Z} + (\Delta + \sqrt{\Delta})/2\,\mathbb{Z}$ is an *imaginary quadratic order* of *discriminant* Δ. Its field of fractions is $\mathbb{Q}(\sqrt{\Delta})$. The discriminant Δ is called *fundamental* if $\Delta/4$ or Δ is square free for $\Delta \equiv 0 \pmod 4$ or $\Delta \equiv 1 \pmod 4$, respectively. (If Δ is fundamental, then $\mathcal{O}_\Delta$ is a *maximal* order.)

The *fractional ideals* of any imaginary quadratic order are of the form $q(a\mathbb{Z} + (b + \sqrt{\Delta})/2\,\mathbb{Z})$ with $q \in \mathbb{Q}$, $a, b \in \mathbb{Z}$, $a > 0$, $4a \mid b^2 - \Delta$, and $\gcd(a, b, c) = 1$ where $c = (b^2 - \Delta)/(4a)$. Hence, they can be represented by triples (q, a, b). If $q = 1$, then the ideal is called *integral.* Two ideals $\mathfrak{a}_1, \mathfrak{a}_2 \subseteq \mathcal{O}_\Delta$ are called *equivalent* if there is a non-zero number $\alpha \in \mathbb{Q}(\sqrt{\Delta})$ such that $\mathfrak{a}_2 = \alpha\mathfrak{a}_1$.

The set of equivalence classes forms an abelian group under ideal multiplication. This group is called *class group* and denoted by $\mathrm{Cl}(\Delta)$. The class group is always finite. Its order is called *class number* and is denoted by $h(\Delta)$. The class number is not efficiently computable, if the discriminant is fundamental, but the even part of $h(\Delta)$ can be efficiently computed, if the prime factorization of Δ is known. For example, if Δ is a negative prime, then $h(\Delta)$ is always odd.

To compute with equivalence classes of class groups, one has to select representatives from each class. From the definition of equivalence it is obvious that any non-zero fractional ideal is equivalent to an integral one. Thus, each equivalence class can be represented by an integral ideal. Additionally, the product of two integral ideals is also an integral ideal. Therefore, we shall deal only with integral ideals, which we shall represent by pairs (a, b). Moreover, class groups of imaginary quadratic orders have the property that each equivalence class of ideals contain exactly one *reduced* integral ideal. An integral ideal (a, b) of a quadratic order $\mathcal{O}_\Delta$ is called reduced, if the following conditions are satisfied, where $c = (b^2 - \Delta)/(4a)$: 1. $a \le c$, 2. $-a < b \le a$, and 3. if $a = c$, then $b > 0$.

There are efficient algorithms to reduce integral ideals (i.e. finding an equivalent reduced ideal), see Section 5. Thus, we shall represent each equivalence class by a unique reduced integral ideal. Computing with reduced ideals is usually optimal in terms of efficiency because the bit sizes of reduced ideals are small. In particular, if (a, b) is a reduced ideal of $\mathcal{O}_\Delta$, then $a \le \sqrt{|\Delta|/3}$. Since $|b| \le a$, the bit size of (a, b) is at most that of Δ. The multiplication of reduced ideals takes $\mathrm{O}(\log^2 |\Delta|)$ bit operations. Since the product of reduced ideals is usually not reduced, we shall reduce any (intermediate) result. A careful analysis of the reduction algorithm presented in Section 5 shows that this algorithm takes also $\mathrm{O}(\log^2 |\Delta|)$ bit operations, and therefore, a group operation (i.e. ideal multiplication with subsequent reduction) takes $\mathrm{O}(\log^2 |\Delta|)$ bit operations.

3. Security of IQC

In this section we shall make some statements on the hardness of some computational problems for class groups and on the selection of IQC parameters.

3.1. Some computational problems

Let G be a finite abelian group. Then we define

Discrete logarithm problem (DLP): given $\alpha, \beta \in G$, find the smallest positive integer x such that $\beta = \alpha^x$ (or decide that no such x exists).

Order problem (OP): given $\alpha \in G$, compute $|\langle\alpha\rangle|$.

Root problem (RP): given $\alpha \in G$ and an integer $x > 1$, compute β such that $\beta^x = \alpha$ (or decide that no such $\beta \in G$ exists).

Since $|\langle\alpha\rangle| = \mathsf{DLP}(\alpha, 1_G)$, we have $\mathsf{OP} \le \mathsf{DLP}$. Conversely, the knowledge of the group order does apparently not help to compute discrete logarithms. For example, the group order of multiplicative groups of any finite field is obviously known, yet the computation of discrete logarithms still appears to be intractable. Thus, it is unlikely that $\mathsf{OP} = \mathsf{DLP}$.

Finally, there are efficient methods to compute an xth root (or to decide that such a root does not exist), if we know the group order. Thus, $\mathsf{RP} \le \mathsf{OP}$. It is an open question, whether $\mathsf{RP} = \mathsf{OP}$ or not, and it is also unknown, whether it is possible to compute roots efficiently without knowing the group order.

In the context of class groups of imaginary quadratic orders we denote these three problems by IQ-DLP, IQ-OP, and IQ-RP. It is known that $\mathsf{IQ\text{-}DLP} \le \mathsf{IFP}$[1)],

---

1) Integer factorization problem

so the complexity to solve the IFP is a lower bound for the complexity to solve the IQ-DLP.

IQ-DLP, IQ-OP, and IQ-RP appear to be hard problems. Despite the fact, that these problems do not appear to be equivalent (with respect to complexity theory), the best known algorithms to solve each of these problems are variants of each other with the same asymptotic running time. More precisely, there is no better method known to compute a solution to the IQ-RP than to compute a solution to the IQ-OP, for which in turn is no better method known than to compute a solution to an instance of the IQ-DLP.

The security of the IQ cryptosystems are based either on the IQ-DLP, or on the IQ-RP. Since instances of the IQ-RP are solved by invoking an algorithm for the IQ-DLP, we shall focus on the IQ-DLP. The following properties of class groups are of major interest for cryptographic applications:

- The class group is large, if the discriminant is large. It was known to Gauß that the average class number $\overline{h(\Delta)} = c_1\sqrt{D}$ for all fundamental discriminants up to $-D$, where $c_1 \approx 0.46$. In fact, from the Brauer–Siegel Theorem [19] follows that
$$\sqrt{|\Delta|}^{1+\epsilon} \leq h(\Delta) \leq \sqrt{|\Delta|}^{1-\epsilon} \tag{1}$$
for any positive real ϵ. Moreover, if one assumes the Extended Riemann Hypothesis, then it is possible to show [21] that
$$\frac{1+\mathrm{o}(1)}{c_2 \ln\ln|\Delta|}\sqrt{|\Delta|} < h(\Delta) < (1+\mathrm{o}(1))c_3\sqrt{|\Delta|}\ln\ln|\Delta| \tag{2}$$
where $c_2 = 12e^{\gamma}/\pi \approx 6.8$ and $c_3 = 2e^{\gamma}/\pi \approx 1.134$.
- The class group of a randomly chosen fundamental discriminant contains a very large cyclic subgroup with very high probability. This follows from [9, conjecture C5].
- The probability that the class number of a randomly chosen fundamental discriminant is very smooth, is negligible. This follows from [9, conjecture C2] and an additional assumption [7], see [14].

This shows, that class groups are suitable for cryptographic purposes. The above algorithms are called *generic*, for they work in any finite abelian group. As we shall see in the next subsection, there are faster algorithms known to compute discrete logarithms in class groups that are not generic.

3.2. Choosing the cryptographic parameter

Since the class group depends only on the discriminant, the discriminant is the main cryptographic parameter. In [14] all known strategies to compute discrete logarithms in class groups have been investigated. These are:

- Reductions to discrete logarithm computations in multiplicative groups of finite fields. Such reduction has been found for *totally non-maximal orders.*

Conversely, there are no such reduction known for maximal orders, hence the discriminant should be chosen to be fundamental. The simplest way to achieve this is to select a (large) prime p and set $\Delta = -p$ or to select two (large) primes p and q and set $\Delta = -pq$ (such that, in both cases, $\Delta \equiv 1 \pmod 4$).

- A method similar to the $(p-1)$-factoring algorithm to compute the class number and the Pohlig–Hellman algorithm. This $(p-1)$-like method can be used to compute the class number, if it is very smooth (a similar algorithm has been used in an other factoring algorithm due to Schnorr and H.W. Lenstra [28]). Then, the Pohlig–Hellman algorithm can be used for discrete logarithm computations. The $(p-1)$ algorithm has exponential running time in the size of the smoothness bound. In [14] it was shown that the probability that the class number of a random fundamental discriminant is very smooth is negligible. Hence, we expect the $(p-1)$ algorithm to have exponential running time on average.
- Square-root algorithms, such as the Baby-Step-Giant-Step method or the ρ and λ algorithms (see [22, Chap. 3] for an overview and further references). The running time of these algorithms is exponential in the size of the class group, and from the previous section follows, that they are also exponential in the size of the discriminant.
- Index-calculus algorithms. These algorithms have subexponential running time (and need subexponential space) in the size of the discriminant (see [17] or [8, Sect. 5.5]). Therefore, in order to protect IQ cryptosystems from attacks by index-calculus algorithms, much larger discriminants are required than for any square-root algorithm or the $(p-1)$-algorithm.

The fastest known algorithm to solve the IQ-DLP is a variant of the MPQS factoring algorithm (IQ-MPQS). It is asymptotically much slower than the GNFS, the fastest known algorithm to factor integers (see Figure 1). If one compares the expected running times of the GNFS and the IQ-MPQS, one gets that factoring 1024-bit integers requires about as much computational work as the computation of class groups with a 687-bit discriminant. Similar results are summarized in Table 1.

Table 1. Estimated expected computational work of the GNFS for factoring integers and the IQ-MPQS for computing discrete logarithms in class groups aligned

magnitude of n	magnitude of $\|\Delta\|$	expected no. of MIPS-years
2^{768}	2^{540}	4.99×10^{7}
2^{1024}	2^{687}	6.01×10^{10}
2^{1536}	2^{958}	5.95×10^{15}
2^{2048}	2^{1208}	7.05×10^{19}
2^{3072}	2^{1665}	2.65×10^{26}
2^{4096}	2^{2084}	5.87×10^{31}

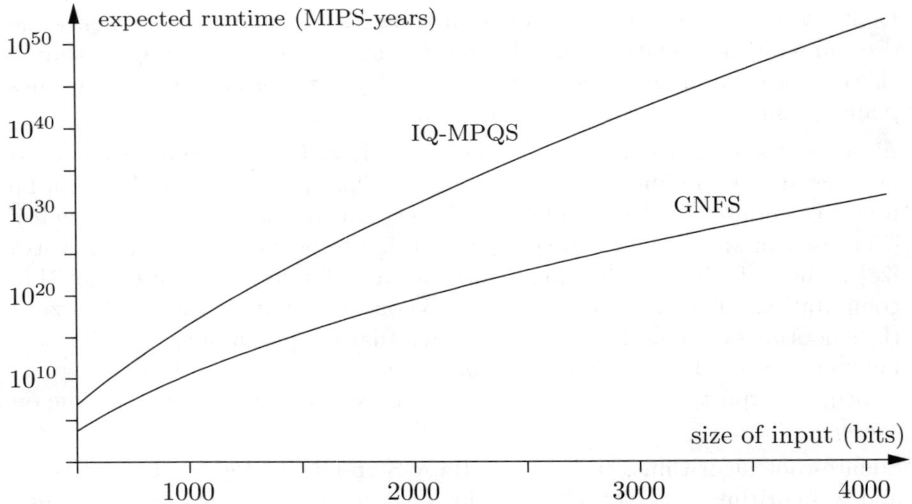

Figure 1. Asymptotic expected running times for the GNFS and the IQ-MPQS. Here we assume the expected running times to be $L_n\left[\frac{1}{3}, \sqrt[3]{64/9}\right]$ for the GNFS and $L_{|\Delta|}\left[\frac{1}{2}, 1\right]$ for the IQ-MPQS. The details can be found in [14].

4. IQC protocols

We shall now describe some protocols for IQC. Since class groups are ordinary finite abelian groups, we could use any scheme that is based on discrete logarithms. However, the class number (i.e. the order of a class group) is usually not efficiently computable. Thus, cryptographic schemes that require the knowledge of the group order (such as DSA or the Schnorr signature scheme) cannot be used in a straightforward manner. But these protocols can modified in such a way that the knowledge of the group order is not required.

From the above introduction it is clear that protocols like the Diffie–Hellman key exchange or the ElGamal encryption scheme can be used in a straightforward manner with class groups.

4.1. The ElGamal encryption

The ElGamal encryption scheme could be used just as described in [11]. However, this would require a user to embed the plain text into a group element. In [26] probabilistic methods have been presented to do this efficiently, but neither of the described methods are very fast in practice. It is also not really necessary to embed a group elements. Instead, one can use the following modified ElGamal signature:

Key Generation: A randomly selects a fundamental discriminant Δ of size according to the desired security level. Then A randomly selects $\gamma \in \mathrm{Cl}(\Delta)$, $a \leq \sqrt{|\Delta|}$, and computes $\alpha = \gamma^a$. A's public key is (Δ, γ, α), the private key is a.

Encryption: To encrypt the plain text m, B randomly selects $k \leq \sqrt{|\Delta|}$ and computes $\kappa = \gamma^k$ and $\beta = \alpha^k$. Then the cipher text is $C = (\kappa, c)$, where $c = m \oplus f(\beta)$, f is a preimage and collision resistant hash function, and $\oplus$ denotes bitwise xor-ing.

Decryption: A computes $\beta = \kappa^a$ and recovers m by computing $m = c \oplus f(\beta)$.

This scheme actually resembles the Diffie–Hellman key exchange, where A enters the protocol at the key generation stage, and B completes the protocol on encrypting the plain text. The same idea has been used for elliptic curves in [1]. This scheme is deterministic and very efficient.

Finally, observe that the Cramer–Shoup encryption scheme [10] can be used in a straightforward way with class groups. Moreover, the same modification we have proposed for the IQ version of the ElGamal encryption scheme above can be applied to the Cramer–Shoup encryption scheme.

4.2. Two DSA variants

As noted before, DSA [12] or the Schnorr [27] signature scheme (and similar signature schemes of ElGamal type) cannot be used in a straightforward way with class groups. This is because the class number, i.e. the order of the class group cannot be computed efficiently, at least if the discriminant is fundamental. But signing a message with a signature scheme of ElGamal type requires a reduction modulo the group order or a divisor thereof. We demonstrate this with a generalized version of DSA:

Key Generation: A randomly selects a group G, such that $|G|$ has a 160-bit prime divisor q (the actual size of G depends on properties of G and the specific security requirements). Then A randomly selects $\gamma_0 \in G$ and computes

$$\gamma = \gamma_0^{|G|/q}. \tag{3}$$

(If $\gamma = 1_G$ A selects another γ_0.) Finally, A randomly selects $a \leq 2^{160}$ and computes $\alpha = \gamma^a$. A's public key is (G, q, γ, α), the private key is a.

Signature: To sign the message m, A randomly selects $k \leq 2^{160}$ and computes $\varrho = \gamma^k$, an integer $r = f(\varrho)$, and an integer

$$s = k^{-1}\big(f(m) + ar\big) \bmod q, \tag{4}$$

where f is a collision and preimage resistant hash function. Then the signature for m is $S = (r, s)$, where f is a preimage and collision resistant hash function.

Verification: B checks that $0 < s < q$. Then B computes $w = s^{-1} \bmod q$, $u_1 = wf(m) \bmod q$, $u_2 = wr \bmod q$, and $v = f(\gamma^{u_1}\alpha^{u_2})$. Finally, B checks that $v = r$.

Without knowledge of $|G|$ the computation of (3) and (4) is impossible.

In the context of class groups (i.e. $G = \mathrm{Cl}(\Delta)$), (3) can be replaced by a random selection of $\gamma \in \mathrm{Cl}(\Delta)$, because $\mathrm{Cl}(\Delta)$ contains large cyclic subgroups with very high probability. Moreover, the probability that $h(\Delta)$ is very smooth is negligible.

Hence, by group theoretic arguments, the probability that $\langle\gamma\rangle$ is large and $|\langle\gamma\rangle|$ has a large prime divisor is very high. It remains to answer the question: what can we do about (4)? There are at least two possible solutions:

1. Replace q by an integer that is not related to $h(\Delta)$. The main motivation for this is the fact that $s = x \bmod q$ is equivalent to $s = x - \ell q$ for some integer ℓ
2. Omit the modular reduction.

First, in order to simplify the matter, we modify the signature scheme such that the modular inversion of (4) disappears. To do this, we pick DSA variant EG II.3 from [15], in which (4) becomes

$$s = -af(m, \varrho) + k \bmod q, \tag{5}$$

with appropriate modification of the signature (which becomes $S = (\varrho, s)$) and the verification procedure. The resulting scheme is actually more similar to the Schnorr signature scheme than to DSA, but for technical reasons we shall call it DSA, too. In the following subsections we shall discuss the two solutions.

4.2.1. Replacing the modulus. This variant of DSA has been described in [3]. It is called RDSA. The security of the resulting protocol is based on the *root problem* (and *not* on the discrete logarithm problem), whence the name.

Key generation: A randomly selects a fundamental discriminant Δ. Then A randomly selects $\gamma \in \mathrm{Cl}(\Delta)$ and a 160-bit prime q. Then A randomly selects $a \leq 2^{160}$ and computes $\alpha = \gamma^a$. A's public key is $(\Delta, q, \gamma, \alpha)$, the private key is a.

Signature: To sign the message m, A randomly selects $k \leq 2^{160}$ and computes $\varrho = \gamma^k$ and an integer $x = -af(m, \varrho) + k$. Then A divides x by q with remainder, i.e. A computes integers s and ℓ such that $x = q\ell + s$ and $0 \leq s < q$. Finally, A computes $\lambda = \gamma^\ell$. Then the signature for m is $S = (s, \varrho, \lambda)$.

Verification: B checks that $0 \leq s < q$ and that $\gamma^s \alpha^{f(m,\varrho)} \lambda^q = \varrho$.

From this description it is obvious that an attacker, who can compute qth roots in $\mathrm{Cl}(\Delta)$, can forge signatures for A: He simply randomly selects $\varrho \in \mathrm{Cl}(\Delta)$ and $s \in \{0, \ldots, q-1\}$, then he sets $\lambda = \left(\varrho\gamma^{-s}\alpha^{-f(m,\varrho)}\right)^{1/q}$. It is easily checked that (s, ϱ, λ) is a valid signature for m under A's public key.

In [3] the converse has been shown in the *random oracle model* [2], i.e. an attacker who can efficiently compute existentially forged signatures in a chosen message attack, can efficiently compute qth roots. Therefore, under the assumption of the hardness of the root problem, RDSA is secure against existential forgeries even in a chosen message attack.

4.2.2. Omitting the modular reduction. This variant of DSA is due to an idea in [25]. Unlike the DSA variant in the previous subsection, this variant is based on the discrete logarithm problem.

Key generation: A randomly selects a fundamental discriminant Δ. Then A randomly selects $\gamma \in \mathrm{Cl}(\Delta)$, an integer $a \leq 2^{160}$, and computes $\alpha = \gamma^a$. A's public key is (Δ, γ, α), the private key is a.

Signature: To sign the message m, A randomly selects $k \leq 2^{160}$ and computes $\varrho = \gamma^k$ and an integer $s = -af(m, \varrho) + k$. Then the signature for m is $S = (s, \varrho)$.

Verification: B checks that $\gamma^s \alpha^{f(m,\varrho)} = \varrho$.

This surprisingly simple DSA variant appears to be superior to RDSA, for it requires apparently less computation and is based on an allegedly harder problem. However, as in [25], this scheme is secure against existential forgery in a chosen message attack only if af/k is negligible. For example, if a and the output of f are 160 bits wide, and if $1/2^{80}$ is negligible, then k has to be selected 400 bits large (note that s will also have 400 bits). Thus, this DSA variant is considerably less efficient than RDSA.

4.3. The Guillou–Quisquater signature scheme

The Guillou–Quisquater signature scheme [13] works unmodified in our context. For convenience, we present a stripped down version:

Key generation: A randomly selects a fundamental discriminant Δ. Then A randomly selects $\alpha \in \mathrm{Cl}(\Delta)$ and an integer $q \leq 2^{160}$. Then A computes $\theta = \alpha^{-q}$. A's public key is (Δ, q, θ), the private key is α.

Signature: To sign the message m, A randomly selects $\kappa \in \mathrm{Cl}(\Delta)$ and computes $\varrho = \kappa^q$. Then A computes $\ell = f(m, \varrho)$ and $\sigma = \kappa\alpha^\ell$. The signature for m is $S = (\ell, \sigma)$.

Verification: B computes $v = f(m, \sigma^q\theta^\ell)$ and checks that $v = \ell$.

It is obvious that this signature scheme is based on the intractability to compute roots in $\mathrm{Cl}(\Delta)$. According to [24], the Guillou–Quisquater signature scheme also can be proven to be secure against existential forgery in a chosen message attack.

5. Efficiency of IQ protocols

In this section, we shall review some algorithms to perform IQ arithmetic, then we shall present some benchmarks for the IQ-RDSA signature scheme and compare these to benchmarks of the RSA signature scheme.

5.1. IQ arithmetic

IQ arithmetic is essentially the same as the composition of binary quadratic forms [8, Sect. 5.2]. The algorithms here are taken from [17]. Recall that integral ideals are represented by pairs (a, b), where a and b are integers, a is positive, and $4a \mid b^2 - \Delta$. For any ideal (a, b) set $c = (b^2 - \Delta)/(4a)$.

First, we present the algorithms for ideal multiplication and ideal squaring. For integers a and b we denote by $(d, x, y) \leftarrow \mathrm{xgcd}(a, b)$ the computation of integers d, x, and y such that d is the greatest common divisor of a and b, and $d = ax + by$.

The hard part of Algorithms 1 and 2 is the extended Euclidean algorithm (line 1 in both algorithms). Therefore, both algorithms have the best performance, if their input ideals are reduced. Note that the second extended Euclidean algorithm in Algorithm 1 (line 5) takes on average only a few steps, because d_1 is very small on average.

Algorithm 1. Compute $\mathfrak{a}_3 \leftarrow \mathfrak{a}_1 \times \mathfrak{a}_2$
Input: Reduced ideals $\mathfrak{a}_1 = (a_1, b_1)$ and $\mathfrak{a}_2 = (a_2, b_2)$, $\mathfrak{a}, \mathfrak{b} \subseteq \mathcal{O}_\Delta$
Output: the ideal $\mathfrak{a}_3 = (a_3, b_3)$

1: $d_1, v, w \leftarrow \mathrm{xgcd}(a_1, a_2)$
2: $a_3 \leftarrow a_1 a_2$
3: $b_3 \leftarrow v a_1 (b_2 - b_1)$
4: **if** $d_1 \neq 1$ **then**
5: $\quad d_2, v, w \leftarrow \mathrm{xgcd}(d_1, (b_1 + b_2)/2)$
6: $\quad a_3 \leftarrow a_3 / d_2^2$
7: $\quad b_3 \leftarrow (b_3 v + w(\Delta - b_1^2)/2)/d_2$
8: **end if**
9: $b_3 \leftarrow b_1 + b_3 \bmod 2a_3$

Algorithm 2. Compute $\mathfrak{a}_3 \leftarrow \mathfrak{a}_1^2$
Input: A reduced ideal $\mathfrak{a}_1 = (a_1, b_1)$, $\mathfrak{a}_1 \subseteq \mathcal{O}_\Delta$
Output: the ideal $\mathfrak{a}_3 = (a_3, b_3)$

1: $d, v, w \leftarrow \mathrm{xgcd}(a_1, b_1)$
2: $a_3 \leftarrow (a_1/d)^2$
3: $b_3 \leftarrow w(\Delta - b_1^2)/(2d)$
4: $b_3 \leftarrow b_1 + b_3 \bmod 2a_3$

Next, we present a reduction algorithm, since it is essential to reduce any (intermediate) result. The reduction is performed by Algorithms 3. Algorithm 3 is similar to the extended Euclidean algorithm, and it can be shown that the loop (lines 13 to 22) is executed at most $2 + \lceil \log_2(a/\sqrt{|\Delta|}) \rceil$ times [8, Proposition 5.4.3]. A run time analysis of Algorithm 3 similar to the analysis of the extended Euclidean algorithm shows that Algorithm 3 performs $\mathrm{O}(\log^2 |\Delta|)$ steps [5, Sect. 5.6], if the input to this algorithm is the output of either algorithm 1 or 2.

Finally, we note that if (a, b) represents an ideal class, then $(a, -b)$ represents the inverse. If (a, b) is reduced, then so is $(a, -b)$, unless $a = b$, which happens with negligible probability. For instance, if Δ is a negative odd prime, then there is only one such reduced ideal, and that is $(1, 1)$. Therefore, the inversion of a group element takes almost always constant time. This can be utilized for fast exponentiations of ideals, because one could use a *signed-digit exponent recoding* [22, Sect. 14.7], which reduces the number of ideal multiplications.

Algorithm 3. Reduce $\mathfrak{a}$
Input: An ideal $\mathfrak{a} = (a, b)$, $\mathfrak{a} \subseteq \mathcal{O}_\Delta$
Output: A reduced ideal equivalent to $\mathfrak{a}$

```
if b < 0 then
    s ← −1
else
    s ← 1
end if
b ← |b|, r ← b mod 2a
if a < r then
    b ← a
else
    b ← r, s ← −s
end if
c ← (b^2 − Δ)/4a
while a > c do
    t_a ← a, a ← c
    q ← ⌊b/2a⌋, r ← b mod 2a
    c ← t_a − q(r + b)/2
    if a < r then
        b ← 2a − r, c ← c + a − r
    else
        b ← r, s ← −s
    end if
end while
if s < 0 then
    b ← −b
end if
if b > a then
    b ← b − 2a, c ← c − b + a
else if b ≤ −a then
    b ← b + 2a, c ← c + b + a
end if
if a = c and b < 0 then
    b ← −b
end if
```

Algorithms 1, 2, and 3 have been implemented using the GNU multiprecision arithmetic library (GMP, version 3.1.1, see `http://www.swox.com/gmp/`). On a SUN machine with Sparc Ultra II processor (333 MHz), we get the benchmarks in Table 2. From this table, one can conclude that on average an ideal squaring is only slightly faster than an ideal multiplication. Furthermore, the greatest amount of work has to be spent for the reduction, thus any optimization efforts should start here. This could be done by replacing Algorithm 3 by Schönhage's reduction algorithm [29, 33]. Another idea is to use the NUCOMP and NUDUPL algorithms of Shanks [32] for ideal multiplication and squaring. Shanks' Algorithms have the advantage, that most intermediate results are of size $O(\sqrt{|\Delta|})$. Moreover, their outputs are already reduced. This is achieved to the expense of more multiprecision multiplications. But if these algorithms are implemented carelessly, they are still slower than multiplication or squaring with subsequent reduction.

Table 2. Performance of IQ arithmetic (in milliseconds)

size of Δ	multiply	square	reduce
512	0.1570	0.1550	0.4455
724	0.2390	0.2295	0.7035
1024	0.3570	0.3380	1.1400
1448	0.5635	0.5230	1.9130
2048	0.9145	0.8340	3.3165
2896	1.4855	1.3475	6.0395
4096	2.4590	2.2270	10.8880

5.2. Efficiency of some IQ cryptosystems

In this subsection we present benchmarks for the IQ-RDSA signature scheme. This have been implemented using the arithmetic modules that were described in the previous subsection. We have used the following optimizations: For the exponentiation we have used signed-digit exponent recoding. For a signature, we have also used Gordon–Brickell precomputation [22, Sect. 14.6.3] (generated in the key generation step). For the verification we have used simultaneous multiple exponentiation. Some averaged timings are given in Table 3.

Table 3. Some IQ-RDSA timings (in milliseconds)

Size of	IQ-RDSA	
Δ	Sig	Ver
687	82.4	245.4
1208	191.4	554.5
2084	411.3	1248.8

For a comparison, we present some benchmarks of the RSA signature scheme in Table 4. We have used the openssl implementation of RSA (openssl version 0.9.6, see `http://www.openssl.org/`). This implementation makes use of Montgomery exponentiations and Chinese remaindering. As public exponent we have selected the Fermat prime 65537. Note that the parameter size for IQ-RDSA and for RSA has been selected according to Table 1.

The key observation is that an IQ-RDSA signature is eventually faster than a RSA signature. The cross over is at a security level that is rather moderate

Table 4. Some RSA timings (in milliseconds)

Size of n	RSA Sig	RSA Ver
1024	50.7	3.0
2048	342.7	10.7
4096	2447.5	39.8

(according to [20]). Since the efficiency of IQ arithmetic received comparably little attention in the past, it is reasonable to expect significant improvements in the future. These improvements will make IQ cryptosystems even more competitive to traditional cryptosystems.

6. Conclusion

We have given a summary over the selection of the cryptographic parameter for IQ cryptosystems. We have presented IQ versions of some well known protocols, and we have presented efficient algorithms to implement IQ cryptosystems. Finally we have seen that they are efficient and practical. The next steps are: Apply the optimizations to IQ arithmetic as outlined at the end of Section 5.1, investigate more protocols (for example, blind signatures, undeniable signatures, signatures with message recovery, and key exchange protocols), and finally, a formal description of IQC similar to [30, 31].

References

[1] ANSI X9.63. Public Key Cryptography for the Financial Services Industry: Key Agreement and Transport Using Elliptic Curve Cryptography. American Bankers Association, 1999. Working Draft.

[2] Bellare, M., Rogaway, P., Random oracles are practical: a paradigm for designing efficient protocols. In: Proceedings of the 1st ACM Conference on Computer and Communication Security (CCS '93), 62–73. ACM Press, New York 1993.

[3] Biehl, I., Buchmann, J., Hamdy, S., Meyer, A., A signature scheme based on the intractability of extracting roots. Des. Codes Cryptogr. (to appear).

[4] Borevich, Z I., Shafarevich, I.R., Number Theory. Academic Press, New York 1966.

[5] Buchmann, J., Algorithms for binary quadratic forms. `http://www.informatik.tu-darmstadt.de/TI/Lehre/WS00_01/Vorlesung/Algorithmische_Zahlentheorie/quf.ps.gz`, TU-Darmstadt, FB Informatik, October 2000. Draft.

[6] Buchmann, J., Williams, H.C., A key-exchange system based on imaginary quadratic fields. J. Cryptology 1 (1988), 107–118.

[7] Buell, D.A., The expectation of success using a Monte Carlo factoring method — some statistics on quadratic class numbers. Math. Comp. 43 (1984), 313–327.

[8] Cohen, H., A Course in Computational Algebraic Number Theory (Grad. Texts in Math. 138). Springer, Berlin 1993.

[9] Cohen, H., Lenstra, Jr., H.W., Heuristics on class groups of number fields. In: Number Theory, Noordwijkerhout 1983 (ed. by H. Jager; Lecture Notes in Math. 1068), 33–62. Springer, Berlin 1984.

[10] Cramer, R., Shoup, V., A practical public key cryptosystem provably secure against adaptive chosen ciphertext attack. In: Advances in Cryptology – CRYPTO '98 (ed. by H. Krawczyk; Lecture Notes in Comput. Sci. 1462), 13–25. Springer, Berlin 1998.

[11] ElGamal, T., A public key cryptosystem and a signature scheme based on discrete logarithms. IEEE Trans. Inform. Theory 31 (1985), 469–472.

[12] FIPS 186-2. Digital Signature Standard (DSS). NIST, 2000. Federal Information Processing Standards Publication 186-2.

[13] Guillou, L.C., Quisquater, J.-J., A practical zero-knowledge protocol fitted to security microprocessors minimizing both transmission and memory. In: Advances in Cryptology – EUROCRYPT '88 (ed. by C.G. Günther; Lecture Notes in Comput. Sci. 330), 123–128. Springer, Berlin 1988.

[14] Hamdy, S., Möller, B., Security of cryptosystems based on class groups of imaginary quadratic orders. In: Advances in Cryptology – ASIACRYPT 2000 (ed. by T. Okamoto; Lecture Notes in Comput. Sci. 1976), 234–247. Springer, Berlin 2000.

[15] Horster, P., Michels, M., Petersen, H., Meta-ElGamal signature schemes. In: 2nd ACM Conference on Computers and Communications Security, 96–107. ACM Press, 1994.

[16] Hua, L.K., Introduction to Number Theory. Springer, Berlin 1982.

[17] Jacobson, Jr., M.J., Subexponential Class Group Computation in Quadratic Orders. Ph.D. Thesis, Technische Universität Darmstadt, Fachbereich Informatik, 1999.

[18] Koblitz, N., Elliptic curve cryptosystems. Math. Comp. 48 (1987), 203–209.

[19] Lang, S., Algebraic Number Theory, 2 ed. (Grad. Texts in Math. 110). Springer, Berlin 1994.

[20] Lenstra, A.K., Verheul, E.R., Selecting cryptographic keysizes. In: Public Key Cryptography (PKC 2000, Melbourne) (ed. by H. Imai, Y. Zhang; Lecture Notes in Comput. Sci. 1751), 446–465. Springer, Berlin 2000. Full version available from `http://www.cryptosavvy.com/`.

[21] Littlewood, J.E., On the class number of the corpus $P(\sqrt{-k})$. Proc. London Math. Soc. (2) 27 (1928), 358–372.

[22] Menezes, A.J., van Oorschot, P.C., Vanstone, S.A., Handbook of Applied Cryptography. CRC Press, Boca Raton 1997.

[23] Miller, V.S., Use of elliptic curves in cryptography. In: Advances in Cryptology – CRYPTO '85 (ed. by H.C. Williams; Lecture Notes in Comput. Sci. 218), 417–426. Springer, Berlin 1986.

[24] Pointcheval, D., Stern, J., Security arguments for digital signatures and blind signatures. J. Cryptology 13 (2000), 361–396.

[25] Poupard, G., Stern, J., Security analysis of a practical "on the fly" authentication and signature generation. In: Advances in Cryptology – EUROCRYPT '98 (ed. by K. Nyberg; Lecture Notes in Comput. Sci. 1403), 422–436. Springer, Berlin 1998.

[26] Schaub, J., Implementierung von Public-Key-Kryptosystemen über imaginär-quadratischen Ordnungen. Master's Thesis, Technische Universität Darmstadt, Fachbereich Informatik, 1999 (German).

[27] Schnorr, C.P., Efficient signature generation by smart cards. J. Cryptology 4 (1991), 161–174.

[28] Schnorr, C.P., Lenstra, Jr., H.W., A Monte Carlo factoring algorithm with linear storage. Math. Comp. 43 (1984), 289–311.

[29] Schönhage, A., Fast reduction and composition of binary quadratic forms. In: Proceedings ISSAC '91 (ed. by S. M. Watt), 128–133. ACM Press, 1991.

[30] SEC 1. Elliptic Curve Cryptography. Standards for Efficient Cryptography Group, September 2000. Working Draft. Available from `http://www.secg.org/`.

[31] SEC 2. Recommended Elliptic Curve Domain Parameters. Standards for Efficient Cryptography Group, September 2000. Working Draft. Available from `http://www.secg.org/`.

[32] Shanks, D., On Gauss and composition I, II. In: Number Theory and Applications (ed. by R.A. Mollin; NATO Adv. Sci. Inst. Ser. C 265), 163–178, 179–204. Kluwer Academic Publ., Dordrecht 1989.

[33] Weilert, A., Effiziente Algorithmen zur Berechnung von Idealsummen in quadratischen Ordnungen. Ph.D. Thesis, Rheinische Friedrich-Wilhelms-Universität Bonn, 2000 (German).

Algebraic groups and discrete logarithm

Jean-Marc Couveignes

Abstract. We prove two theorems and raise a few questions concerning discrete logarithms and algebraic groups.

1. Introduction

> Je n'aurais jamais pensé que mes pièces dussent s'attirer l'immortalité; mais depuis que quelques poètes fameux m'ont fait l'honneur de les parodier, ce choix de préférences pourrait bien, dans les temps à venir, leur faire partager une réputation qu'elles ne devront originairement qu'aux charmantes parodies qu'elles auront inspirées. Aussi marquai-je d'avance à mes associés bénévoles, dans ce nouveau livre, toute la reconnaissance que m'inspire une société aussi flatteuse, en leur fournissant, dans ce troisième ouvrage, un vaste champ pour exercer leur Minerve.
>
> François Couperin, *Préface au troisième livre*

This paper, based on two talks given in Bordeaux (*Three days in computational analytic and elementary number theory*, May 1998) and Durham (*Computational number theory*, July 2000) is devoted to the study of discrete logarithm in a slightly more general context than usual.

In Section 3 we give the weakest conditions we know for the existence of a *proven* sub-exponential algorithm computing discrete logarithms in the jacobians of curves with increasing genera over a fixed field. The first work in that context is [2] and was repeated in a series of publications all assuming more less implicitly the curves to have very special geometric properties (e.g. being super-elliptic). We show this is an inessential property. We replace it by a more arithmetic one concerning the asymptotic behavior of the size of the jacobian. We believe it has not been stressed enough that the genus does not suffice to control this size (e.g. up to a sub-exponential factor). So it is not reasonable to expect discrete logarithm algorithms to exist that would not take this problem into account. I would like to thank Christophe Betard with whom I had many interesting discussions on that topic. The algorithm we give relies on no heuristic and therefore gives a theorem on the (randomized) complexity of discrete logarithms in the Picard group of curves over a fixed field.

Public-Key Cryptography and Computational Number Theory
K. Alster, J. Urbanowicz and H. C. Williams (Eds.)

Conversely, in Section 4 we define the category of easy algebraic groups and study the problem of discrete logarithm in the groups of points of a fixed algebraic group, rational over increasing finite fields. We show that this category has quite nice properties.

This leads to very natural questions of algorithmic and geometric nature.

Section 2 sets or recalls a few natural and simple definitions from complexity theory to ensure the accuracy of the statements in that work.

2. Discrete logarithm problems and their complexity

Since we shall be concerned with the difficulty of computing discrete logarithms we first give the following

Definition 1. A *discrete logarithm problem* $(\mathcal{G}, \alpha, \beta)$ consists of a finite computational commutative group $\mathcal{G}$, an element α in $\mathcal{G}$, and an element β in $\mathcal{G}$. A *solution* to this problem consists of an integer k such that $\beta = k\alpha$.

We do not assume the order of the group $\mathcal{G}$ to be known. We may slightly generalize the previous definition with

Definition 2. An *extended discrete logarithm problem* $(\mathcal{G}, (\alpha_a)_{1\leq a\leq A}, \beta)$ consists of a finite computational commutative group $\mathcal{G}$, a family $(\alpha_a)_{1\leq a\leq A}$ of elements in $\mathcal{G}$, and an element β in $\mathcal{G}$. A *solution* to this problem consists of A integers $(k_a)_{1\leq a\leq A}$ such that $\beta = \sum_a k_a\alpha_a$.

Although the extended logarithm problem seems to be more difficult than the ordinary one, we notice that all known algorithms for the second one straightfully extend to the first one.

In the sequel we shall concentrate on plain discrete logarithm problem. An algorithm for solving discrete logarithm will be a *randomized* algorithm that gives a solution when it exists but may run indefinitely when there is no solution. Such an algorithm with known complexity may be used to solve the decision problem (existence or not of a solution) in a *probabilistic* way. This means that we may bound the probability of giving a wrong (negative) answer by an arbitrary small constant.

Another use of a generalized discrete logarithm random algorithm is to compute the module of relations between elements $(\alpha_a)_{1\leq a\leq A}$ of a commutative group $\mathcal{G}$ with given cardinality I, that is the kernel $\mathcal{R}$ of the map ξ in

$$0 \to \mathcal{R} \to \mathbb{Z}^A \xrightarrow{\xi} \mathcal{G}.$$

This problem is solved, at least in a *probabilistic* way, by first factoring (with a sieving method like in [8]) the cardinality I. One then computes the order $h_{1,1}$

of α_1 which is a divisor of I. Then, one looks for the smallest divisor $h_{2,2}$ of I such that $h_{2,2}\alpha_2$ is in the subgroup $\langle\alpha_1\rangle$ generated by α_1. This is done using the generalized discrete logarithm algorithm. We find $h_{2,2}\alpha_2+h_{2,1}\alpha_1 = 0$. We continue this way and find the Hermite normal form of $\mathcal{R}$.

An application is to compute a generating set for the intersection

$$\mathcal{H} \cap \langle\alpha_1, \ldots, \alpha_A\rangle$$

where $\mathcal{H}$ is a subgroup of $\mathcal{G}$, provide we are given the cardinality of $\mathcal{G}/\mathcal{H}$ and a generating set for $\mathcal{H}$.

We are interested in the randomized complexity of discrete logarithms. We say that a family of discrete logarithm problems is *easy* if there exists a randomized algorithm that solves these problems in sub-exponential time in the order of the group $\mathcal{G}$ that is in time $\exp(O(1)\log^\alpha|\mathcal{G}|)$ for some α in $]0,1[$. Basic operations in the group are assumed to require polynomial time in the logarithm of the size of the group.

We say that a family of groups is *easy* if the family of all discrete logarithm problems involving any of these groups is solved in time $\exp(O(1)\log^\alpha|\mathcal{G}|)$ for some α in $]0,1[$.

It is a simple consequence of Kraitchik's index calculus method that the family of all multiplicative groups of prime finite fields is easy (we can take any $\alpha > 1/2$). Similarly, the family of all multiplicative groups of finite fields of given characteristic is easy.

It is shown in [1] that the family of multiplicative groups of all finite fields is easy under some reasonable though unproven conjectures.

3. Index calculus and Picard groups

In this section we prove the following

Theorem 1. *Let $(\mathcal{C}_i)_{i\geq 1}$ be a family of curves over $\mathbb{F}_q$ with genus $\gamma_i = g(\mathcal{C}_i)$ tending to infinity and assume the number $\#\mathcal{J}_{\mathcal{C}_i}(\mathbb{F}_q)$ of $\mathbb{F}_q$-rational points on the jacobian of $\mathcal{C}_i$ is bounded by $q^{\gamma_i+O(\sqrt{\gamma_i})}$. The curves $\mathcal{C}_i$ are given as plane curves with polynomial degree in the genus. Assume further we are given an $\mathbb{F}_q$-rational point O_i on the smooth model of $\mathcal{C}_i$ for every i. Then there exists an algorithm that computes discrete logarithms in the groups $\mathcal{H}_i = \mathcal{J}_{\mathcal{C}_i}(\mathbb{F}_q)$ in random sub-exponential time $q^{O(\gamma_i^{0.5+\epsilon})}$ for any positive ϵ.*

Subsection 3.1 recalls a few basic facts about the algorithmics of algebraic curves. We find it necessary since our experience is that the amazing recent development of this topic tends to hide the existence of quite old and useful results (after all, the theory of algebraic curves was very much algorithmic in its origins). The proof for Theorem 1 is given in Subsection 3.2.

3.1. Basic algorithms for algebraic curves

In this subsection we review a few basic facts about the algorithmics of algebraic curves. We shall represent any curve over $\mathbb{F}_q$ as a plane curve $\mathcal{C}$ with degree d and genus g and will assume that the degree is bounded polynomially in the genus. This is just intended to avoid artificial examples like a series of plane genus zero curves with degree tending to infinity. *Computing a smooth model* from the plane model is then done in polynomial time in the genus. For most purposes, it suffices to have an explicit description of the different places above any singular point. These places may be given by the Puiseux expansion of one coordinate in terms of the other.

By a divisor we always mean a divisor on the smooth model of the curve although the plane model might be more convenient for practical computation. Similarly, the Picard group and the jacobian $\mathcal{J}_{\mathcal{C}}$ of $\mathcal{C}$ always are the ones of the smooth model.

Computation of the linear space of a divisor $A - B$ where A and B are effective divisors is then done in time polynomial in the product $\deg(A) \times \deg(B) \times g$.

The *enumeration of all prime divisors* of degree bounded by a constant c is done in polynomial time in $q^c \times g$.

Elements in the Picard group of degree 0 are represented by divisors of degree zero and computation with such divisors relies on *reducing divisors* that is finding an equivalent divisor of the form $A - gO$ with A effective which can be done in polynomial time. Here, a rational origin O is assumed to exist and to be given.

The techniques described above are very classical. See [5] for a classical review of this folklore and [9] for an updated survey taking into account the progress of computer algebra.

We now come to the problem of *choosing a random element in the Picard group* of $\mathcal{C}$ over $\mathbb{F}_q$. We do not assume the size of this group to be known. We just assume we are given a set $\mathcal{B}$ of divisor classes that generate the Picard group. We assume that the size of $\mathcal{B}$ is polynomial in the genus. Let $\Upsilon > 1$ be a real number. We want to choose a random combination of elements in $\mathcal{B}$ in such a way that the corresponding sum in the Picard group is random with Υ-*almost uniform probability*. This means that every element occurs with probability at least Υ^{-1} and at most Υ times the inverse of the size of the Picard group $\mathcal{G}$. This is achieved in the following way. Let d be the size of $\mathcal{B}$ and I the size of the Picard group. We assume $d \geq 2$, otherwise the problem is trivial. Let $\mathcal{R}$ be the kernel of the generating map ξ

$$0 \to \mathcal{R} \to \mathbb{Z}^d \xrightarrow{\xi} \mathcal{G} \to 0$$

and set $\rho = d4^d(d!)^2 I$. From Minkowski's theorems [6, II.6.8, II.6.9, II.7.4] we deduce that $\mathcal{R}$ has a fundamental parallelogram $\mathcal{P}$ of radius bounded by ρ.

Let $\epsilon = 1 - \Upsilon^{-1}$. Let R be an even integer bigger than $\epsilon^{-1} d\rho$. Let C_R be the hypercube in $\mathbb{R}^d$ with center at the origin and edge R defined to be the set of points with integral coordinates in the interval $[-R/2, R/2[$. It is of course very easy to pick an element in C_R with random uniform probability. The map

ξ pushes the uniform probability on C_R to a probability on the set $\mathcal{G}$ which is Υ-almost uniform thus solving the problem of choosing random elements in the Picard group in polynomial time provide we use the fast exponentiation algorithm. To prove that this probability is Υ-uniform we set $r = R - \rho$ and consider the union

$$A = \bigcup_{\substack{z\in\mathcal{R}\\ z+\mathcal{P}\cap C_r\neq\emptyset}} z + \mathcal{P}$$

of $\mathcal{R}$-translates of the fundamental parallelogram $\mathcal{P}$ that meet C_r. We easily check that $C_r \subset A \subset C_R$ and $|C_R - A|/|C_R| \leq |C_R - C_r|/|C_R| \leq d\rho/R$. We similarly set $s = R + \rho$ and

$$B = \bigcup_{\substack{z\in\mathcal{R}\\ z+\mathcal{P}\cap C_R\neq\emptyset}} z + \mathcal{P}$$

the union of $\mathcal{R}$-translates of the fundamental parallelogram $\mathcal{P}$ that meet C_R. We have $C_R \subset B \subset C_s$ and $|B - C_R|/|C_R| \leq |C_s - C_R|/|C_R| \leq d\rho/R$. The result follows. On the way, we have proven the following useful

Lemma 1. *Let $\mathcal{R}$ be a sub-lattice of $\mathbb{Z}^d$ with index I and set $\kappa_d = d^2 4^d (d!)^2$. Let R be an even integer bigger than $\kappa_d I$. Let a be an element in the quotient group $\mathbb{Z}^d/\mathcal{R}$. Then the proportion of points in C_R that are congruent to a modulo $\mathcal{R}$ is in $[(1-\kappa_d I/R)/I, (1+\kappa_d I/R)/I]$.*

In case $\mathcal{R}$ is a subgroup of $\mathbb{Z}^d$ of rank smaller than d we have the even simpler

Lemma 2. *Let $\mathcal{R}$ be a subgroup of $\mathbb{Z}^d$ with infinite index. Let R be an even positive integer. Let a be an element in the quotient group $\mathbb{Z}^d/\mathcal{R}$. Then the proportion of points in C_R that are congruent to a modulo $\mathcal{R}$ is bounded by $1/R$.*

Remark 1. The considerations above apply to the more general situation of a finite commutative group given by generators (relations are unknown) and allow to pick random elements with almost uniform probability as soon as an estimate is given for the size of the group.

3.2. Proof of Theorem 1

In order to prove Theorem 1 we shall first describe an algorithm and then the complexity analysis for it.

Let $\mathcal{C}_i$ be any curve in the family. We denote by I the cardinality of its Picard group over $\mathbb{F}_q$. Since it is not known, we shall replace it by a majoration of it in all the estimates bellow. A discrete logarithm problem consists of two effective divisors A and B of degree γ_i. The solution is an integer k such that $[B-\gamma_i O_i] = k[A-\gamma_i O_i]$ whenever it exists.

We denote by $\mathcal{B}_i$ the set of divisors on $\mathcal{C}_i$ of degree lower than or equal to $7\sqrt{\gamma}_i$. We denote by $\bar{\mathcal{B}}_i$ the set of divisor classes of the form $[D - \deg(D)O_i]$ where D is in $\mathcal{B}_i$. One easily checks (using Riemann hypothesis for function fields) that these divisor classes generate the full Picard group of $\mathcal{C}_i$ over $\mathbb{F}_q$. We say that a divisor is $\mathcal{B}_i$-smooth or just smooth if it is a linear combination of elements in $\mathcal{B}_i$. We denote by d the cardinality of $\mathcal{B}_i$. From [12, Corollary V.2.10] we have $d = q^{7\sqrt{\gamma_i}+O(\log \gamma_i)}$. The number of effective divisors with degree $2\gamma_i$ whose all prime divisors are in $\mathcal{B}_i$ is at least

$$\frac{d^{\lfloor \frac{2}{7}\sqrt{\gamma_i}\rfloor}}{\lfloor \frac{2}{7}\sqrt{\gamma_i}\rfloor !}.$$

Indeed, to any unordered family of $\lfloor \frac{2}{7}\sqrt{\gamma_i}\rfloor$ elements in $\mathcal{B}_i$ we associate their product and thus form the expected number of smooth effective divisors. In this expression, the numerator is $q^{2\gamma_i+O(\sqrt{\gamma_i}\log\gamma_i)}$ and the denominator is $q^{O(\sqrt{\gamma_i}\log\gamma_i)}$ so we have at least $q^{2\gamma_i+O(\sqrt{\gamma_i}\log\gamma_i)}$ effective divisors of degree $2\gamma_i$ that are $\mathcal{B}_i$-smooth.

As expected the algorithm goes in several steps.

Step 1. We first express $\alpha = [A - \gamma_i O_i]$ and $\beta = [B - \gamma_i O_i]$ as linear combinations of elements in $\bar{\mathcal{B}}_i$. To this end, we choose a linear combination of elements in $\bar{\mathcal{B}}_i$ such that the corresponding sum in the Picard group of $\mathcal{C}_i$ over $\mathbb{F}_q$ is random with 10/9-almost uniform probability (we take random coordinates in C_R with $R = 10d\rho$ where $\rho = d4^d(d!)^2 I$). We call $D = D^+ - \gamma_i O_i$ the corresponding reduced divisor and compute the linear space $\mathcal{L}(A + D + \gamma_i O_i)$ and pick a random function ϕ in it. We set

$$(\phi) = E - A - D^+$$

where E is an effective divisor of degree $2\gamma_i$. If E is smooth, we thus get an expression of $\alpha = [A - \gamma_i O_i]$ as a linear combination of smooth divisor classes. We repeat *the whole process* (taking another random element in the jacobian) until we have found a relation.

We do the same with β.

Step 2. We then look for linear relations between elements in $\bar{\mathcal{B}}_i$. Again we choose a linear combination of elements in $\bar{\mathcal{B}}_i$ such that the corresponding sum in the Picard group of $\mathcal{C}_i$ over $\mathbb{F}_q$ is random with 10/9-almost uniform probability. We call $D = D^+ - \gamma_i O_i$ the corresponding reduced divisor and compute the linear space $\mathcal{L}(2\gamma_i O_i + D)$ and pick a random function ϕ in it. If $(\phi) = E - D^+ - \gamma_i O_i$ with E a $\mathcal{B}_i$-smooth effective divisor, we get a linear relation. We repeat the process until we have found d independent relations.

Step 3. We now have a lattice $\mathcal{T} \subset \mathbb{Z}^d$ of relations and we compute its volume K which is a multiple of I bounded by $(R + 2\gamma_i)^d d^d$. We look for an integer k such that $\beta = k\alpha$ in $\mathbb{Z}^d/\mathcal{T}$. This is done by mere linear algebra. If such a k exists, we output it and stop. If such a k does not exist, we try to enlarge $\mathcal{T}$ by looking for extra linear relations. We proceed as in Step 2 but this time allow coefficients in C_Q with $Q = 10d\rho'$ with $\rho' = d4^d(d!)^2 K$. We go on enlarging $\mathcal{T}$ until we find some k.

Remark 2. The above algorithm solves discrete logarithm in random sub-exponential time which means that if some k exists, it shall be found in random sub-exponential time. However, if k does not exist, the algorithm will run indefinitely. This algorithm may also be used to compute the cardinality and the structure of the Picard group in probabilistic sub-exponential time as explained in Section 2.

Remark 3. It may look strange that we use so big coefficients in the linear combinations. This is to ensure the almost uniformity of the probability distributions. In practice, one may rely on heuristics and take smaller coefficients. The necessity of translating the divisor by a random element in the jacobian instead of looking for smooth functions in a fixed linear space is also quite theoretical. However one may suspect that heuristics concerning the density of smooth functions in a given linear space (like the ones in [2]) turn to be false for some very special infinite series of curves. By contrast we notice that our algorithm makes no use of smooth functions (just smooth divisors).

As for the complexity analysis of Step 1 of the algorithm, we observe that this first step consists in taking a random element in the set $\mathcal{S}$ with cardinality $R^d q^{\gamma_i+1}$ of pairs consisting of a point in C_R and a function in the corresponding linear space. The lucky elements in this set are made of in the following way: take a $\mathcal{B}_i$-smooth divisor E of degree $2\gamma_i$. Then to any point in C_R that is mapped onto $[E - A - \gamma_i O_i]$ by ξ there corresponds a lucky element in $\mathcal{S}$. The number of lucky elements is thus at least the number of smooth divisors of degree $2\gamma_i$ times the minimum number of elements in C_R in a given class modulo $\mathcal{R}$. We thus have at least $0.9R^d q^{2\gamma_i+O(\sqrt{\gamma}_i \log \gamma_i)}/I$ lucky elements. Since $I = q^{\gamma_i+O(\sqrt{\gamma}_i)}$, the proportion of lucky elements is $q^{O(\sqrt{\gamma}_i \log \gamma_i)}$.

For Step 2 we have to remove from the subset of lucky elements those which lead to relations that are in the sub-lattice generated by the relations we already found. But this lattice is of infinite index and from Lemma 2 we loose at most $1/R$ of the lucky elements.

For Step 3 we again have to remove from the subset of lucky elements those which lead to relations that are in the sub-lattice generated by the relations we already found. This time the lattice $\mathcal{T}$ has finite index at least two in $\mathcal{R}$ but since we allow coefficients of size Q the proportion of lucky elements to remove is at most 0.5×1.1. This finishes the proof of Theorem 1. □

Remark 4. If $(\mathcal{C}_i)_{i\geq 1}$ is an infinite family of curves that are all coverings of the projective line of degree bounded by a constant n, then from [3, Th. 3] the class number is $O(q^g(g \log q)^{n-1})$ thus the conditions of Theorem 1 are met. This is the case in particular for super-elliptic curves.

If one takes arbitrary curves of increasing genera γ_i there is no evidence that they form an easy family of discrete logarithm problems although their jacobians are quite computational groups as soon as we avoid fancy models. This point seems to be ignored in the literature where people tend to assume (often implicitly) the existence of a map of fixed degree to the projective line. A geometric condition which is useless and quite stronger than the arithmetic condition we assume.

4. Easy algebraic groups

Que si l'on te reproche que tu ne parles pas le langage des villageois, et que toi ni ta troupe ne sentez guère les brebis ni les chèvres, réponds leur, ma bergère, que pour peu qu'ils aient connaissance de toi, ils sauront que tu n'es pas, ni celles aussi qui te suivent, de ces bergères nécessiteuses, qui, pour gagner leur vie, conduisent les troupeaux aux pâturages, mais que vous n'avez toutes pris cette condition que pour vivre plus doucement et sans contrainte. Que si vos conceptions et paroles étaient véritablement telles que celles des bergers ordinaires, ils auraient aussi peu de plaisir de vous écouter, que vous auriez beaucoup de honte à les redire.

Honoré d'Urfé, *Epître-Préface de l'auteur à la bergère Astrée*

In the previous section we studied the asymptotic complexity of discrete logarithms in Picard groups of algebraic curves over a fixed field. In this section, on the contrary, we study discrete logarithms in the group of points of a fixed algebraic group over increasing extensions of the base field.

An algebraic group G over $\mathbb{F}_q$ will be given as a finite collection of disjoint regular varieties all defined over $\mathbb{F}_q$ with multiplication and inversion morphisms also defined over $\mathbb{F}_q$ (see [10]). In some cases (e.g. jacobian varieties) one may prefer to work with more natural models. The choice of a model is not decisive as far as pushing a point from one model to the other one requires a constant number of operations in the field of definition of that point.

In this paper we only consider *commutative* algebraic groups.

Definition 3. Let p be a prime and $\mathbb{F}_q$ the field with $q = p^d$ elements. Let G be a commutative algebraic group over $\mathbb{F}_q$. We say that G is an *easy algebraic group* over $\mathbb{F}_q$ if there exists an algorithm that solves discrete logarithm problems in the groups $\mathcal{G}_k = G(\mathbb{F}_{q^k})$ in sub-exponential random time in the size $|\mathcal{G}_k|$ of the groups. An operation in the base field $\mathbb{F}_q$ is assumed to require one unit of time. So we say that G is *easy* if and only if there exists a real α in $]0,1[$ and a randomized algorithm that computes discrete logarithms in the groups $\mathcal{G}_k$ in time $\exp(O(1)\log^\alpha |\mathcal{G}_k|)$.

We notice that an algebraic group over $\mathbb{F}_q$ is easy if an only if the family of groups $G(\mathbb{F}_{q^r})$ for $r \geq 1$ is easy.

Remark 5. We may give a more accurate definition and specify a value for α. For example we may say that a group is γ-easy if one can take in the previous definition $\alpha = \gamma + \epsilon$ for any positive ϵ. However, all the proven sub-exponential algorithm for discrete logarithms and factoring integers correspond to the case $\gamma = 1/2$. See Remark 6.

We start with a few lemmata.

Lemma 3. *Let G be a commutative algebraic group over $\mathbb{F}_q$, of dimension d. For any positive integer r, the size of $G(\mathbb{F}_{q^r})$ is equivalent to a constant c times q^{dr} when r tends to infinity*:

$$|G(\mathbb{F}_{q^r})| = q^{dr}(c + o(1)).$$

The cardinality of $G(\mathbb{F}_{q^r})$ is equal to the number of connected components times the cardinality of $G_0(\mathbb{F}_{q^r})$ where G_0 is the identity component (recall G/G_0 is assumed to decompose over $\mathbb{F}_q$). The lemma then follows from some weak form of Weil's conjectures (for a non-complete variety) like in [4, Corollary 5]. □

Lemma 4. *Let G be a commutative connected algebraic group over a finite field and let $1 \to L \to G \to A \to 1$ be the strict exact sequence of connected commutative algebraic groups from Chevalley's theorem. There exists an isogeny defined over $\mathbb{F}_q$ from G to the direct product $L \times A$.*

This results from the properties of the Ext functor for algebraic groups ([11, Proposition VII.6] and [11, Théorème VII.12]), the solvability of commutative linear groups (Rosenlicht [10]) and the computations for G_a and G_m ([11, Théorème VII.6, Théorème VII.7]). □

We list natural properties of easy groups in the following

Theorem 2. *The following are true.*

(i) *If $Q = q^a$ is a power of q and $\mathbb{F}_Q \supset \mathbb{F}_q$ the corresponding field extension and G an algebraic group over $\mathbb{F}_q$ then G is an easy group over $\mathbb{F}_q$ if and only if $G \otimes_{\mathbb{F}_q} \mathbb{F}_Q$ is an easy group over $\mathbb{F}_Q$.*

(ii) *Algebraic groups of dimension zero are easy. A group G is easy if and only if the component of the identity G_0 is easy.*

(iii) *Subgroups and direct product of easy groups are easy. An isogenous group to an easy group is easy.*

(iv) *If A, B and C are algebraic groups over $\mathbb{F}_q$ with a strict exact sequence*

$$1 \to A \to B \to C \to 1$$

then B is easy if and only if A and C are.

(v) *The multiplicative group G_m and the additive group G_a are easy over $\mathbb{F}_p$. Linear groups are easy. A connected algebraic group is easy if and only if its maximal complete quotient is an easy abelian variety.*

(vi) *Any super-singular elliptic curve is easy over its field of definition.*

Proof. Points (i), (ii) and (iii) are trivial.

If $1 \to A \to B \to C \to 1$ is a strict exact sequence of commutative algebraic groups and A and C are easy then B is easy since we can push a discrete logarithm problem (α, β) from B to C and find k such that $\beta\alpha^k$ is in A. Because we know the order of the group of rational points of C (from Weil's conjectures) we may factor it using a sieving method and compute the smaller positive power of α lying in A. We thus reduce to a discrete logarithm problem in A. This proves half of point (iv).

The additive group is trivially easy and also the multiplicative group by the index calculus method. Linear commutative groups are solvable (Rosenlicht [10]) therefore easy. If $1 \to L \to G \to A \to 1$ is the exact sequence from Chevalley's theorem then A is isogenous to a subgroup of G by Lemma 4 and is thus easy if G is. The converse follows from what we just proved of point (iv). Therefore quotients of easy groups are easy. This finishes the proof of point (iv).

Assertion (vi) is the result in [7]. □

Remark 6. The theorem above remains true if we replace everywhere the word easy by α-easy (following Remark 5) provide $\alpha \geq 1/2$. This condition is imposed by the use of a proven factoring algorithm in proving point (iv). One may of course wonder if there are provable factoring algorithms for a smaller α (see related Question 4 below).

We denote by AG_q the category of algebraic groups over $\mathbb{F}_q$ and AG_0 the category of algebraic groups over $\bar{\mathbb{F}}_q$. We denote by EAG_q the full sub-category of AG_q consisting of easy algebraic groups over $\mathbb{F}_q$ and similarly for EAG_0.

We ask how large are EAG_q and EAG_0. In view of Chevalley's theorem [10] and Theorem 2 we may restrict our attention to simple abelian varieties over $\bar{\mathbb{F}}_q$. It is remarkable that there has been no result in the sense of enlarging these categories since the proof by Menezes, Okamoto and Vanstone [7] of statement (vi) in Theorem 2.

From the properties listed above we can expect the category EAG_0 to be characterized by simple geometric ways. This raises the following

Question 1. *How big is* EAG_0*?*

More accurate questions

Question 2. *Are all abelian varieties easy?*

Question 3. *Are all abelian varieties with zero p-rank easy?*

Question 4. *Can one prove a theorem like Theorem* 1 *with a sharper estimate for the running time namely* $q^{O(\gamma_i^a)}$ *with a smaller than* $1/2$*?*

References

[1] Adleman, L.M., DeMarrais, J., A subexponential algorithm for discrete logarithms over all prime fields. Math. Comp. 61 (1993), 1–15.

[2] Adleman, L.M., DeMarrais, J., Huang, M.-D., A subexponential algorithm for discrete logarithms over the rational subgroup of the jacobians of large genus hyperelliptic curves over finite fields. In: Algorithmic Number Theory – ANTS I (ed. by L.M. Adleman and M.-D. Huang; Lecture Notes in Comput. Sci. 877), 28–40. Springer, Berlin 1994.

[3] Lachaud, G., Martin-Deschamps, M., Nombre de points des jacobiennes sur un corps fini. Acta Arith. 56 (1990), 329–340.

[4] Lang, S., Weil, A., Number of points of varieties in finite fields. Amer. J. Math. 76 (1954), 819–827.

[5] Lebesgue, H., Leçons sur les constructions géométriques. Gauthier-Villars, Paris 1950.

[6] Martinet, J., Les réseaux parfaits des espaces euclidiens (Mathématiques). Masson, Paris 1996.

[7] Menezes, A., Okamoto, T., Vanstone, S.A., Reducing elliptic curves logarithms to logarithms in a finite field. IEEE Trans. Inform. Theory 39 (1993), 1639–1646.

[8] Pomerance, C., Fast, rigorous factorization and discrete logarithm algorithms. In: Discrete Algorithms and Complexity (ed. by D.S. Johnson et al.), 119–143. Academic Press, 1987.

[9] Poonen, B., Computational aspects of curves of genus at least 2. In: Algorithmic Number Theory – ANTS II (ed. by H. Cohen; Lecture Notes in Comput. Sci. 1122), 283–306. Springer, Berlin 1996.

[10] Rosenlicht, M., Some basic theorems on algebraic groups. Amer. J. Math. 78 (1956), 401–443.

[11] Serre, J.-P., Groupes algébriques et corps de classes (Publications de l'institut de mathématiques de l'université de Nancago VII). Hermann, Paris 1959.

[12] Stichtenoth, H., Algebraic Function Fields and Codes. Springer, Berlin 1993.

Fermat numbers, Wieferich and Wilson primes: computations and generalizations

Karl Dilcher

Abstract. In this survey of recent and ongoing computational and theoretical work I report on the following interrelated topics, all dealing with very large integers: Fermat numbers, generalized Fermat numbers, the search for Wieferich and Wilson primes, and Fermat and Wilson quotients for composite moduli. A number of related topics will also be briefly discussed.

1. Introduction

This paper is a survey of classical and recent results on several interrelated topics in computational number theory, all of them dealing with computations on large integers. Only the most important definitions and results will be given for each of the topics; for further information, the reader is encouraged to consult the original papers or other references provided.

We begin with the question of how we can most easily obtain large integers. Powers of fixed integers, especially 2^n, will probably first come to mind, along with $n!$. Both are not very interesting, as far as factorizations are concerned; therefore we consider the nearest neighbours $2^n \pm 1$ and $n! \pm 1$. At the same time this will remind us of two fundamental theorems of modular arithmetic, namely Fermat's "little" theorem

$$2^{p-1} \equiv 1 \pmod{p}, \quad p \neq 2 \tag{1.1}$$

(here in the special case with base 2), and Wilson's theorem

$$(p-1)! \equiv -1 \pmod{p}. \tag{1.2}$$

Here and in the remainder of the paper p always denotes a prime.

Starting with $2^n \pm 1$, we have on the one hand the Mersenne numbers and primes $2^p - 1$, and on the other hand Fermat numbers and primes, $2^{2^n} + 1$. The latter can be generalized to numbers of the form $a^{2^n} + b^{2^n}$. One can try to find factors by computation, or by more theoretical means.

Public-Key Cryptography and Computational Number Theory
K. Alster, J. Urbanowicz and H. C. Williams (Eds.)

By Fermat's little theorem (1.1), the quotient

$$q_p := \frac{2^{p-1} - 1}{p} \tag{1.3}$$

is an integer. It is an interesting and important problem to determine those p for which $q_p \equiv 0 \pmod{p}$. Next, the quotient (1.3) can be generalized in two steps. First, the base 2 can be replaced by any integer base a not divisible by p. Then Euler's generalization of Fermat's little theorem can be used to define the generalized Fermat quotient

$$q(a, m) := \frac{a^{\varphi(m)} - 1}{m}, \quad (a, m) = 1, \tag{1.4}$$

and it becomes again an interesting question to determine, for a given a, those positive integers m for which $q(a, m) \equiv 0 \pmod{m}$.

Moving on to $n!$, we note that not much is known about the numbers $n! \pm 1$, except for some calculations and primality testing. Primes of this form are sometimes called "factorial primes". The latest numerical and heuristic results can be found in [12]; see also [36], p. 7 or [56], p. 26. By Wilson's theorem (1.2), the quotient

$$w_p := \frac{(p-1)! + 1}{p} \tag{1.5}$$

is an integer, and again we may ask for which p the congruence $w_p \equiv 0 \pmod{p}$ holds. Wilson's theorem also has a generalization, due to Gauss, and this leads to a generalization of the Wilson quotient (1.5) for composite moduli m. Finally, we may ask again for which positive integers m this quotient vanishes modulo m.

For a schematic representation of these topics and their interrelations, see Figure 1. Among the various topics mentioned above, I will restrict my attention in this survey to the following ones:

- Fermat numbers
- Generalized Fermat numbers
- The search for Wieferich primes
- Fermat quotients for composite moduli
- The search for Wilson primes
- Wilson quotients for composite moduli

These topics correspond to the following six sections of this paper.

2. Fermat numbers

Because of the polynomial factorization

$$x^{ab} + 1 = (x^b + 1)(x^{(a-1)b} - x^{(a-2)b} + \cdots - x^b + 1),$$

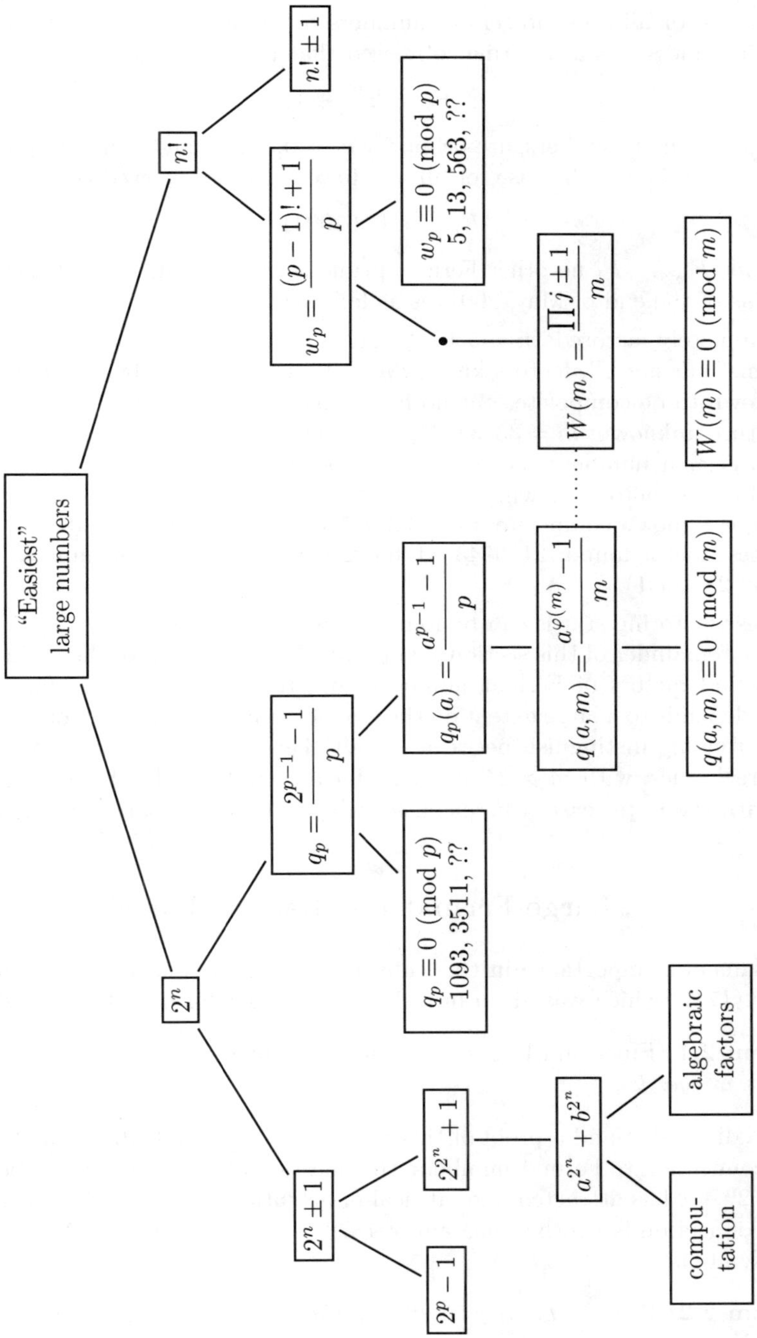

Figure 1. Logical connection of topics

which holds for all odd integers a, numbers of the form $2^n + 1$ will certainly be composite unless $n = 2^m$. Fermat observed that the numbers

$$F_m = 2^{2^m} + 1, \tag{2.1}$$

now called Fermat numbers, are prime for $m = 0, 1, 2, 3$, and 4, and he conjectured in 1637 that this was the case for all m. However, Euler showed that

$$F_5 = 641 \cdot 6700417.$$

Except for $F_0, \ldots, F_4$ no other Fermat prime is known. The current factorization status (as of the end of May, 2001) is as follows:

- Completely factored: $m = 5, 6, \ldots, 11$.
- Some, but not all, factors known: $m = 12, 13, 15, 16, 17, 18, 19, 21, 23, \ldots$
- Known to be composite, but no factor known: $m = 14, 20, 22, 24$.
- Nature unknown: $m = 33, 34, 35, 40, 41, 44, \ldots$
- 190 Fermat numbers known to be composite.
- 223 prime factors known.
- Largest known composite: $m = 382\,447$ (Cosgrave & Gallot, 1999)
- Latest factor found: $2139543641769 \cdot 2^{50} + 1 \mid F_{48}$ (Bodschwinna & Durman, May 28, 2001).

The latest factoring status can be found on Keller's website [39].

In the remainder of this section I will give a brief overview of the main methods used to find factors of Fermat numbers, and tests for primality. The choice of method depends to a large extent on the size of the number in question. At present, one can roughly distinguish between "small" Fermat numbers F_m, with $m \leq 24$, and "large" ones, with $m > 24$, allowing for a certain overlap between the two. I begin with the latter case, where considerably more factors are being found.

2.1. Large Fermat numbers and sieving

Of fundamental importance in this connection is the following well-known result of Euler (1747) which was given in a slightly stronger form by Lucas (1879):

Theorem 2.1 (Euler and Lucas). *If a prime p divides F_m, then $p = k \cdot 2^{m+2} + 1$ for some integer k.*

For a discussion and a proof of this result, see [69], p. 38, 101 f. Partly because of this connection to Fermat numbers, the question of primality of numbers of the form $k \cdot 2^n + 1$ has attracted a great deal of attention. The most important result in this connection is Proth's theorem of 1878. In what follows, (a/N) denotes the Jacobi symbol.

Theorem 2.2 (Proth). *Let $N = k \cdot 2^n + 1$, where k is odd and $2^n > k$, and suppose that $(a/N) = -1$ for some integer a. Then N is a prime if and only if*

$$a^{(N-1)/2} \equiv -1 \pmod{N}.$$

For a proof and a discussion of Proth's work, see [69], p. 122. In practice, one can dispose of the Jacobi symbol condition and use a slightly weaker version with the conclusion "If $a^{(N-1)/2} \equiv -1 \pmod{N}$ for some integer a, then N is prime."

Once such a number $k \cdot 2^n + 1$ has been identified as a prime p, one uses the quadratic iteration

$$f_{m+1} = f_m^2, \quad \text{where} \quad f_m = F_m - 1, \tag{2.2}$$

which follows immediately from (2.1). Starting with $f_0 = 2$, reduce modulo $p = k \cdot 2^n + 1$ after each squaring. By Theorem 2.1 it is enough to compute the $f_m \pmod{p}$ for $m \leq n-2$. Then p is a factor of F_m when $f_m \equiv -1 \pmod{p}$. For computational details see, e.g., [70].

Although Proth's theorem is indeed used in practice, it was observed that a test for divisibility of F_m by $N = k \cdot 2^n + 1$ is essentially the same as a test for primality of N. Indeed, Suyama proved the following result:

Theorem 2.3 (Suyama). *If $N = k \cdot 2^n + 1$ and $N \mid F_m$ for some m, then N is a prime if $k \cdot 2^{n-m-2} < 9 \cdot 2^{m+2} + 6$.*

A proof can be found in [69], p. 322. At the time of writing, the most recent factors of Fermat numbers were found based on this approach; a program and further information on a coordinated search can be found in [28]. Similarly, a program based on Proth's theorem is available at [11].

2.2. Small Fermat numbers and primality testing

All Fermat numbers, including those with $m \leq 24$, have been subject to the sieving process described above. However, the "smaller" numbers are also amenable to various methods of factoring and primality testing. In the case of those Fermat numbers for which no factors have been found by sieving, it is of primary interest to determine whether they are prime or (more likely) composite. The main tool for this purpose is the following theorem of Pepin (1877):

Theorem 2.4 (Pepin). *Let $m \geq 2$ and $k \geq 2$. Then the following are equivalent:*
(i) *F_m is prime and $(k/F_m) = -1$;*
(ii) *$k^{(F_m-1)/2} \equiv -1 \pmod{F_m}$.*

A proof can be found, e.g., in [56]. It is easy to see, using the reciprocity law for Jacobi symbols, that possible choices for k include $k = 3, 5$, or 10. The most usual choice in recent times has been $k = 3$. Pepin's test then involves repeated squaring (in fact, $2^m - 1$ times), starting with 3 and reducing modulo F_m after each step. Then F_m is prime if and only if the highest (the 2^{2^m}-th) binary bit position for the residue is 1.

The arithmetic for the latest, very large-scale, computations on F_{20}, F_{22}, and F_{24} was done using fast Fourier transforms (FFT) and appropriate variants, such as the discrete weighted transform (DWT) of Crandall and Fagin [20]. For a brief discussion of such methods, see [69], p. 326 ff.; more practical aspects and implementations are discussed in [22] and in [16]. A popular exposition can be found in [17].

The Fermat number F_{20} was shown to be composite by Young and Buell [71]. The calculations for F_{22} were done by Crandall, Doenias, Norrie, and Young [19] and independently by Trevisan and Carvalho [65]. Finally, the result of the proof of compositeness of F_{24} by Crandall, Mayer, and Papadopoulos was reported in [21]. This last paper also contains detailed information on the computational and algorithmic issues involved in such massive computations, as well as some interesting "cultural history" of these computations.

2.3. Small Fermat numbers and factoring

While new factors of large Fermat numbers are found on a regular basis by sieving (see [39]), factors of smaller Fermat numbers are found considerably less frequently, by "genuine" factorization methods. Given the size of even the smaller Fermat numbers, the method of choice in recent years has been the elliptic curve method (ECM) which was invented by H.W. Lenstra, Jr., and first published in 1987; for relevant references, see, e.g., [10].

For several years, a special piece of hardware, the "Dubner Cruncher", provided a very efficient way of factoring small Fermat numbers (up to about F_{15}) on a PC, and indeed, a number of new factors were found this way. However, for F_m with about $m \geq 15$ it was found that it is more efficient to use fast DWT-based multiplication algorithms on standard workstations or PCs. For a detailed discussion of these issues, see [10] or [16]. The relevant software by Crandall can be obtained through `http://www.perfsci.com/`. The last three factors, all found with this software, were a 27-digit factor of F_{16} (December, 1996), a 33-digit factor of F_{15} (July, 1997), and a 23-digit factor of F_{18} (April, 1999). For details, see [10].

3. Generalized Fermat numbers

While the Fermat numbers F_m have long been of great interest to mathematicians, the generalized Fermat numbers

$$F_m(a,b) = a^{2^m} + b^{2^m}, \quad \gcd(a,b) = 1,$$

and the more special case $b = 1$, were not seriously studied until the 1960s; see [9], [27], [57], [58], [60], and most recently [26].

Computational results were published in [27], [60], and [9] (and in some earlier papers), and some theoretical work appeared in [58] and [27]. In particular, Riesel [58] explains a connection between factors of ordinary Fermat numbers and generalized Fermat numbers $F_m(a,1)$ with even bases a, and Dubner and Keller [27] observe and prove that a given prime p of the form $p = k \cdot 2^n + 1$ divides a surprisingly large number of generalized Fermat numbers $F_m(a,1)$. To be more exact, they showed that given such a prime p, close to $1/k$ of all bases a, $1 < a < p$, are such that p divides $F_m(a,1)$ for some m.

If a is odd, then $F_m(a,1)$ is clearly even. But one may ask whether $F_m(a,1)/2$ can be a prime. While most computational work until the most recent project by Björn and Riesel [60], [9] was done primarily for even bases a, Dubner and Keller [27] observed that of the 26 primes of the form $3 \cdot 2^n + 1$ that were known at that time, 19 divide some $F_m(3,1)$, and that 8 of 18 primes of the form $5 \cdot 2^n + 1$ divide some $F_m(5,1)$. In this connection, Dubner and Keller suggested that some attention be given to odd-based generalized Fermat numbers.

Returning to the special case of $F_m(3,1)$, one can observe that all primes of the form $3 \cdot 2^n + 1$ with $3 \mid n$ divide some $F_m(3,1)$ with $m < n$, and that all those primes of the form $3 \cdot 2^n + 1$ that divide an ordinary Fermat number, also divide an $F_m(3,1)$ with $m < n$. These two cases cover the 19 factors mentioned in [27], as well as all the relevant primes discovered since then. Similar observations can be made for prime factors of the form $5 \cdot 2^n + 1$, and indeed for any prime $k \cdot 2^n + 1$ with odd k.

3.1. Results

The observations above are explained by the results in this subsection; they were recently published by I. Jiménez Calvo [38].

Theorem 3.1 (Jiménez Calvo). *Let $p = k \cdot 2^n + 1$ be a prime, where k is odd and $n = n'2^l$, with $n' \geq 3$ odd. If p divides the Fermat number $F_m = 2^{2^m} + 1$, then it also divides the generalized Fermat number*

$$F_{m-l}(k,1) = k^{2^{m-l}} + 1.$$

To put the next result of Jiménez Calvo into perspective, we quote from [9] the following generalization of the Euler-Lucas theorem (Theorem 2.1):

Theorem 3.2 (Björn and Riesel). *Let $p = k \cdot 2^n + 1$ be a prime, where k is odd. Suppose that $p \mid F_m(a,b)$ and $u \equiv a/b \pmod{p}$ is a 2^t-th power residue but not a 2^{t+1}-th power residue $\pmod{p}$. Then $n = m + t + 1$.*

For a proof, see [9]. A partial converse is given by the following

Theorem 3.3 (Jiménez Calvo). *Let $p = k \cdot 2^n + 1$ be a prime, with k odd. Let $v := \operatorname{ord}_2(n)$, and u be such that 2 is a 2^u-th power residue but not a 2^{u+1}-th power residue $\bmod p$. Furthermore, suppose that n, k have a common divisor $d \geq 3$, and if $k' := k/d$, then 2 is a k'-th power residue $\bmod p$.*

Then $p \mid k^{2^m} + 1$, with $m = n - u - v - 1$.

As an immediate consequence (in the case $k' = 1$) we get the following result. By the supplementary laws of quadratic reciprocity we always have $u \geq 1$.

Corollary 3.4. *Let* $p = k \cdot 2^n + 1$ *be a prime, with* k *odd and* $k \,|\, n$. *Then*

$$p \,|\, k^{2^m} + 1 \quad \textit{for some} \quad m \le n - 2.$$

We illustrate these results with a few examples.

Example 1. The first 36 primes of the form $p = 3 \cdot 2^n + 1$ are given by $n =$ 1, 2, 5, 6, 8, 12, 18, 30, 36, 41, 66, 189, 201, 209, 276, 353, 408, 438, 534, 2208, 2816, 3168, 3189, 3912, 20909, 34350, 42294, 42665, 44685, 48150, 55182, 59973, 80190, 157169, 213321, and 303093. (For another one, see Example 2).

All but the ones with $n = 1, 2, 5, 8, 353, 2816, 20909, 42665$ divide a generalized Fermat number $3^{2^m} + 1$.

- In the cases $n = 41, 209, 157169, 213321, 303093$: p divides a Fermat number (Theorem 3.1).
- In all other cases: $3 \,|\, n$ (Corollary 3.4).

I am grateful to W. Keller for pointing out that for $n = 213321$ and 303093 both of the above conditions are satisfied. (This is also true for the prime below).

Example 2. $p = 3 \cdot 2^{382449} + 1$ is to date the largest known prime factor of any Fermat number (namely of F_{382447}; see Section 2). Then by Theorem 3.1 we have

$$p \,|\, 3^{2^{382447}} + 1.$$

Example 3. Consider the prime $p = 9 \cdot 2^{33} + 1$. Hence $k' = d = 3$ in the notation of Theorem 3.3. To decide whether 2 is a cubic residue, we use the following criterion due to Jacobi (see, e.g., [46]):

Write $4p = L^2 + 27M^2$, $L \equiv 1 \pmod 3$; *this is a unique representation in positive integers* L, M. *If* L *and* M *are both even, then* 2 *is a cubic residue modulo* p.

In our specific example, we have $L = 550\,078$ and $M = 15\,696$. Hence 2 is indeed a cubic residue, so by Theorem 3.3, p divides a generalized Fermat number $9^{2^m} + 1$, with $m \le 31$. (In fact, $m = 28$).

3.2. A divisibility result for $F_m(a, b)$

Among numerous factors of $F_m(a, b)$ found by Björn and Riesel [9] there is, e.g.,

$$5 \cdot 2^{127} + 1 \,|\, 5^{2^{118}} + 2^{2^{118}}.$$

The following unpublished result of the author will explain this factor.

Theorem 3.5. *Suppose that* $N = k \cdot 2^n + 1 \,|\, F_m$. *Write* $n = 2^s - r$, $0 \le r \le 2^{s-1} - 1$. *Then* N *also divides*

$$F_{m-s}(k, 2^r) = k^{2^{m-s}} + (2^r)^{2^{m-s}}.$$

Example 4. Let $k = 5$, $n = 127$. It is known that $5 \cdot 2^{127} + 1 \,|\, F_{125}$; see [39]. Theorem 3.5, with $s = 7$, $r = 1$, then explains the factor given above.

Example 5. Let $k = 3$, $n = 382449$ (see Example 2). Then by Theorem 3.5,

$$p = 3 \cdot 2^{382449} + 1 \,|\, 3^{2^{382428}} + (2^{141839})^{2^{382428}}.$$

4. The search for Wieferich primes

One of the easiest and most pleasing criteria in the classical theory of Fermat's last theorem is the following celebrated result of Wieferich [67] of 1909:

Theorem 4.1 (Wieferich). *If the first case of Fermat's last theorem is false for the exponent p, i.e., if there exist integers x, y, z with $x^p + y^p + z^p = 0$, p an odd prime, and $p \nmid xyz$, then*

$$2^{p-1} \equiv 1 \pmod{p^2}. \tag{4.1}$$

By Fermat's "little" theorem, the congruence (4.1) always holds modulo p. This means that the *Fermat quotient* defined by

$$q_p(a) := \frac{a^{p-1} - 1}{p}, \quad p \nmid a, \tag{4.2}$$

is always an integer, and Wieferich's theorem can be reformulated as follows:

If the first case of Fermat's last theorem is false for the exponent p, then $q_p(2)$ vanishes modulo p.

Primes for which $q_p(2) \equiv 0 \pmod{p}$ are called *Wieferich primes.* Because of their connection to Fermat's last theorem, Fermat quotients and Wieferich primes have been studied quite intensively; see, e.g., [56] and the references provided there. Even before Wieferich's theorem, Lerch [47] published an important paper in which he referred to earlier results by other mathematicians. More recently, Wieferich primes have occurred in the theory of primality proving; see [13], p. 451.

In this section I will concentrate on recent computations, in particular on the search for Wieferich primes; later in the section I will mention some related topics. I begin with a brief summary of the main results in the search for Wieferich primes:

- $p = 1093$ is a Wieferich prime (Meissner [50], 1913)
- $p = 3511$ is a Wieferich prime (Beeger [7], 1922)
- No others $\leq 6 \cdot 10^9$ (D.H. Lehmer [44], 1981)
- No others $\leq 4 \cdot 10^{12}$ (Crandall et al. [18], 1997)
- No others $\leq 4.6 \cdot 10^{13}$ (Brown and McIntosh [49], 2001).

4.1. Recent computations

In this subsection I will briefly mention some of the computational enhancements used to facilitate the recent large-scale searches for Wieferich primes; for a detailed discussion, see [18] or [16], p. 97 ff.

The power of 2 $(\bmod\, p^2)$ in (4.1) is fairly easily computed by way of a "binary powering ladder" (see, e.g., [43], p. 441 ff.) which involves the binary expansion of $p - 1$. Furthermore, the following refinements have been used to speed up the process:

1. It is easy to see that if p is a Wieferich prime then

$$2^{(p-1)/2} \equiv \pm 1 \pmod{p^2}.$$

This reduces the size of the exponent.

2. Since all arithmetic is done modulo p^2, some intermediate results may be as large as p^4. If p is large, the standard precision multiplication in most computing machinery will not be able to handle this. In this case one can use the following base-p representation to split the multiplication into smaller ones:

If $\{a, b\}$ is the representation of $x \pmod{p^2}$, namely $x = a + bp$, $0 \le a, b \le p-1$, then

$$x^2 \equiv a^2 + 2abp \pmod{p^2},$$

and consequently x^2 has representation

$$\left\{a^2 \pmod{p},\ 2ab + \left\lfloor \frac{a^2}{p} \right\rfloor \pmod{p}\right\}.$$

As reported in [18], on some machines this base-p scheme is about twice as fast as the straightforward scheme.

3. In cases where many divisions are carried out with the same denominator (such as p or p^2), a "steady state division" leads to considerable savings. Brief discussions of this can be found in [18] or [16], p. 9 f.; see also [20], p. 320. This scheme allows for a divide operation in roughly 3/2 of the time of a single multiply.

4.2. Related matters

A. *Wieferich's theorem generalized.* Soon after Wieferich published his famous criterion, other mathematicians proved analogous criteria for bases other than 2. The first such criterion, for base 3, is due to Mirimanoff (1910), and the analogous result for base 5 was obtained by Vandiver in 1914. For details and further advances, see [54]. The first real progress in decades was made in [35] (1988):

Theorem 4.2 (Granville, Monagan). *If the first case of Fermat's last theorem is false for the exponent p, then*

$$q_p(q) \equiv 0 \pmod{p}$$

for all primes $q \le 89$.

This was extended by Suzuki [63] to $q \le 113$. Because of these results (and for other reasons—see below), it was of interest to compute generalized Wieferich primes to an integer base $a \ge 2$, i.e., primes which satisfy $q_p(a) \equiv 0 \pmod{p}$. For example, 11 and 1006003 are Wieferich primes to base 3. The results of recent computations are reported in [53], [29], [41], and [40]; for shorter tables, see [56], p. 347, or [55], p. 235 ff.

B. *Sums of reciprocals.* Let p be an odd prime, and for $1 \le N \le p-1$, $0 \le k \le N-1$, define

$$s(k, N) \equiv \sum_{j=\lfloor \frac{kp}{N} \rfloor + 1}^{\lfloor \frac{(k+1)p}{N} \rfloor} \frac{1}{j} \pmod{p}. \tag{4.3}$$

Lerch [47] proved the following connection between these sums of reciprocals and Fermat quotients:

Theorem 4.3 (Lerch). *For any integer $N \geq 1$,*

$$Nq_p(N) \equiv \sum_{k=0}^{N-1} ks(k,N) \pmod{p}. \tag{4.4}$$

By Theorem 4.2 and Suzuki's result, as well as by the "logarithmic property" $q_p(ab) \equiv q_p(a) + q_p(b) \pmod{p}$ for integers a, b we see that if the first case of Fermat's last theorem is false for the exponent p, then the left-hand side of (4.4) vanishes modulo p for all integers N whose prime divisors are at most 113. In view of congruence (4.4) it is surprising that given this hypothetical assumption, each of the summands on the right-hand side of (4.4) also vanishes modulo p, for $N \leq 46$. In fact, we have the following result [25]:

Theorem 4.4 (Dilcher, Skula). *If the first case of Fermat's last theorem is false for the exponent p, then*

$$s(k,N) \equiv 0 \pmod{p}$$

for all $1 \leq N \leq 46$ and $0 \leq k \leq N-1$.

Even though Fermat's last theorem has been proven, results such as Theorems 4.2 and 4.4 are of independent interest, for instance because of their connections with "Kummer systems of congruences". Also, it is conceivable that a proof of the first case of Fermat's last theorem may be found by means of classical methods going back to Kummer.

C. *Connections with Fibonacci numbers.* Let U_n denote the n-th Fibonacci number ($U_0 = 0$, $U_1 = 1$, $U_{n+1} = U_n + U_{n-1}$); let $(\frac{a}{p})$ be the Legendre symbol. It is a well-known fact that for any prime p,

$$U_{p-(\frac{5}{p})} \equiv 0 \pmod{p}.$$

In [23], p. 393, this result is attributed to Lagrange (1775) and Gauss [30], Art. 123 (1801). In [66], Wall raises the question whether there exist primes such that $U_{p-(5/p)} \equiv 0 \pmod{p^2}$. The following result, apparently due to Sun and Sun [62], establishes a connection to the sums of reciprocals defined above, and thus also to Fermat's last theorem.

Theorem 4.5 (Sun and Sun). *For primes $p > 5$ we have*

$$\frac{2}{5}\, s(1,5) \equiv \frac{1}{p}\, U_{p-(\frac{5}{p})} \pmod{p}.$$

A result equivalent to this was obtained earlier by Williams [68], and another one of a similar nature was obtained earlier still by Andrews [4]. As an immediate consequence we see with Vandiver's result mentioned above (or with Theorem 4.4) that if the first case of Fermat's last theorem is false for the exponent $p > 5$, then

$$U_{p-(\frac{5}{p})} \equiv 0 \pmod{p^2}. \tag{4.5}$$

It is therefore of interest to find those primes, called *Wall–Sun–Sun primes* in [16], that satisfy the congruence (4.5). So far, however, no such primes have been found. The following is a summary of some major search efforts.

- $p < 10^9$ (Williams [68], 1982).
- $p < 2^{32}$ (Montgomery [53], 1993).
- $p < 1.6 \cdot 10^{13}$ (McIntosh [49], 2001).

For some remarks on the computational issues involved, see also [16], p. 110 ff.

D. *Catalan's conjecture* states that the only integer solution of the equation

$$x^M - y^N = 1 \quad (xy \neq 0, M > 1, N > 1) \tag{4.6}$$

is $3^2 - 2^3 = 1$. After this was settled by Tijdeman [64] in 1976 except for possibly finitely many cases, a great deal of further progress was made very recently. Conditions involving Fermat quotients have been know for some time; see, e.g., [55]. All these also involved various class number conditions. However, Mihăilescu [52] proved the following result.

Theorem 4.6 (Mihăilescu). *If Catalan's equation $x^p - y^q = 1$ (p, q odd primes) has a nontrivial solution, then*

$$q^{p-1} \equiv 1 \pmod{p^2} \quad \text{and} \quad p^{q-1} \equiv 1 \pmod{q^2}. \tag{4.7}$$

Six pairs (p, q) are known for which both congruences in (4.7) hold: $(83, 4871)$, $(2903, 18787)$, $(911, 318917)$, $(3, 1006003)$, $(5, 1645333507)$, $(5, 188748146801)$; for these cases, different criteria must be used. These "double Wieferich pairs" were discovered by various researchers, from Aaltonen in the late 1980s to Keller in the 1990s. There are no further such pairs in the range $p < 10^6$, $q < 10^{10}$; see [41]. More details and related results can be found in [51]; for computations concerning the above pairs, see [40], [41], and [29]. Although Mihăilescu's result has drastically increased the speed of computations, the remaining gap is still too large to be filled by computations at the present time. The current state of computations (as of March, 2001) is as follows: If the equation (4.6) has a solution for odd prime exponents p and q, then

- $p, q > 39 \times 2^{23} \simeq 3.27 \times 10^8$ (Grantham and Wheeler [33], 2001);
- $\max\{p, q\} < 7.78 \times 10^{16}$ and $\min\{p, q\} < 7.15 \times 10^{11}$ (Mignotte [51], 2001).

5. Fermat quotients for composite moduli

As already mentioned in Section 4, the Fermat quotients defined by (4.2) were studied quite intensively by Lerch [47], as well as by researchers before and after. In his paper [47], and then later in greater generality in [48], Lerch defined the quotient

$$q(a, m) = \frac{a^{\phi(m)} - 1}{m}, \quad (a, m) = 1,\ m \geq 2.$$

By Euler's generalization of Fermat's little theorem, these quotients are integers. They were also studied by Baker in [5], but apart from Lerch's and Baker's work, the first detailed study was done in [2], where the $q(a,m)$ were called "generalized Fermat quotients" (or "Euler quotients").

In the papers mentioned above it was shown that almost all properties of the Fermat quotients (4.2) carry over to the generalized Fermat quotients. I will mention only the "logarithmic property" (see the remark following Theorem 4.3) which here takes the form $q(ab,m) \equiv q(a,m) + q(b,m) \pmod{m}$, and the following representation in terms of Bernoulli numbers and polynomials, which was proved in [2]: If $a \geq 1$ and $m \geq 2$ are relatively prime integers, then

$$aq(a,m) = -\frac{a^{\phi(m)}}{mB_{\phi(m)}} \sum_{j=1}^{a-1} \Big(B_{\phi(m)}\left(\frac{j}{a}\right) - B_{\phi(m)} \Big).$$

Further work on generalized Fermat quotients appeared in [1] and [61].

5.1. Wieferich numbers

In analogy to the Wieferich primes one can now ask, given a base a, for which integers m do we have $q(a,m) \equiv 0 \pmod{m}$.

An integer m with this property will be called a *Wieferich number* with base a. It turns out that these numbers can be completely characterized in terms of Wieferich *primes* with base a. For reasons of clarity I will quote only the special case for *odd* moduli m; the general case and the proof can be found in [2].

Theorem 5.1 (Agoh, Dilcher, Skula). *Let* $m = p_1^{\alpha_1} \cdots p_k^{\alpha_k}, p_j \neq 2$ $(j = 1, \dots, k)$, $(m,a) = 1$. *Then m is a base-a Wieferich number if and only if*

$$\alpha_j \leq \operatorname{ord}_{p_j} \prod_{i=1}^{k} (p_i - 1) + \operatorname{ord}_{p_j} q(a, p_j), \quad 1 \leq j \leq k.$$

Here $\operatorname{ord}_p(a)$ stands for the highest power of the prime p dividing the integer a. I conclude this brief section with an immediate consequence and two examples.

Corollary 5.2. *The largest factor of a Wieferich number with base a has to be a Wieferich prime with base a.*

Example 1. Let $a = 42$. The only known base-42 Wieferich prime up to $5 \cdot 10^8$ is $p = 23$; see [40]. This prime has the unusual property $\operatorname{ord}_p q(42, P) = 2$. Since $p - 1 = 2 \cdot 11$, $11 - 1 = 2 \cdot 5$, Theorem 5.1 says that the only Wieferich *numbers* less than $5 \cdot 10^8$ are 23, 23^2, $11 \cdot 23$, $11 \cdot 23^2$, $5 \cdot 11 \cdot 23$, $5 \cdot 11 \cdot 23^2$.

Example 2. We consider now the classical Wieferich case $a = 2$; see Section 4. Since $1093 - 1 = 2^2 \cdot 3 \cdot 7 \cdot 13$ and $3511 - 1 = 2 \cdot 3^3 \cdot 5 \cdot 13$, we expect considerably more base-2 Wieferich numbers. Indeed, there are 104 of them less than $4 \cdot 10^{13}$.

- Smallest one: 1093 (see Corollary);
- largest one: $3^6 \cdot 5 \cdot 7 \cdot 13^2 \cdot 1093 \cdot 3511$.

6. The search for Wilson primes

Closely related to the Fermat quotients are the *Wilson quotients* defined by

$$w_p = \frac{(p-1)!+1}{p}. \tag{6.1}$$

By the well-known theorem of Wilson, this is an integer. The Wilson quotients were studied in depth by Lerch [47] and E. Lehmer [45]; see also [54], p. 155. The following connection with the Fermat quotients was proved by Lerch [47]:

$$w_p \equiv \sum_{j=1}^{p-1} q_p(j) \pmod{p}, \tag{6.2}$$

and Beeger [6] proved the remarkable congruence

$$pw_p \equiv pB_{p-1} - p + 1 \pmod{p^2} \tag{6.3}$$

involving a Bernoulli number.

Again, it will be interesting to know for which primes p the Wilson quotient is divisible by p or, in other words, $w_p \equiv 0 \pmod{p}$. Such primes are called *Wilson primes*. Before I make some remarks on recent computations, here is a brief summary of what is known:

- $p = 5$, 13 are Wilson primes (obvious).
- No others < 113 (Beeger [6], 1913, using congruence (6.3)).
- $p = 563$ is a Wilson prime (Goldberg [31], 1953).
- No others $< 1.88 \cdot 10^7$ (Gonter and Kundert [32], 1994).
- No others $< 5 \cdot 10^8$ (Crandall et al. [18], 1997).

6.1. Recent computations

As I did in Section 4, I will now briefly discuss the more important computational enhancements used in the latest efforts to search for further Wilson primes. Details can be found in [18] and in [16], p. 102 ff.

1. Probably the most important enhancement, already used in earlier calculations, is a congruence sometimes known as "Morley's theorem"; for a proof see, e.g., [37], p. 105.

Theorem 6.1 (Morley). *If p is an odd prime, then*

$$(p-1)! \equiv \left(\left(\frac{p-1}{2}\right)!\right)^2 (1-2^p) \pmod{p^2}. \tag{6.4}$$

Other congruences of this type are known and can be found in [34] or [8], p. 280 ff. For example, suppose that $p \equiv 5 \pmod{12}$, and write $p = a^2 + b^2$, $a \equiv 1 \pmod{4}$. Then

$$(p-1)! \equiv \left(\left(\frac{p-1}{4}\right)!\right)^4 (3 \cdot 2^p - 4)(2a^2 - p) \pmod{p^2}. \tag{6.5}$$

A similar formula with $(\frac{p-1}{6})!$ holds for $p \equiv 1, 7 \pmod{12}$. Altogether, on average the size of the factorial is cut by a factor of $48/13$.

2. By a method called "factorial sieving" in [18] and [16], the number of multiplications required to compute the already reduced factorials can be further cut. This is best explained by way of an example (see [16], p. 105, or [18]): Suppose that $6 \mid n$. By factoring out appropriate terms we easily obtain

$$n! = \frac{2^{n/3}3^{n/6}(\frac{n}{2})!(\frac{n}{3})!}{(\frac{n}{6})!} \prod_{\substack{1 \le k < n \\ (k,6)=1}} k.$$

Thus $n!$ can be evaluated with $\sim 2n/3$ multiplications. By using numerous other cases, the number of multiplications in n! has been reduced to $\sim 283n/768$; however, it is not known how far this scheme can be taken.

3. After using the enhancements described above we will still be left with relatively large factorials and products of integers in arithmetic progressions. These can be computed by way of existing fast algorithms for polynomial evaluations. In practice, variants of an algorithm described in [43], p. 469, can be used to evaluate the polynomial

$$f(x) = x(x+d)(x+2d)\cdots(x+(n-1)d)$$

at appropriate integers x. For details and algorithms, see [18] or [16], p. 105 f.

6.2. Related results

A. The schemes mentioned above were optimized by Crandall who calculated, for $p = 2^{40} + 5^4 = 1\,099\,511\,628\,401$,

$$(p-1)! \equiv -1 - 533091778023\,p \pmod{p^2}.$$

This may well be the largest number proven prime by Lagrange's converse of Wilson's Theorem. For details, see again [18] or [16], p. 107.

B. Crandall in 1995 observed the "nested squares" identity

$$\left((x^2-85)^2-4176\right)^2-2880^2 = (x^2-1^2)(x^2-7^2)(x^2-11^2)(x^2-13^2).$$

(see [16], p. 109). Note that the product of eight integers on the right can be evaluated as a succession of three squares and three subtractions. The existence of longer identities of this kind would have profound effects on factorial evaluations. However, in [24] the following was shown:

Theorem 6.2 (Dilcher).
(a) *There are no such identities of length equal to or greater than* 4.
(b) *There are infinitely many such identities of length* 3, *the "smallest" one being*

$$\left((x^2-65)^2-1696\right)^2-1440^2 = (x^2-3^2)(x^2-7^2)(x^2-9^2)(x^2-11^2).$$

7. Wilson quotients for composite moduli

Just as in the case of Fermat quotients, we can extend the definition of Wilson quotients to composite moduli. The basis for this is the following generalization of Wilson's theorem due to Gauss [30], Art. 78; see also [23], p. 65.

Theorem 7.1 (Gauss). *Let $m \geq 2$ be an integer, and set $\epsilon_m = -1$ when $m = 2$, 4, p^α, or $2p^\alpha$, where p is an odd prime and α a positive integer, and $\epsilon_m = 1$ otherwise. Then*

$$\prod_{\substack{j=1\\(j,m)=1}}^{m} j \equiv \epsilon_m \pmod{m}.$$

This allows us to define the *generalized Wilson quotient* of m,

$$W(m) = \frac{P(m) - \epsilon_m}{m},$$

and we say that m is a *Wilson number* if $W(m) \equiv 0 \pmod{m}$. Although generalized Wilson quotients did appear in the literature before, the first detailed study was done in [3]. As in the case of the generalized Fermat quotients, most of the properties of the Wilson quotients extend quite naturally to their generalized analogues. For instance, the congruence

$$\epsilon_m \varphi(m) W(m) \equiv \sum_{\substack{1 \leq a \leq m\\(a,m)=1}} q(a,m) \pmod{m}$$

generalizes Lerch's congruence (6.2). However, in contrast to the generalized Fermat case, no simple characterization of Wilson numbers has been found.

7.1. Computational results

It turns out that there are classes of composite numbers for which $W(m)$ cannot be divisible by m. In particular, the search for Wilson numbers can essentially be restricted to squarefree integers m; see [3]. The following result also helps reduce the amount of computations necessary:

Theorem 7.2 (Agoh, Dilcher, Skula). *If $p \nmid m$, $m \geq 2$, $p \geq 3$, then*

$$-mW(pm) \equiv w_p \varphi(m) + \sum_{r^j \mid m} q_p(r) \varphi\Big(\frac{m}{r^j}\Big) \pmod{p}.$$

Thus, the search for Wilson numbers basically reduces to evaluating Wilson quotients and Fermat quotients, both of which have been discussed before.

It is worth mentioning the following curious consequence: If $p \equiv 1$ or $3 \pmod{8}$, then $4p$ is a Wilson number if and only if

$$w_p \equiv -q_p(2) \pmod{p}.$$

This connects two of the more basic objects discussed in this paper; it follows from Theorem 7.2 and another result in [3] not mentioned here.

Further work on generalized Wilson quotients can be found in [61]. I will close this section with a list of all known Wilson numbers, including the three Wilson primes, up to $5 \cdot 10^8$; this is the same search limit as that for the Wilson primes. The first and second composite Wieferich numbers below are due to Kloss [42] (1965) and Crandall [15] (1994), respectively.

$$
\begin{aligned}
&5,\ 13,\ 563\\
5\,971 &= 7 \cdot 853\\
558\,771 &= 3 \cdot 19 \cdot 9803\\
1\,964\,215 &= 5 \cdot 11 \cdot 71 \cdot 503\\
8\,121\,909 &= 3 \cdot 139 \cdot 19477\\
12\,326\,713 &= 7 \cdot 1760959\\
23\,025\,711 &= 3 \cdot 1867 \cdot 4111\\
26\,921\,605 &= 5 \cdot 67 \cdot 80363\\
341\,569\,806 &= 2 \cdot 3 \cdot 181 \cdot 409 \cdot 769\\
399\,292\,158 &= 2 \cdot 3 \cdot 17 \cdot 97 \cdot 40357
\end{aligned}
$$

Note added in proof. I was informed by Wilfrid Keller that on May 29, 2001, John Cosgrove found the new very large prime $3 \cdot 2^{916773} + 1$ and verified that this prime divides the generalized Fermat numbers $3^{2^{916771}} + 1$ (see Corollary 3.4) and $10^{2^{916772}} + 1$.

Acknowledgment. I wish to thank Wilfrid Keller, Tauno Metsänkylä, Jörg Richstein, and Gary Walsh for many helpful remarks and corrections concerning various topics in this paper.

References

[1] Agoh, T., Fermat and Euler type quotients. C. R. Math. Acad. Sci. Soc. R. Can. 17 (1995), 159–164.

[2] Agoh, T., Dilcher, K., Skula, L., Fermat quotients for composite moduli. J. Number Theory 66 (1997), 29–50.

[3] —, —, —, Wilson quotients for composite moduli. Math. Comp. 67 (1998), 843–861.

[4] Andrews, G.E., Some formulae for the Fibonacci sequence with generalizations. Fibonacci Quart. 7 (1969), 113–130.

[5] Baker, H.F., Remark on the Eisenstein-Sylvester extension of Fermat's theorem. Proc. London Math. Soc. (2) 4 (1906), 131–135.

[6] Beeger, N.G.W.H., Quelques remarques sur les congruences $r^{p-1} \equiv 1 \pmod{p^2}$ et $(p-1)! \equiv -1 \pmod{p^2}$. Messenger of Math. 43 (1913–1914), 72–84.

[7] —, On a new case of the congruence $2^{p-1} \equiv 1 (p^2)$. Messenger of Math. 51 (1922), 149–150.

[8] Berndt, B.C., Evans, R.J., Williams, K.S., Gauss and Jacobi Sums. Wiley, New York 1998.

[9] Björn, A., Riesel, H., Factors of generalized Fermat numbers. Math. Comp. 67 (1998), 441–446.

[10] Brent, R.P., Crandall, R.E., Dilcher, K., van Halewyn, C., Three new factors of Fermat numbers. Math. Comp. 69 (2000), 1297–1304.

[11] Caldwell, C.K., Proth.exe by Yves Gallot. A Windows program for finding very large primes. `http://www.utm.edu/research/primes/programs/gallot/`.

[12] Caldwell, C.K., Gallot, Y., On the primality of $n! \pm 1$ and $2 \times 3 \times 5 \times \cdots \times p \pm 1$. Math. Comp., posted on 11 May 2001, PII S 0025-5718(01)01315-1 (to appear in print).

[13] Cohen, H., A Course in Computational Algebraic Number Theory. Springer, New York 1993.

[14] Crandall, R.E., Projects in Scientific Computation. Springer, New York 1994.

[15] —, personal communication. 1994.

[16] —, Topics in Advanced Scientific Computation. Springer, New York 1996.

[17] —, The challenge of large numbers. Scientific American, February 1997, 74–78.

[18] Crandall, R.E., Dilcher, K., Pomerance, C., A search for Wieferich and Wilson primes. Math. Comp. 66 (1997), 433–449.

[19] Crandall, R.E., Doenias, J., Norrie, C., Young, J., The twenty-second Fermat number is composite. Math. Comp. 64 (1995), 863–868.

[20] Crandall, R.E., Fagin, B., Discrete weighted transforms and large-integer arithmetic. Math. Comp. 62 (1994), 305–324.

[21] Crandall, R.E., Mayer, E.W., Papadopoulos, J.S., The twenty-fourth Fermat number is composite. Math. Comp. (to appear).
`http://www.perfsci.com/free/techpapers/freepapers.html`.

[22] Crandall, R.E., Pomerance, C., Prime Numbers. A Computational Perspective. Springer, New York 2001.

[23] Dickson, L.E., History of the Theory of Numbers. Volume 1: Divisibility and Primality. Chelsea, New York 1952.

[24] Dilcher, K., Nested squares and evaluations of integer products. Experiment. Math. 9 (2000), 369–372.

[25] Dilcher, K., Skula, L., A new criterion for the first case of Fermat's last theorem. Math. Comp. 64 (1995), 363–392.

[26] Dubner, H., Gallot, Y., Distribution of generalized Fermat prime numbers. Math. Comp., posted on 17 May 2001, PII S 0025-5718(01)01350-3 (to appear in print).

[27] Dubner, H., Keller, W., Factors of generalized Fermat numbers. Math. Comp. 64 (1995), 397–405.

[28] Durman, L., Distributed search for Fermat number divisors.
`http://www.fermatsearch.org/`.

[29] Ernvall, R., Metsänkylä, T., On the p-divisibility of Fermat quotients. Math. Comp. 66 (1997), 1353–1365.

[30] Gauss, C.F., Disquisitiones arithmeticae. Translated and with a preface by Arthur A. Clarke. Springer, New York 1986.

[31] Goldberg, K., A table of Wilson quotients and the third Wilson prime. J. London Math. Soc. 28 (1953), 252–256.

[32] Gonter, R.H., Kundert, E.G., All prime numbers up to 18,876,041 have been tested without finding a new Wilson prime. Preprint, 1994.

[33] Grantham, J., personal communication. March 5, 2001.

[34] Granville, A., Arithmetic properties of binomial coefficients. I: Binomial coefficients modulo prime powers. In: Organic Mathematics (ed. by J. Borwein et al.; CMS Conf. Proc. 20), 253–276. Amer. Math. Soc., Providence 1997.

[35] Granville, A., Monagan, M.B., The first case of Fermat's last theorem is true for all prime exponents up to 714,591,416,091,389. Trans. Amer. Math. Soc. 306 (1988), 329–359.

[36] Guy, R., Unsolved Problems in Number Theory, 2nd ed. Springer, New York 1994.

[37] Hardy, G.H., Wright, E.M., An Introduction to the Theory of Numbers, fifth ed. Oxford University Press, 1979.

[38] Jiménez Calvo, I., A note on factors of generalized Fermat numbers. Appl. Math. Lett. 13.6 (2000), 1–5.

[39] Keller, W., Prime factors $k \cdot 2^n + 1$ of Fermat numbers F_m and complete factoring status. `http://www.prothsearch.net/fermat.html`.

[40] Keller, W., Richstein, J., Fermat quotients $q_p(a)$ that are divisible by p. `http://www.informatik.uni-giessen.de/cnth/FermatQuotient.html`.

[41] —, —, Solutions of the congruence $a^{p-1} \equiv 1 \pmod{p^r}$. Preprint, March 2001.

[42] Kloss, K.E., Some number-theoretic calculations. J. Res. Nat. Bur. Standards Sect. B 69 (1965), 335–339.

[43] Knuth, D.E., The Art of Computer Programming, vol. 2. Addison-Wesley, Reading 1973.

[44] Lehmer, D.H., On Fermat's quotient, base two. Math. Comp. 36 (1981), 289–290.

[45] Lehmer, E., On congruences involving Bernoulli numbers and the quotients of Fermat and Wilson. Ann. of Math. (2) 39 (1938), 350–360.

[46] —, Criteria for cubic and quartic residuacity. Mathematika 5 (1958), 20–29.

[47] Lerch, M., Zur Theorie des Fermatschen Quotienten $(a^{p-1} - 1)/p = q(a)$. Math. Ann. 60 (1905), 471–490.

[48] —, Sur les théorèmes de Sylvester concernant le quotient de Fermat. C. R. Acad. Sci. Paris 142 (1906), 35–38.

[49] McIntosh, R., personal communication. March 4, 2001.

[50] Meissner, W., Über die Teilbarkeit von $2^p - 2$ durch das Quadrat der Primzahl $p = 1093$. Sitzungsber. Akad. d. Wiss. Berlin (1913), 663–667.

[51] Mignotte, M., Catalan's equation just before 2000. In: Number Theory (Turku, 1999) (ed. by M. Jutila and T. Metsänkylä), 247–254. Walter de Gruyter, Berlin 2001.

[52] Mihăilescu, P., A class number free criterion for Catalan's conjecture. J. Number Theory (to appear).

[53] Montgomery, P., New solutions of $a^{p-1} \equiv 1 \pmod{p^2}$. Math. Comp. 61 (1993), 361–363.

[54] Ribenboim, P., 13 Lectures on Fermat's Last Theorem. Springer, New York 1979.

[55] —, Catalan's Conjecture. Academic Press, Boston 1994.

[56] —, The New Book of Prime Number Records. Springer, New York 1996. Additions and errata:
`http://www.utm.edu/research/primes/notes/errata/index.html`.

[57] Riesel, H., Some factors of the numbers $G_n = 6^{2n} + 1$ and $H_n = 10^{2n} + 1$. Math. Comp. 23 (1969), 413–415. Corrigenda: Math. Comp. 24 (1970), 243.

[58] —, Common prime factors of the numbers $A_n = a^{2^n} + 1$, Nordisk Tidskr. Informationsbehandling (BIT) 9 (1969), 264–269.

[59] —, Prime Numbers and Computer Methods for Factorization. Birkhäuser, Boston 1985.

[60] Riesel, H., Björn, A., Generalized Fermat numbers. In: Mathematics of Computation 1943–1993: a Half-Century of Computational Mathematics (Vancouver, 1993) (ed. by W. Gautschi; Proc. Sympos. Appl. Math. 48), 583–587. Amer. Math. Soc., Providence 1994.

[61] Skula, L., Fermat and Wilson quotients for p-adic integers. Acta Math. Inform. Univ. Ostraviensis 6 (1998), 167–181.

[62] Sun, Zhi Hong, Sun, Zhi Wei, Fibonacci numbers and Fermat's last theorem. Acta Arith. 60 (1992), 371–388.

[63] Suzuki, J., On the generalized Wieferich criteria. Proc. Japan Acad. Ser. A Math. Sci. 70 (1994), 230–234.

[64] Tijdeman, R., On the equation of Catalan. Acta Arith. 29 (1976), 197–209.

[65] Trevisan, V., Carvalho, J.B., The composite character of the twenty-second Fermat number. J. Supercomputing 9 (1995), 179–182.

[66] Wall, D.D., Fibonacci series modulo m. Amer. Math. Monthly 67 (1960), 525–532.

[67] Wieferich, A., Zum letzten Fermat'schen Theorem. J. Reine Angew. Math. 136 (1909), 293–302.

[68] Williams, H.C., A note on the Fibonacci quotient $F_{p-\varepsilon}/p$. Canad. Math. Bull. 25 (1982), 366–370.

[69] —, Édouard Lucas and Primality Testing. Wiley, New York 1998.

[70] Young, J., Large primes and Fermat factors. Math. Comp. 67 (1998), 1735–1738.

[71] Young, J., Buell, D., The twentieth Fermat number is composite. Math. Comp. 50 (1988), 261–263.

How to distinguish hyperelliptic curves in even characteristic

Andreas Enge

Abstract. We provide a complete algorithmic characterisation of hyperelliptic curves over perfect fields of characteristic 2. Given a quadratic equation, we show how to decide whether it is hyperelliptic and how to distinguish between imaginary, real and inert quadratic curves. If the curve is indeed hyperelliptic, we compute its genus. Furthermore, we show that any hyperelliptic function field may be represented by a smooth affine hyperelliptic curve. By developing normal forms for hyperelliptic curves we also provide a new normal form for non-supersingular elliptic curves.

1. Hyperelliptic curves and cryptography

Public key cryptosystems related to hyperelliptic curves over finite fields have been suggested as a possible replacement for elliptic curve cryptosystems with two aims in mind: First, the additional flexibility introduced might lead to more efficient systems (cf. [6]), second, variety is desired to be able to fall back upon a replacement in case elliptic curves turn out to be insecure. Hyperelliptic cryptosystems come in two flavours, both of which are based on some kind of discrete logarithm problem; either in the Jacobian or equivalently ideal class group of an imaginary quadratic curve ([8]) or in the infrastructure of a real quadratic curve ([11, 9]).

Hyperelliptic curves over fields of odd characteristic are treated in much detail in Artin's doctoral thesis (see [1, 2]); hyperelliptic curves in characteristic 2, however, have received much less attention. The cryptographic literature is usually concerned with some specific type of equation for hyperelliptic curves suited for the given application, and no attempt at an exhaustive classification has been made so far. In this paper, we provide a complete theoretical and algorithmic classification of hyperelliptic curves over perfect fields of characteristic 2, with special regards to finite fields. For any quadratic equation, we show how to determine whether the equation defines a hyperelliptic curve, and if it does, whether the curve is imaginary or real quadratic and what is its genus. Notice that Weil's theorem, which gives an estimate on the size of Jacobians over finite fields in terms of the field size and the curve genus, implies that the genus is important for the security of a hyperelliptic cryptosystem.

Furthermore, we provide normal forms of hyperelliptic curves, which, similarly to those of elliptic curves, may be used to obtain a faster arithmetic.

Public-Key Cryptography and Computational Number Theory
K. Alster, J. Urbanowicz and H. C. Williams (Eds.)

2. Definitions

Before examining hyperelliptic curves, we introduce the notion of a hyperelliptic function field, which varies slightly in the literature. Its most general definition is as follows.

Definition 2.1. A *hyperelliptic function field* F is an extension of degree 2 and genus $g \geq 1$ of a function field F_0 of genus zero.

There is an equivalent, but more involved definition via differentials, see [4]. In principle, F_0 need not be rational. Since a function field of genus zero is rational if and only if it has a divisor of degree 1, and this is always the case over finite fields ([12], § 8), which are of biggest interest in the cryptographic context, we assume from now on that F_0 is indeed rational. In geometric language, we consider only function fields corresponding to degree 2 covers of the projective line. For the more general case, i.e. for covers of twisted projective lines, see [10].

Whether the case $g = 1$ of *elliptic function fields* should be included into the definition is essentially a matter of taste. Usually, as the morpheme "hyper" indicates, one assumes $g \geq 2$. Since all results of this paper equally hold for elliptic function fields and curves, we simply include this case.

The genera of F and F_0 being distinct and the extension degree being prime, both function fields must have the same constant field, which we denote by K and assume to be perfect; the most interesting case for cryptography is that in which K is finite, but all theoretical results hold in general.

Of course, a hyperelliptic function field is the function field of some smooth projective curve, and such a curve is often called hyperelliptic. However, it need not be plane, and it is an interesting question to ask whether a hyperelliptic function field admits a plane smooth projective model. While this is in general not the case, it will turn out that it is always possible to represent a hyperelliptic function field by a plane smooth affine curve which is allowed to have singularities at infinity. This leads to the following definition.

Definition 2.2. Let K be a perfect field. By a *hyperelliptic curve* over K we understand a plane, smooth, absolutely irreducible affine curve $H \in K[X, Y]$ whose function field $K(H)$ is a hyperelliptic extension of $K(X)$. The *genus* of H is that of $K(H)$. An absolutely irreducible curve which is quadratic in Y is called *imaginary*, *real* resp. inert depending on whether the infinite prime divisor of $K(X)$ is ramified, split resp. inert in $K(H)$.

By a prime divisor we understand a discrete valuation of $K(H)$ resp. $K(X)$, and the infinite prime divisor of $K(X)$ is given by the negative degree valuation.

The requirement that the curve be absolutely irreducible is not a serious restriction. By its definition, a hyperelliptic curve is quadratic in Y; hence, it is either absolutely irreducible or splits as a product of two lines over K or over a quadratic extension K' of K. Often, inert quadratic curves are counted among the imaginary quadratic ones because in both cases there is a unique infinite prime divisor; more often yet they are not even taken into account at all because it can be argued that they arise only when considered over the "wrong" constant field. In fact, the same curve becomes real quadratic over K', see Corollary 3.4.

Natural questions are now how to determine, given a polynomial quadratic in Y, whether it defines a hyperelliptic curve, if so, whether it is real, imaginary or inert quadratic and which is its genus. These questions are answered in Section 3. Finally, we show in Section 4 that each hyperelliptic function field may be represented by a hyperelliptic curve in the sense of Definition 2.2, which can moreover be assumed to be in a certain normal form. Since all these problems are solved in odd characteristic, we suppose from now on that K is of characteristic 2.

3. Classification of hyperelliptic curves

Our aim in this section is to classify general quadratic equations in even characteristic into hyperelliptic curves and others and to distinguish the different types of hyperelliptic curves. Special equations have been treated in [8] and [14, 15]. Koblitz determines curves of genus g of the form $Y^2 + vY + u$ with $u, v \in K[X]$, u monic of degree $2g + 1$ and v of degree at most g as imaginary quadratic and of genus g. Zuccherato is interested in real quadratic curves of the special form $Y^2 + vY + u$ with $u, v \in K[X]$ and *both* monic. He gives the necessary condition for this polynomial to be real quadratic that $\deg v \geq 1$ and the necessary and sufficient condition that the equation has a root for Y in the field of *Puiseux expansions* $K((\frac{1}{X}))$. If it exists, such a root may be found algorithmically.

Example. In [14], p. 32, Zuccherato claims that

$$C = Y^2 + (X+1)Y + (X^6 + X^2 + X + 1)$$

is a real quadratic hyperelliptic curve over $\mathbb{F}_2$. Letting $y = \sum_{i=-\infty}^{N} c_i X^i$ with $c_i \in \mathbb{F}_2$ be a hypothetical root of C in $\mathbb{F}_2((\frac{1}{X}))$, plugging it into C and equating the coefficients of powers of X with zero, one finds $N = 3$ and the quadratic equations

$$\begin{aligned} c_3^2 + 1 &= 0 \\ c_2^2 + c_3 &= 0 \\ c_2 + c_3 &= 0 \\ c_1^2 + c_1 + c_2 + 1 &= 0 \\ c_0 + c_1 + 1 &= 0 \\ c_0^2 + c_{-1} + c_0 + 1 &= 0 \\ c_{-2i} + c_{-2i+1} &= 0 \quad \text{for } i \geq 1 \\ c_{-i}^2 + c_{-2i-1} + c_{-2i} &= 0 \quad \text{for } i \geq 1. \end{aligned}$$

The first equations have two distinct solutions for $(c_3, c_2, c_1, c_0, c_{-1})$, namely $(1, 1, 0, 1, 1)$ and $(1, 1, 1, 0, 1)$; the following equations are linear in the variables with smallest index they contain and can thus be solved recursively.

This example can be turned into a general polynomial time algorithm which *decides* whether a root exists or not.

Theorem 3.1. *Let $K = \mathbb{F}_{2^m}$ be finite. Given an equation of the form $C = Y^2 + vY + u$ with $u, v \in K[X]$, not necessarily monic, it can be decided in deterministic polynomial time whether it has a root in $K((\frac{1}{X}))$. If so, there is a deterministic, successively polynomial algorithm for computing the root, i.e., each further coefficient can be computed in polynomial time.*

Proof. If $v = 0$, then C has a root in $K((\frac{1}{X}))$ if and only if u is a square already in $K[X]$, which can be decided trivially.

Otherwise, if $y = \sum_{i=-\infty}^{N} c_i X^i$ with $c_i \in K$ is to be a root, then plugging it into C and equating the coefficients of X^i with zero implies that $N = \max\bigl(\deg v, \lfloor \frac{1}{2} \deg u \rfloor\bigr)$ and that the c_i satisfy an infinite system of quadratic equations.

Consider first the system of equations obtained from the coefficients of X^i with $i \geq 0$, which contains the variables $c_N, \ldots, c_{-\deg v}$. Notice that this system does not involve mixed terms $c_i c_j$ so that it can be solved in deterministic polynomial time. Rewrite $c_i = \sum_{j=1}^{m} c_{ij} e_j$ for a fixed basis $\{e_1, \ldots, e_m\}$ of $\mathbb{F}_{2^m}$ over $\mathbb{F}_2$ with new variables c_{ij} taking values in $\mathbb{F}_2$. Then each quadratic equation of the type

$$c_r^2 + \sum_i a_i c_i + b = 0$$

over $\mathbb{F}_{2^m}$ with $a_i = \sum_{j=1}^{m} a_{ij} e_j \in \mathbb{F}_{2^m}$ and $b = \sum_{j=1}^{m} b_j e_j \in \mathbb{F}_{2^m}$ can be written as

$$\sum_{j=1}^{m} c_{rj}^2 e_j^2 + \sum_i \sum_{j,k=1}^{m} a_{ij} c_{ik} e_j e_k + \sum_{j=1}^{m} b_j e_j = 0.$$

Taking into account that $c_{rj}^2 = c_{rj}$ since $c_{rj} \in \mathbb{F}_2$, expressing e_j^2 and $e_j e_k$ as linear combinations of the basis elements $e_1, \ldots, e_m$ and equating coefficients, we obtain a set of m linear equations over $\mathbb{F}_2$. Likewise, any linear equation over $\mathbb{F}_{2^m}$ translates into m linear equations over $\mathbb{F}_2$. Thus, we obtain a linear system over $\mathbb{F}_2$ with m times as many variables and equations as before, which can be solved in deterministic polynomial time.

If this finite system has no solution, then C has no root. Otherwise, for any $i < -\deg v$ there is a unique equation among the remaining ones in which all occurring variables c_j satisfy $j \geq i$. In this equation, c_i occurs linearly. Thus, the set of infinitely many equations has a solution, and c_i can be found in polynomial time once the c_j for $j > i$ are known. In fact, in the latter case the system has two distinct solutions with the same values of c_i for $i < 0$ since the two roots of C have to sum up to v. These two solutions correspond to the two prime divisors of $K(C)$ above infinity and the curve is indeed real quadratic. □

Unfortunately, the algorithm does not distinguish between imaginary and inert quadratic curves and does not allow to determine the genus of the curve; in fact, we show below that the curve of the example above is not hyperelliptic, but rational. We propose a different algorithm to answer these questions.

Let $C = wY^2 + vY + u$ with u, v, $w \in K[X]$ be a (not necessarily irreducible) polynomial. To check if C is singular, we may clearly assume that $\gcd(w, v, u) = 1$, since otherwise C would be trivially reducible. Assume that a point $(x, y) \in \overline{K} \times \overline{K}$ with $\overline{K}$ being the algebraic closure of K is singular on C, i.e. $0 = \frac{\partial C}{\partial Y}(x, y) = v(x)$, $0 = C(x, y) = w(x)y^2 + u(x)$ and $0 = \frac{\partial C}{\partial X}(x, y) = \frac{\partial w}{\partial X}(x)y^2 + \frac{\partial v}{\partial X}(x)y + \frac{\partial u}{\partial X}(x)$. If $w(x) = 0$, then $u(x) = 0$, and the minimal polynomial of x over K divides u, v and w, a contradiction. Thus to test for singularity, we may compute the roots of v over $\overline{K}$ which are not roots of w and check for the multiple roots x whether $\frac{\partial w}{\partial X}(x)\frac{u(x)}{w(x)} + \frac{\partial u}{\partial X}(x) = 0$ and for the simple roots x whether

$$\left(\frac{\frac{\partial w}{\partial X}(x)u(x)}{w(x)\frac{\partial v}{\partial X}(x)} + \frac{\frac{\partial u}{\partial X}(x)}{\frac{\partial v}{\partial X}(x)}\right)^2 = \frac{u(x)}{w(x)}.$$

Notice that in fact these conditions can be tested by computing a number of greatest common divisors over K.

Let now $F_0 = K(X)$ and denote by y the image of Y in the function field $F = K(C) = F_0[Y]/(C)$ of C, so that $F = F_0(y)$. Then $y' = \frac{wy}{v}$ is another generating element of F over F_0 and a zero of $C' = Y^2 + Y + u'$ with $u' = \frac{uw}{v^2} \in F_0[Y]$, which is in the Artin–Schreier form (cf. [3]). Clearly, the (absolute) irreducibility of C is equivalent with that of C'. Function field extensions of Artin–Schreier (and Kummer) type have been studied by Hasse in [7], and we follow his argumentation to derive the splitting behaviour of the infinite prime divisor of F_0 and the genus of F.

Let $p_1, \ldots, p_t$ be the irreducible factors of the denominator of u'. Determine the partial fraction decomposition

$$u' = \sum_{i=1}^{t} u_i + u_0 + r_0$$

with

$$u_i = \sum_{j=1}^{\mu_i} \frac{r_{ij}}{p_i^j}, \qquad r_{ij} \in K[X] \text{ of degree less than } \deg p_i \text{ for } i = 1, \ldots, t;$$

$$u_\infty = \sum_{j=1}^{\mu_\infty} r_{\infty j} X^j, \quad r_{\infty j} \in K;$$

$$r_0 \in K.$$

Notice that $t = 0$ is possible if u' is a polynomial and $\mu_\infty = 0$ if $\deg u' \leq 0$. This step involves the factorisation of the denominator of u' and several applications of the extended Euclidean algorithm.

If one of the μ_i is even for $1 \leq i \leq t$, then find a polynomial r of degree less than $\deg p_i$ such that $r^2 \equiv r_{i,\mu_i} \pmod{p_i}$, say $r^2 = qp_i + r_{i,\mu_i}$. This is possible since $K[X]/(p_i)$ is a finite extension of K and thus perfect. Replacing y' by the generator $y' + \frac{r}{p_i^{\mu_i/2}}$ of F/F_0, which is a zero of $C' + \left(\frac{r_{i,\mu_i}}{p_i^{\mu_i}} + \frac{q}{p_i^{\mu_i - 1}} + \frac{r}{p_i^{\mu_i/2}}\right)$, we reduce μ_i in the partial fraction decomposition. If μ_∞ is even, we find $r \in K$ such that $r^2 = r_{\infty,\mu_\infty}$ and replace y' by $y' + rX^{\mu_\infty/2}$. By induction, we obtain a polynomial $C'' = Y^2 + Y + u''$ with $u'' \in K[X]$ whose root y'' is a generator of F/F_0 and which is (absolutely) irreducible if and only if C is. Moreover, it is in the following special form.

Definition 3.2. A polynomial

$$C'' = Y^2 + Y + u'' \in K(X)[Y]$$

is in *special Artin–Schreier form* if all irreducible polynomials in the denominator of u'' occur to an odd multiplicity, and the degree of u'' is either positive and odd, or non-positive.

If u'' is non-constant, then C'' (and C) are absolutely irreducible. If it were reducible over $\overline{K}(X)$, it would have two roots y'' and $y'' + 1$ in $\overline{K}(X)$ and $u'' = (y'')^2 + y''$. Then, if ord is a discrete valuation of $\overline{K}(X)$ with $\operatorname{ord}(y'') < 0$, the strict triangle inequality yields that $\operatorname{ord}(u'') = 2\operatorname{ord}(y'')$ is even and negative, contradicting that C'' is in special Artin–Schreier form. Thus, $\operatorname{ord}(u'') \geq 0$ for all discrete valuations ord of $\overline{K}(X)$ and u'' is a constant, a contradiction.

On the other hand, if $u'' \in K$, then C'' has a root either in K itself or in its quadratic extension K', and C'' (and C) are not absolutely irreducible. This solves the question of irreducibility in an algorithmic way, and we assume from now on that u'' is not constant.

Let $\mathfrak{p}_i$ be the prime divisor of $K(X)$ with local parameter p_i and ∞ the infinite prime divisor with local parameter $\frac{1}{X}$. Then the pole divisor of u'' is

$$\operatorname{div}_\infty(u'') = \sum_{i=1}^{t} \mu_i \mathfrak{p}_i + \mu_\infty \infty.$$

Any prime divisor $\mathfrak{p}$ occurring in this sum is ramified. Any prime divisor $\mathfrak{p}$ not occurring in this sum is unramified, and it is split resp. inert depending on whether or not C'' has a root in the residue class field of $\mathfrak{p}$. Thus, ∞ is ramified and the curve is imaginary quadratic if and only if $\deg u'' > 0$, i.e. u'' is a polynomial. Otherwise, ∞ is unramified and the curve is real quadratic if and only if there is a rational function $z \in K(X)$ of non-positive degree (with the convention $\deg 0 = -\infty$) such that $\deg(z^2 + z + u'') < 0$. If $\deg u'' < 0$, then $z = 0$ is such a function. If $\deg u'' = 0$, a suitable element z exists if and only if the leading coefficient of u'' can be written as $z^2 + z$ with some $z \in K$; for $K = \mathbb{F}_{2^m}$ finite, this can be decided in deterministic polynomial time by solving a system of linear equations over $\mathbb{F}_2$ as in the proof of Theorem 3.1.

By [7], the genus of F and thus of C is given by

$$\frac{1}{2}\Big(\sum_{i=1}^{t}(\mu_i + 1)\deg p_i + \mu'_\infty\Big) - 1. \tag{1}$$

with $\mu'_\infty = \mu_\infty + 1$ for C imaginary and $\mu'_\infty = \mu_\infty = 0$ for C real or inert. Hence, we have shown the following result.

Theorem 3.3. *Given a polynomial $C \in K[X,Y]$ with K perfect of characteristic 2 which is quadratic in Y, there is an algorithm which decides whether C is non-singular and absolutely irreducible. If so, it determines the genus of C and decides whether C is a real, imaginary or inert hyperelliptic curve.*

The algorithm performs a polynomial number of field operations in K and of computations of square roots in extensions of K of polynomially bounded degree. Furthermore, it factors a polynomial over K and possibly solves a quadratic equation of the form $Y^2 + Y = \gamma$ with $\gamma \in K$. Over a finite field K, all these steps are of deterministic polynomial complexity except for the polynomial factorisation, which to our current knowledge is of only probabilistic polynomial complexity.

Example. Consider again the curve C with $w = 1$, $v = X+1$ and $u = X^6 + X^2 + X + 1 = X^2(X^4+1) + X + 1$ from above. Then

$$u' = \frac{uw}{v^2} = X^4 + X^2 + \frac{1}{X+1}$$

and

$$u'' = u' + (X^4 + X^2) = \frac{1}{X+1}.$$

Hence, the genus of the curve is

$$g = \frac{1}{2}(1+1)\cdot 1 - 1 = 0,$$

and the curve is not hyperelliptic, but rational. Indeed, if y'' is the image of Y in $K(C'') = K(X)[Y]/(C'')$, then $K(C) = K(C'') = K(X, y'') = K(y'')$ since

$$X = \frac{1}{(y'')^2 + y''} - 1.$$

The algorithm also proves the aforementioned result on inert quadratic curves.

Corollary 3.4. *An absolutely irreducible inert quadratic curve over K becomes real quadratic over a quadratic extension K' of K.*

4. Normal forms

In this section we show that any hyperelliptic function field is represented by some hyperelliptic curve in the sense of Definition 2.2 and that furthermore, hyperelliptic curves may be turned algorithmically into certain normal forms. We have already seen in the previous section how to transform a hyperelliptic curve into special Artin–Schreier form while preserving the function field. Conversely, since a hyperelliptic function field is a cyclic quadratic extension of genus at least 1 of the rational function field, it has a generating element whose minimal polynomial is absolutely irreducible and in Artin–Schreier form, which can be transformed into a special Artin–Schreier polynomial as discussed above. We show now that the absolutely irreducible special Artin–Schreier polynomials representing a hyperelliptic function field are in bijection with certain hyperelliptic curves.

Let Y^2+Y+w with $w=\frac{r}{s}\in K(X)\backslash K$, r, $s\in K[X]$ be such a polynomial. To turn it into a curve, we must get rid of the denominator s of w. We might replace a root y of the polynomial by the element ys and still generate the same function field. The minimal polynomial of ys is the plane curve $Y^2+sY+sr$; however, it may be singular since each multiple root x of s leads to the singularity $(x,0)$. To avoid these singularities, we split off the square part of the denominator of w and write $w=a\frac{r}{s^2t}$ with $a\in K^\times$, r, s, $t\in K[X]$ monic, $\gcd(r,st)=1$ and t square-free. Since K is perfect, a has a root in K. We replace s by $\frac{s}{\sqrt{a}}$ and assume $a=1$ while sacrificing the monicity of s. The generating element yst has the minimal polynomial

$$C=Y^2+stY+rt=Y^2+vY+u$$

with $u=rt$ monic and $v=st$.

Since C is in special Artin–Schreier form, the prime divisors in the denominator of w occur to an odd power and any irreducible polynomial dividing s also divides t, i.e., t is the square-free part of st. Hence, any irreducible divisor of v is a simple divisor of u. Using this property, it is easily seen that C has no affine singularity.

Concerning the ramification of the infinite prime divisor, consider first the case that w is of odd positive degree, i.e., $\deg r>2\deg s+\deg t$ and $\deg(rt)$ is odd, so that ∞ is ramified. For an irreducible polynomial $p\in K[X]$ denote by ord_p the discrete valuation attached to it. Then by (1), the genus of C is

$$\begin{aligned} g &= \frac{1}{2}\Big(\sum_{p\,|\,s^2t,\ p \text{ irreducible}} (\mathrm{ord}_p(s^2t)+1)\deg p+(\deg w+1)\Big)-1\\ &= \frac{1}{2}\big(\deg(s^2t)+\deg t+\deg r-\deg(s^2t)+1-2\big)=\frac{1}{2}\big(\deg(rt)-1\big)\\ &= \frac{1}{2}(\deg u-1). \end{aligned}$$

Thus, $\deg u=2g+1$ and $\deg v=\frac{1}{2}\deg(s^2t^2)\le\left\lfloor\frac{1}{2}\deg(rt)\right\rfloor=\left\lfloor\frac{1}{2}\deg u\right\rfloor=g$.

Now, let $\deg w\le 0$. Then ∞ is unramified, and the genus of C is

$$g=\frac{1}{2}\sum_{p\,|\,s^2t,\ p \text{ irreducible}}(\mathrm{ord}_p(s^2t)+1)\deg p-1=\deg(st)-1=\deg v-1.$$

Thus, $\deg v=g+1$ and $\deg u=\deg(rt)\le\deg(s^2t^2)=2g+2$ since $\deg w\le 0$. More precisely, ∞ splits if and only if $\deg w<0$, i.e. $\deg u<2g+2$; or $\deg w=0$, i.e. $\deg u=2g+2$, and the leading coefficient of w equals z^2+z for some $z\in K$. Denoting the leading coefficient of a polynomial p by $l(p)$, we have $l(w)=l(s)^{-2}=l(v)^{-2}$. Notice that $l(w)=z^2+z$ is equivalent with $l(v)^{-1}=(\sqrt{z})^2+\sqrt{z}$.

We have shown the following result.

Theorem 4.1. *Let K be perfect of characteristic 2. Consider absolutely irreducible curves $C = Y^2 + vY + u$ in the following normal forms: u, $v \in K[X]$, u monic and any irreducible polynomial dividing v is a simple divisor of u.*

- *If $g \geq 1$, $\deg u = 2g+1$ and $\deg v \leq g$, then C is an imaginary quadratic hyperelliptic curve of genus g.*
- *If $g \geq 1$, $\deg v = g+1$ and $\deg u < 2g+2$, or $\deg u = 2g+2$ and the inverse of the leading coefficient of v equals z^2+z for some element $z \in K$, then C is a real quadratic hyperelliptic curve of genus g.*
- *If $g \geq 1$, $\deg v = g+1$, $\deg u = 2g+2$ and the inverse of the leading coefficient of v cannot be written as z^2+z for any element $z \in K$, then C is an inert hyperelliptic curve of genus g.*

Moreover, given a hyperelliptic curve, it can be transformed into one of these normal forms with the same complexity as in Theorem 3.3.

Conversely, any hyperelliptic function field over K has an affine model of this type.

As different special Artin–Schreier polynomials lead to different curves in normal form, we have indeed constructed a bijection between them. We remark that the monicity of u can be traded for that of v by starting with a special Artin–Schreier polynomial with $w = \frac{r}{s^2 t}$ and s and t monic, but r in general not.

The property that any hyperelliptic function field in even characteristic has an affine model in which the irreducible factors of v are simple divisors of u has not been observed before. It is an interesting open question whether this property can be used to speed up the arithmetic of the Jacobian. Of special interest is the case of elliptic curves, which is obtained from Theorem 4.1 by letting $g=1$.

Corollary 4.2. *Over a perfect field K of characteristic 2, any imaginary quadratic non-supersingular elliptic curve may be effectively transformed into*

$$Y^2 + XY + (X^3 + a_2X^2 + a_4X)$$

with $a_2 \in K$ and $a_4 \in K^\times$. The complexity of the algorithm is the same as in Theorem 3.3.

Proof. By Theorem 4.1, we obtain either a curve $Y^2 + a_3Y + u$ with $a_3 \in K^\times$ and $u \in K[X]$ cubic, which is supersingular (cf. [13, 5]), or

$$E = Y^2 + \alpha(X+\beta)Y + (X+\beta)(X^2 + \gamma X + \delta)$$

with α, β, γ, $\delta \in K$, $\alpha \neq 0$ and $\beta^2 + \beta\gamma + \delta \neq 0$. Let y be the image of Y in $K(E)$, $y' = \frac{y}{\alpha^3}$ and $X' = \frac{X+\beta}{\alpha^2}$. Then $K(E) = K(X,y) = K(X',y') = K(E')$ with

$$\begin{aligned} E'(X',Y) &= \alpha^6\, E\big(\alpha^2X' + \beta, \alpha^3 y\big) \\ &= Y^2 + X'Y + \big(X'^3 + \gamma' X'^2 + \delta' X'\big), \\ \gamma' &= \alpha^{-2}\gamma, \\ \delta' &= \alpha^{-4}(\beta^2 + \beta\gamma + \delta) \neq 0. \end{aligned}$$

□

Acknowledgements. I thank Dirk Hachenberger for fruitful discussions concerning the proof of Theorem 3.1 and Steven Galbraith for helpful comments.

References

[1] Artin, E., Quadratische Körper im Gebiete der höheren Kongruenzen I. Math. Z. 19 (1924), 153–206.

[2] —, Quadratische Körper im Gebiete der höheren Kongruenzen II. Math. Z. 19 (1924), 207–246.

[3] Artin, E., Schreier, O., Eine Kennzeichnung der reell abgeschlossenen Körper. Abh. Math. Sem. Univ. Hamburg 5 (1927), 225–231.

[4] Chevalley, C., Introduction to the Theory of Algebraic Functions of one Variable. Amer. Math. Soc., New York 1951.

[5] Enge, A., Elliptic Curves and Their Applications to Cryptography — An Introduction. Kluwer Academic Publishers, Dordrecht 1999.

[6] —, The extended Euclidian algorithm on polynomials, and the computational efficiency of hyperelliptic cryptosystems. Des. Codes Cryptogr. 23 (2001), 53–74.

[7] Hasse, H., Theorie der relativ-zyklischen algebraischen Funktionenkörper, insbesondere bei endlichem Konstantenkörper. J. Reine Angew. Math. 172 (1935), 37–54.

[8] Koblitz, N., Hyperelliptic cryptosystems. J. Cryptology 1 (1989), 139–150.

[9] Müller, V., Vanstone, S., Zuccherato, R., Discrete logarithm based cryptosystems in quadratic function fields of characteristic 2. Des. Codes Cryptogr. 14 (1998), 159–178.

[10] Poonen, B., Computational aspects of curves of genus at least 2. In: Algorithmic Number Theory — ANTS II (ed. by H. Cohen; Lecture Notes in Comput. Sci. 1122), 283–306. Springer, Berlin 1996.

[11] Scheidler, R., Stein, A., Williams, H.C., Key-exchange in real quadratic congruence function fields. Des. Codes Cryptogr. 7 (1996), 153–174.

[12] Schmidt, F.K., Analytische Zahlentheorie in Körpern der Charakteristik p. Math. Z. 33 (1931), 1–32.

[13] Silverman, J.H., The Arithmetic of Elliptic Curves (Grad. Texts in Math. 106). Springer, New York 1986.

[14] Zuccherato, R., New Applications of Elliptic Curves and Function Fields in Cryptography. Ph.D. Thesis, University of Waterloo, 1997.

[15] —, The continued fraction algorithm and regulator for quadratic function fields of characteristic 2. J. Algebra 190 (1997), 563–587.

Limitations of constructive Weil descent

*Steven D. Galbraith**

Abstract. Weil restriction of scalars can be used to construct curves suitable for cryptography whose Jacobian has known group order. The description given by Hess, Seroussi and Smart shows how to construct hyperelliptic curves of genus two or three over finite fields of characteristic two. One drawback of their method is discussed, namely that only a small proportion of the set of all curves can be constructed in this way. We also indicate that the method cannot be generalised to overcome this drawback.

1991 Mathematics Subject Classification: 11G20, 14G15, 14G50.

1. Introduction

Jacobians of hyperelliptic curves (and also some more general higher genus curves) have been proposed for use in public key cryptography. However, an impediment to their use has been the problem of computing the number of points on the Jacobian variety of a curve. The currently available methods for finding a curve of genus $2 \leq g \leq 4$ over a finite field whose Jacobian has known order are:

- Using subfield curves (i.e., curves C defined over a small field $\mathbb{F}_q$ but considering their Jacobian as a group over some larger extension field $\mathbb{F}_{q^l}$).
- Using curves with known endomorphism structure.
- Choosing curves at random and counting points using baby-step-giant-step, Pollard or Schoof style methods.
- Using modular curves.
- Constructive Weil descent.

The first method was initially described by Koblitz [12] and has since been used in several other papers. As a means of computing the number of points on a Jacobian it is very effective.

The second method appears in several forms. The most advanced is the method of Spallek [14] (also see Weng [17]) which is an analogue of the CM method for elliptic curves. This method has only been developed over fields of odd characteristic. Nevertheless, it has been used to generate curves suitable for cryptography.

* This research was supported by the NRW-Initiative für Wissenschaft und Wirtschaft "Innovationscluster für Neue Medien" and cv cryptovision gmbh (Gelsenkirchen).

Public-Key Cryptography and Computational Number Theory
K. Alster, J. Urbanowicz and H. C. Williams (Eds.)

There are many other papers which give methods for counting points on certain curves which have a special endomorphism structure.

The main drawback of both of these methods is that the curves have special properties and are thus treated with some suspicion by the community. Admittedly the only known weakness of these curves is that their automorphism structures allow the use of equivalence classes to speed up the Pollard methods (see [2]).

The third method would allow the choice of truly random curves. In the elliptic curve case there are efficient methods for counting points and so this is the method which is used in most applications. Unfortunately, despite the best efforts of Gaudry and Harley [9], this method is not practical when the genus is two or more[1)].

Modular curves have a lot of structure which can be used to help count their points (see [4]). The algorithms for computing the coefficients of weight two modular forms have exponential complexity in terms of q so, in the case of genus less than or equal to four, modularity alone is not enough for efficient point counting. Gaudry (see Section 6.2 of [8]) has used the real multiplication structure of modular curves to accelerate point counting. Nevertheless, modular curves are special and so these methods cannot be used to produce random curves.

The fifth method, that of constructive Weil descent, seems to be an excellent solution to the problem. The seeds for this idea were sown by Frey [3]. The paper of Hess, Seroussi and Smart [11] gives a construction for genus two and three curves over fields of characteristic two and this is the only known method for generating curves over such fields (i.e., not subfield curves) with known group order. Furthermore, constructive Weil descent can produce $O(q^{g/2})$ curves of genus $g \in \{2, 3\}$ whereas the CM and subfield curve methods produce what is, in practice, essentially a constant number of curves.

Unfortunately, the curves which arise from constructive Weil descent do have special properties and are not 'random hyperelliptic curves' in the most general sense. The goal of this paper to give evidence that only a small proportion of all isogeny classes of Jacobians of hyperelliptic curves can be constructed using Weil descent.

Furthermore, we hope to convince the reader that any generalisation of constructive Weil descent to produce curves of any form (i.e., not necessarily hyperelliptic) will also only produce a small proportion of the available isogeny classes of curves. Indeed, our results suggest that most isogeny classes of Jacobians of curves do not arise as a sub-abelian-variety of the Weil restriction of an elliptic curve.

In conclusion, though constructive Weil descent is an interesting new technique, it cannot provide a general solution to the problem of constructing curves of genus $g \geq 2$ suitable for cryptography.

1) *Note added in proof*. New work of Kedlaya, Gaudry, Harley and Mestre means that there are now efficient algorithms for counting points on curves over finite fields of small characteristic.

2. Summary of constructive Weil descent

Frey [3] was the first to observe that Weil restriction of scalars could be used as a tool to find curves whose Jacobian has known order (for details about Weil descent see [3], [4], [7], [10]). From a practical point of view the decisive step is the paper of Hess, Seroussi and Smart [11], which we now outline.

In [11] the goal is to find a hyperelliptic curve C over a finite field $\mathbb{F}_q$ of genus $g \in \{2,3\}$ where $q = 2^l$. The method proceeds as follows:

- Choose non-supersingular elliptic curves E over $\mathbb{F}_{q^g}$ at random (though there are some restrictions on E in the case of $g = 3$) and count their points (using the Schoof–Elkies–Atkin algorithm or Satoh's algorithm) until one is found such that the number of points on E is a product of a small number and a large prime.
- Construct a curve C of genus g on the Weil restriction of scalars A of E with respect to the extension $\mathbb{F}_{q^g}/\mathbb{F}_q$ using the method of Gaudry, Hess and Smart [10].

One then knows the order of $\mathrm{Jac}(C)(\mathbb{F}_q)$ as it will be the same as the number of points on $E(\mathbb{F}_{q^g})$.

The main drawback of this method is that the Jacobians produced are isogenous to the Weil restriction of the elliptic curve E. In other words, they are essentially a different way of writing down the elliptic curve group $E(\mathbb{F}_{q^g})$ and their security is no more than the security of the original elliptic curve discrete logarithm problem.

3. The curves produced by constructive Weil descent

We now give several reasons why constructive Weil descent as above produces curves which are very special compared with a general hyperelliptic curve. These ideas are well known to researchers in algebraic geometry, but we feel it is worth explaining them to the wider cryptographic community.

First consider the number of isomorphism classes of curves produced by this method. We get only one isomorphism class of curves C from each isomorphism class of elliptic curves E, since the function field of C is determined up to isomorphism. The moduli space of hyperelliptic curves of genus g has dimension $2g-1$. It is difficult to derive precise estimates about the number of points on a space solely by considering its dimension. We estimate the number of hyperelliptic curves of genus g over a field $\mathbb{F}_q$ to be $\Omega(q^{2g-1})$. There are roughly $2q^n$ isomorphism classes of elliptic curves over $\mathbb{F}_{q^n}$. We therefore conclude that the proportion of isomorphism classes of genus two curves which can arise from constructive Weil descent is at most c/q for some constant c. For genus three curves we expect that at most c'/q^2 of all genus three curves can arise in this way for some constant c'.

The difficulty of the discrete logarithm problem seems to be more closely related to the isogeny class than the isomorphism class, therefore it makes more sense to estimate the proportion of isogeny classes which can be produced by constructive Weil descent. We study these questions via the characteristic polynomial of Frobenius.

Let A be an abelian variety of dimension g over a finite field $\mathbb{F}_q$. Associated to A is the characteristic polynomial $P_A(T)$ of the Frobenius endomorphism. Tate's isogeny theorem [15] implies that two abelian varieties A and B over a finite field $\mathbb{F}_q$ are isogenous over $\mathbb{F}_q$ if and only if the polynomials $P_A(T)$ and $P_B(T)$ are equal.

Our first step is to understand the abelian varieties which arise from taking Weil restriction of scalars.

Proposition 3.1. *Let $A/\mathbb{F}_{q^n}$ be an abelian variety with characteristic polynomial of Frobenius equal to $P_A(T)$. Let $B/\mathbb{F}_q$ be the Weil restriction of A with respect to the extension $\mathbb{F}_{q^n}/\mathbb{F}_q$. Then $P_B(T) = P_A(T^n)$.*

Proof. The Frobenius endomorphism on A is just the nth power of the Frobenius endomorphism on B. □

We observe the following easy corollary to the above result.

Corollary 3.2. *Let the notation be as above. If* $\gcd(m,n) = 1$ *then* $\#A(\mathbb{F}_{q^{mn}}) = \#B(\mathbb{F}_{q^m})$. *If $n|m$ then $\#B(\mathbb{F}_{q^m}) = (\#A(\mathbb{F}_{q^{m/n}}))^n$.*

Proof. Let $\zeta_n = \exp(2\pi\sqrt{-1}/n)$ and let $P_A(T) = \prod_{i=1}^{2g}(1-\alpha_i)$. Then $P_B(T) = \prod_i \prod_{j=1}^{n}(1-\zeta_n^j \alpha_i^{1/n})$ and $\#B(\mathbb{F}_{q^m}) = \prod_i \prod_j (1-\zeta_n^{mj}\alpha_i^{m/n})$. □

When $\gcd(m,n) = 1$ then it is the case that the groups $A(\mathbb{F}_{q^{nm}})$ and $B(\mathbb{F}_{q^m})$ are isomorphic.

We return to the curves produced by the method of [11]. The original elliptic curve E over $\mathbb{F}_{q^g}$ has $P_E(T) = T^2 + aT + q^n$ (with $|a| < 2\sqrt{q^g}$) and so, by Proposition 3.1, the abelian variety A has $P_A(T) = T^{2g} + aT^g + q^g$. If E is non-supersingular then a is coprime to q and so A is also non-supersingular (see [5]). In the construction of [11] we have $\mathrm{Jac}(C)$ isogenous to A and so it has the same polynomial $P(T)$. This proves the following result.

Proposition 3.3. *Let $g \in \{2,3\}$ and let q be a power of* 2. *Then constructive Weil descent as in* [11] *produces at most $2\sqrt{q^g}$ isogeny classes of Jacobian varieties of hyperelliptic curves of genus g over $\mathbb{F}_q$.*

A general abelian variety A of dimension n over $\mathbb{F}_q$ has characteristic polynomial of Frobenius $P_A(T)$ equal to

$$T^{2n} + a_1 T^{2n-1} + a_2 T^{2n-2} + \cdots + a_n T^n + q a_{n-1} T^{n-1} + \cdots + a_1 q^{n-1} T + q^n$$

where the coefficients $a_i \in \mathbb{Z}$ are constrained by various complicated conditions. There are $O(\binom{2g}{i} q^{i/2})$ possibilities for each a_i. Thus there are $O(q^{3/2})$ isogeny classes of abelian varieties of dimension two over $\mathbb{F}_q$ and $O(q^3)$ isogeny classes of abelian varieties of dimension three over $\mathbb{F}_q$.

For cryptography, groups whose order has a very large prime factor are the most important. Thus the abelian varieties we consider in this paper are simple, or in other words, the characteristic polynomial of Frobenius is irreducible. Of course, a simple abelian variety does not necessarily have an 'almost prime' order. In the analysis below we pay attention to simplicity but disregard the issue of whether the order has a large prime factor.

Rück [13] has shown, for abelian varieties of dimension two over prime fields $\mathbb{F}_p$, that 'most' are simple and that 'most' are isogenous over $\mathbb{F}_p$ to the Jacobian of a hyperelliptic curve $C/\mathbb{F}_p$. We expect that similar results hold for more general finite fields $\mathbb{F}_q$, but a proof of this is beyond the scope of the present work. Therefore we expect $O(q^{3/2})$ isogeny classes of simple Jacobians of genus two curves over $\mathbb{F}_q$. On the other hand, the constructive Weil descent method can only produce $O(q)$ such curves. This is an exponentially small proportion of all isogeny classes.

The majority of abelian varieties of dimension three are simple. In the case of dimension three we also expect that most abelian varieties over $\mathbb{F}_q$ are isogenous over $\mathbb{F}_q$ to the Jacobian of a curve. Hence we expect $O(q^3)$ isogeny classes of Jacobians of genus three curves over $\mathbb{F}_q$. However, a genus three curve can be a plane quartic or a hyperelliptic curve. It is not clear what is the proportion of isogeny classes of hyperelliptic genus three curves. In any case, constructive Weil descent only produces $O(q^{3/2})$ isogeny classes of hyperelliptic curves, and we expect this to be an exponentially small proportion of all isogeny classes of hyperelliptic curves.

4. Experimental evidence

We now give some data for small fields $\mathbb{F}_{2^n}$ (essentially obtained by a brute-force search). Here $N_1(q)$ is the number of isogeny classes of Jacobians of genus two curves over $\mathbb{F}_q$. $N_2(q)$ is the number of classes as in N_1 for which the Jacobian is non-supersingular and simple (the abelian variety is simple if and only if the characteristic polynomial of Frobenius is irreducible). $N_3(q)$ is the number of classes as in N_2 but with the further condition that there is a curve representing the class which can be written in the form with a single point at infinity. Finally, $N_4(q)$ is the number of classes as above which can be constructed from the Weil restriction of an elliptic curve over $\mathbb{F}_{q^2}$.

q	$N_1(q)$	$N_2(q)$	$N_3(q)$	$N_4(q)$
2	20	12	6	2
4	70	46	21	4
8	≥ 160	≥ 121	56	10
16	≥ 415	≥ 366	≥ 168	24

The most important columns for the present discussion are $N_3(q)$ and $N_4(q)$, since the method of [11] produces curves which satisfy the conditions imposed in the definition of $N_3(q)$. It is seen that $N_3(q)$ is growing significantly faster than $N_4(q)$. This suggests that constructive Weil descent only produces a small proportion of the possible isogeny classes of curves.

Here is the analogous data for hyperelliptic genus three curves over $\mathbb{F}_{2^n}$. In this case $N_1(q)$ is the number of isogeny classes of Jacobians of hyperelliptic genus three curves over $\mathbb{F}_q$ and the other $N_i(q)$ are defined analogously.

q	$N_1(q)$	$N_2(q)$	$N_3(q)$	$N_4(q)$
2	59	32	24	2
4	≥ 420	≥ 200	≥ 145	6

In conclusion we see that a rather small proportion of isogeny classes of simple non-supersingular Jacobians of hyperelliptic curves with one point at infinity of genus two or three arise as the Weil restriction of an elliptic curve.

5. Generalising the construction of Hess, Seroussi and Smart

The natural strategy to overcome the problem mentioned above is the following: Rather than constructing the Weil restriction A of an elliptic curve and asking for it to be isogenous to the Jacobian of a curve C, we could ask for $\mathrm{Jac}(C)$ to be a sub-abelian-variety of A.

It is conceivable that the construction of [11] could be modified to give such a method. One key modification would be to search for elliptic curves $E/\mathbb{F}_{q^n}$ such that the characteristic polynomial of Frobenius of the Weil restriction has a factor corresponding to a g-dimensional sub-abelian-variety which has suitable order. It is then plausible that methods related to those of [10] would allow the construction of a genus g curve whose Jacobian is isogenous to that factor. One might hope that this strategy would give a much larger class of suitable curves.

For example, over $\mathbb{F}_2$ we find four new isogeny classes of two-dimensional abelian varieties. These classes correspond to the degree four factors in the following factorisations:

$$\begin{aligned}
T^6 - 5T^3 + 2^3 &= (T^4 - T^3 - T^2 - 2T + 4)(T^2 + T + 2)\\
T^6 + 5T^3 + 2^3 &= (T^4 + T^3 - T^2 + 2T + 4)(T^2 - T + 2)\\
T^{12} + 11T^6 + 2^6 &= (T^4 - 3T^3 + 5T^2 - 6T + 4)(T^4 - T^2 + 4)\\
&\quad (T^4 + 3T^3 + 5T^2 + 6T + 4).
\end{aligned}$$

We now give a small table of results for genus 2 curves over $\mathbb{F}_{2^n}$. Here, as before $N_4(q)$ is the number of non-supersingular, simple Jacobians of curves of genus two over $\mathbb{F}_q$ which can be constructed from Weil descent of an elliptic curve with respect to $\mathbb{F}_{q^2}/\mathbb{F}_q$. The quantity $N_5(q)$ is the number of isogeny classes of non-supersingular, simple abelian varieties of dimension two over $\mathbb{F}_q$ which are isogenous to a factor of the Weil descent of an elliptic curve with respect to $\mathbb{F}_{q^n}/\mathbb{F}_q$ with $n \leq 6$. We do not know if all such abelian varieties are isogenous to the

Jacobian of a curve over $\mathbb{F}_q$. Experimental evidence suggests that taking $n > 6$ yields no new examples.

This data indicates that we do not obtain a significant number of new isogeny classes from this generalised approach. This phenomenon is also observed in the genus three case. If the generalised construction is used to produce curves which are not hyperelliptic (for instance, superelliptic curves [6] or more general curves) then it will still be true that only a small proportion of the available isogeny classes of curves will appear.

q	$N_4(q)$	$N_5(q)$
2	2	4
4	4	6
8	10	10
16	24	12
32	52	18

6. The curves produced by the generalised method

In this section we prove some results about the genus two curves which would arise from the generalised method outlined in Section 5. All results in this section apply only to the case of characteristic two and genus two. The main result is that the genus two curves which would be produced by the generalised method (and which would not have been produced from the original method of [11]) do not have an equation with a single point at infinity.

Lemma 6.1. *Let $q = 2^l$ and let C be a hyperelliptic curve of genus 2 over $\mathbb{F}_q$ whose Jacobian has characteristic polynomial equal to $P(T) = T^4 + a_1T^3 + a_2T^2 + qa_1T + q^2$. If C can be given a hyperelliptic model over $\mathbb{F}_q$ with a single point at infinity then at least one of a_1 or a_2 is even. If all hyperelliptic models of C over $\mathbb{F}_q$ have two points at infinity then both a_1 and a_2 are odd.*

Proof. Let C have equation $y^2+h(x)y = f(x)$ where $\deg h(x) \leq 3$ and $\deg f(x) \leq 6$. Then for each point (x_0, y_0) on C there is the conjugate under the hyperelliptic involution given by $(x_0, y_0+h(x_0))$. The fixed points of the hyperelliptic involution arise from the solutions to $h(x) = 0$.

The set $C(\mathbb{F}_q)$ therefore consists of the $\mathbb{F}_q$-points at infinity, the $\mathbb{F}_q$-points fixed by the involution, and the remaining $\mathbb{F}_q$-points which all occur in pairs.

We now consider the possible equations for genus two curves over $\mathbb{F}_{2^l}$. The cases where there is one point at infinity are when $\deg(h(x)) \leq 2$ and so we can have the four subcases: $\deg(h(x)) = 0$, $\deg(h(x)) = 1$, $\deg(h(x)) = 2$ and $h(x)$ splits into linear factors, $\deg(h(x)) = 2$ and $h(x)$ has no roots in $\mathbb{F}_{2^l}$. The number of points on C over $\mathbb{F}_q$ in the first, third and fourth subcases is therefore odd, and so $a_1 = q + 1 - \#C(\mathbb{F}_q)$ is even. In the second subcase we have $\#C(\mathbb{F}_q)$ even so we must consider $\#C(\mathbb{F}_{q^2})$. Write x_0 for the root of $h(x)$. Every $x_1 \in \mathbb{F}_q\backslash\{x_0\}$

gives rise to two points $(x_1, y_1), (x_1, y_1 + h(x_1))$ in $C(\mathbb{F}_{q^2})$. Furthermore, points $(x_2, y_2) \in C(\mathbb{F}_{q^2})$ for which $x_2 \notin \mathbb{F}_q$ come in groups of four conjugates with $(x_2, y_2 + h(x_2)), (x_2^\sigma, y_2^\sigma), (x_2^\sigma, y_2^\sigma + h(x_2^\sigma))$. Thus $\#C(\mathbb{F}_{q^2}) = 1 + 1 + 2(q-1) + 4m \equiv 0 \pmod 4$ and so $a_2 = \frac{1}{2}((\#C(\mathbb{F}_{q^2}) - q^2 - 1) + a_1^2) \equiv 0 \pmod 2$.

The case with two points at infinity occurs only when $\deg(h(x)) = 3$. The equation is $\mathbb{F}_q$-isomorphic to one with a single point at infinity if and only if $h(x)$ has a root in $\mathbb{F}_q$. Thus, assume that $h(x)$ has no roots in the ground field. In this situation there can be zero or two $\mathbb{F}_q$-points at infinity and always two $\mathbb{F}_{q^2}$-points at infinity. Since the polynomial $h(x)$ has no root in $\mathbb{F}_q$ and has degree three it must in fact be irreducible over $\mathbb{F}_{q^2}$. Once again we easily see that $\#C(\mathbb{F}_q) \equiv 0 \pmod 2$ and so a_1 is odd. Now, we obtain two points of $C(\mathbb{F}_{q^2})$ for every element $x \in \mathbb{F}_q$ and we also obtain sets of 4 conjugate points for elements of $x \in \mathbb{F}_{q^2} \backslash \mathbb{F}_q$. It follows that $\#C(\mathbb{F}_{q^2}) \equiv 2 \pmod 4$ and that a_2 is odd. □

We observe the following trivial but interesting corollary of the above result.

Corollary 6.2. *Let $q = 2^l$ and let C_1 and C_2 be hyperelliptic curves of genus 2 over $\mathbb{F}_q$. Suppose that C_1 has one point at infinity and that C_2 has two points at infinity and is not isomorphic over $\mathbb{F}_q$ to a hyperelliptic model with one point at infinity. Then* $\mathrm{Jac}(C_1)$ *is not isogenous to* $\mathrm{Jac}(C_2)$ *over $\mathbb{F}_q$.*

We note that these results are not true for genus three curves, for example $y^2 + (x^3+1)y = x^7 + x^5 + x^4$ and $y^2 + (x^4 + x + 1)y = x^7$ have isogenous Jacobians but the first has one point at infinity while the second is not isomorphic over $\mathbb{F}_2$ to a curve with one point at infinity.

Suppose that C is a genus two curve whose Jacobian is a factor of the Weil restriction of some elliptic curve with characteristic polynomial of Frobenius equal to $T^2 + aT + q^n$. We make the assumption that a is odd (i.e., that our process starts with an ordinary elliptic curve). There are at most 5 even values for a which can arise, and so this assumption only excludes a very small number of possible isogeny classes of dimension two abelian varieties.

Write $P(T) = T^4 + a_1T^3 + a_2T^2 + qa_1T + q^2$ for the characteristic polynomial of Frobenius for $\mathrm{Jac}(C)$ over $\mathbb{F}_q$, define $m = n - 2$ and write

$$Q(T) = T^{2m} + \alpha_1 T^{2m-1} + \alpha_2 T^{2m-2} + \cdots + \alpha_m T^m + q\alpha_{m-1}T^{m-1} + \cdots + q^m.$$

Then we must have $P(T)Q(T) = T^{2n} + aT^n + q^n$.

Proposition 6.3. *Let $q = 2^l$ and let C be a genus 2 curve over $\mathbb{F}_q$ with characteristic polynomial $P(T)$ as above such that $P(T)Q(T) = T^{2n} + aT^n + q^n$ for some integer $n \geq 3$. If a is odd then a_2 is odd and α_m is odd.*

Proof. The $n = 3$ case is trivial so we assume $n \geq 4$. With notation as above we find that the coefficient of T^n in the product $P(T)Q(T)$ has the form

$$q^2\alpha_{m-2} + qa_1\alpha_{m-1} + a_2\alpha_m + a_1q\alpha_{m-1} + q^2\alpha_{m-2} = a$$

from which we immediately deduce that $a_2 \equiv \alpha_m \equiv 1 \pmod 2$. □

The fact that $P(T)Q(T) = T^{2n} + aT^n + q^n$ imposes various conditions on the α_i as follows:

$$\begin{aligned} \alpha_1 &= -a_1 \\ \alpha_2 &= -a_1\alpha_1 - a_2 \\ \alpha_3 &= -a_1\alpha_2 - a_2\alpha_1 - qa_1 \\ \alpha_4 &= -a_1\alpha_3 - a_2\alpha_2 - qa_1\alpha_1 - q^2 \end{aligned} \tag{1}$$

and so on. For $i \geq 5$ (and thus $n \geq 7$) the recurrence is always of the form

$$\alpha_i = -a_1\alpha_{i-1} - a_2\alpha_{i-2} - qa_1\alpha_{i-3} - q^2\alpha_{i-4}. \tag{2}$$

By substituting the earlier values of α_i we obtain

$$\begin{aligned} \alpha_1 &= -a_1 \\ \alpha_2 &= a_1^2 - a_2 \\ \alpha_3 &= -a_1^3 + 2a_1a_2 - a_1q \\ \alpha_4 &= a_1^4 + a_1^2(2q - 3a_2) + (a_2^2 - q^2) \end{aligned} \tag{3}$$

and so on.

Lemma 6.4. *Let $q = 2^l$. The α_i can be represented as polynomials in a_1, a_2 and q. These polynomials have the following properties:*

1. *α_{2i+1} is always divisible by a_1 for $i = 0, 1, \ldots$.*
2. *The only pure power of a_1 in α_i (i.e., monomial with no a_2 or q terms) is $\pm a_1^i$.*
3. *The following congruences on $\alpha_i \pmod 2$ hold.*

$i \pmod 6$	$\alpha_i \pmod 2$
1	a_1
2	$a_1 + a_2$
3	a_1
4	$a_1 + a_1a_2 + a_2$
5	$a_1 + a_1a_2$
6	$a_1 + a_1a_2 + a_2$

Proof. The statements are easily proved by induction using the initial cases (3) and the recurrence formula (2). For the third part we use the fact that $x^n \equiv x \pmod 2$ to simplify the expressions. □

Proposition 6.5. *Let $q = 2^l$ and suppose C is a hyperelliptic curve of genus two whose Jacobian is a simple factor of the Weil restriction of an elliptic curve E over $\mathbb{F}_{q^n}$ where $n > 2$. Then one of the following two cases holds*

1. *A curve in the isogeny class of C would have already arisen from the Weil descent of an elliptic curve over the quadratic extension $\mathbb{F}_{q^2}$. This case can arise when n is even.*
2. *C has two points at infinity. This case can arise when $n \equiv 0, 2, 3 \pmod 6$.*

Proof. Consider first the case $n \equiv 3 \pmod 6$ (i.e., $m \equiv 1 \pmod 6$). Proposition 6.3 implies that $a_2 \equiv \alpha_m \equiv 1 \pmod 2$ and part 3 of Lemma 6.4 implies that $a_1 \equiv \alpha_m \pmod 2$. Lemma 6.1 implies that such a curve has two points at infinity.

For the case $n = 4$ write $P(T)(T^4 + \alpha_1 T^3 + \alpha_2 T^2 + q\alpha_1 T + q^2) = T^8 + aT^4 + q^4$ so that $a_1 = -\alpha_1$, $\alpha_2 + a_2 = \alpha_1^2$, $\alpha_1(a_2 - \alpha_2) = 0$ and $\alpha_2 \equiv a_2 \equiv 1 \pmod 2$. The case $\alpha_1 = 0$ is the case where $P(T)$ already arises from an elliptic curve over $\mathbb{F}_{q^2}$. The case $\alpha_1 \neq 0$ implies $a_2 = \alpha_2$ and so $\alpha_1^2 = 2a_2$ which implies that α_1 and a_1 are even but this in turn implies that a_2 is even which is a contradiction.

The cases $n \geq 5$ will be handled using general arguments. The basic approach is to consider the coefficient of the term T^{n+1} in $P(T)Q(T)$, which must be zero. This term is congruent modulo 2 to $a_2\alpha_{m-1} + a_1\alpha_m$. For the case $n \equiv 5 \pmod 6$ (i.e., $m \equiv 3 \pmod 6$) we argue as follows: From Lemma 6.4 and the fact that $\alpha_m \equiv a_2 \equiv 1 \pmod 2$ we have $a_1 \equiv 1 \pmod 2$ and $\alpha_{m-1} \equiv 0 \pmod 2$. But then $a_2\alpha_{m-1} + a_1\alpha_m \equiv 1 \pmod 2$ which is a contradiction. Hence when $n \equiv 5 \pmod 6$ there can be no degree 4 factors of $T^{2n} + aT^n + q^n$.

When $m \equiv 5 \pmod 6$ (i.e., $n \geq 7$ and $n \equiv 1 \pmod 6$) then the facts $a_2 \equiv \alpha_m \equiv 1 \pmod 2$ are incompatible with $\alpha_m \equiv a_1 + a_1 a_2 \pmod 2$ and so again this case cannot arise.

When m is even ($m \geq 4$) we must consider the coefficient of T^{n+1} more closely. Using part 1 of Lemma 6.4 we see that this coefficient is $a_1(q^2\alpha_{m-3}/a_1 + q\alpha_{m-2} + a_2\alpha_{m-1}/a_1 + \alpha_m + q\alpha_{m-1}/a_1)$ from which we see that either $a_1 = 0$ (and the curve would have arisen from a Weil descent over a quadratic extension) or else the second factor is zero. Now consider this factor modulo 2. Recalling that $a_2 \equiv \alpha_m \equiv 1 \pmod 2$ it turns out that $\alpha_{m-1}/a_1 \equiv 1 \pmod 2$ must be satisfied. In the case $m \equiv 0 \pmod 6$ it follows that $a_1 \equiv 1 \pmod 2$ and so there are two points at infinity. In the case $m \equiv 2 \pmod 6$ the condition $\alpha_m \equiv 1 \pmod 2$ implies $a_1 \equiv 0 \pmod 2$ but then $\alpha_{m-1}/a_1 \equiv 0 \pmod 2$ which is a contradiction. In the case $m \equiv 4 \pmod 6$ it follows that $a_1 \equiv 1 \pmod 2$ and so the curve has two points at infinity. □

The meaning of this result is that the only way to get something 'new' results in a curve with 2 points at infinity. Experimental evidence suggests that all the new isogeny classes already appear using $n \leq 6$ in the above construction.

7. Odd characteristic

It would be very useful if one could perform constructive Weil descent in odd characteristic, such as to construct a genus 2 curve over a prime field $\mathbb{F}_p$ from the Weil descent of an elliptic curve over $\mathbb{F}_{p^2}$.

Unfortunately, it seems to be rather difficult to achieve this in practice. Suppose E is an elliptic curve over $\mathbb{F}_{p^n}$ and that C is a curve of genus g on the Weil restriction of E. If $g > n$ then one only knows that $P_E(T^n) | P_{\mathrm{Jac}(C)}(T)$ and this information alone is not enough to determine the number of points on $\mathrm{Jac}(C)(\mathbb{F}_p)$.

Even the knowledge that there is a large prime factor is useless in practice if one does not know the cofactor, since it is not possible to force a random element to have prime order. Thus we are forced to seek curves of genus $g \leq n$. In odd characteristic this condition is not usually satisfied for curves on the Weil restriction.

One case where this is possible has been given by Claus Diem: given an elliptic curve over $\mathbb{F}_{3^m}$ there is an explicitly given genus two curve C over $\mathbb{F}_{3^m}$ such that $\#\mathrm{Jac}(C)(\mathbb{F}_{3^m}) = \#E(\mathbb{F}_{3^{3m}})/\#E(\mathbb{F}_{3^m})$. Details of this construction are given in Section 3.2 of [4].

Of course, in odd characteristic the results of this article still apply: only a small proportion of the isogeny classes of curves can be constructed using this method.

Acknowledgement. I would like to thank the referees for pointing out some errors in an earlier draft of the paper. I would also like to thank Jerzy Urbanowicz and the rest of the organising committee for all their hard work in putting the conference together.

References

[1] Cantor D.G., Computing in the Jacobian of a hyperelliptic curve. Math. Comp. 48 (1987), 95–101.

[2] Duursma, I., Gaudry, P., Morain, F., Speeding up the discrete log computation on curves with automorphisms. In: Advances in Cryptology – ASIACRYPT '99 (ed. by K.Y. Lam, E. Okamoto, C. Xing; Lecture Notes in Comput. Sci. 1716), 103–121. Springer, Berlin 1999.

[3] Frey, G., How to disguise an elliptic curve. Talk at ECC'98, Waterloo, Canada, 1998.
`http://cacr.math.uwaterloo.ca/conferences/1998/ecc98/slides.html`

[4] —, Applications of arithmetical geometry to cryptographic constructions. Preprint, 2000.
`http://www.exp-math.uni-essen.de/preprints/index00.html`

[5] Galbraith, S.D., Supersingular curves in cryptography. Preprint, 2000.

[6] Galbraith, S.D., Paulus, S.M., Smart, N.P., Arithmetic on superelliptic curves. Math. Comp., posted on 26 October 2000, PII S 0025-5718(00)01297-7 (to appear in print).

[7] Galbraith, S.D., Smart, N.P., A cryptographic application of Weil descent. In: Cryptography and Coding (Cirencester, 1999) (ed. by M. Walker; Lecture Notes in Comput. Sci. 1746), 191–200. Springer, Berlin 1999.

[8] Gaudry, P., Algorithmique des courbes hyperelliptiques et applications à la cryptologie. L'École Polytechnique thesis, 2000.
`http://ultralix.polytechnique.fr/~gaudry/`

[9] Gaudry, P., Harley, R., Counting points on hyperelliptic curves over finite fields. In: Algorithmic Number Theory – ANTS IV (ed. by W. Bosma; Lecture Notes in Comput. Sci. 1838), 313–332. Springer, Berlin 2000.

[10] Gaudry, P., Hess, F., Smart, N.P., Constructive and destructive facets of Weil descent on elliptic curves. J. Cryptology (to appear).

[11] Hess, F., Seroussi, G., Smart, N.P., Two topics in hyperelliptic cryptography. Preprint, 2000.

[12] Koblitz, N., Hyperelliptic cryptosystems. J. Cryptology 1 (1989), 139–150.

[13] Rück, H.-G., Abelian surfaces and Jacobian varieties over finite fields. Compositio Math. 76 (1990), 351–366.

[14] Spallek, A.-M., Kurven vom Geschlecht 2 und ihre Anwendung in Public-Key-Kryptosystemen. Ph.D. Thesis, IEM Essen, 1994.

[15] Tate, J., Endomorphisms of abelian varieties over finite fields. Invent. Math. 2 (1966), 134–144.

[16] Waterhouse, W.C., Abelian varieties over finite fields. Ann. Sci. École Norm. Sup. (4) 2 (1969), 521–560.

[17] Weng, A., Constructing hyperelliptic curves of genus 2 suitable for cryptography. Preprint, 2000.
`http://www.exp-math.uni-essen.de/preprints/index00.html`

On the security of a public-key cryptosystem

Otokar Grošek, Spyros S. Magliveras and Wandi Wei

Abstract. In 1992, W. Webb proposed a public-key cryptosystem by using *complementing subsets* of polynomials over finite fields. Two years later, Qu and Vanstone showed that Webb's system is not secure by proving that any factorization of the elementary abelian group $G = \mathbb{Z}_p \oplus \cdots \oplus \mathbb{Z}_p$, must be a translation of a transversal factorization of G. In this paper we present a result that is stronger than the Qu–Vanstone result with a simpler proof.

Key words and phrases: cryptology, transversal, factorizations, elementary abelian groups, PGM, Webb's cryptosystem.

1. Introduction

In 1992, W. Webb [10] proposed a public-key cryptosystem, which can be described as follows.

Let $GF(q)$ be the finite field with $q = p^r$ elements, and P_d the set of polynomials of degree not greater than d over $GF(q)$. Webb calls a collection of subsets $A_1, A_2, \ldots, A_k$ of P_d *complementing* sets if each element $g \in P_d$ can be uniquely represented as a sum of form

$$g = a_1 + a_2 + \cdots + a_k, \quad a_i \in A_i.$$

A collection of complementing sets is called *easy* if for any $g \in P_d$ the representation of g of the above form can be obtained in time polynomial in $\log q$. Otherwise, the collection is called *hard*. Let $(A_1, A_2, \ldots, A_k)$ be easy, and f and h be two polynomials over $GF(q)$ with $\gcd(f, h) = 1$ and $\deg(f) = d + 1$. Then Webb suggests that the collection $(B_1, \ldots, B_k)$ with $B_i = hA_i \pmod f$ is hard. The easy collection $(A_1, A_2, \ldots, A_k)$ together with h and f is used as a private key, while the hard collection $(B_1, B_2, \ldots, B_k)$ is used as a public key, where the A_i and B_j are treated as ordered sets. Messages are the strings $(x_1, x_2, \ldots, x_k)$ with $x_i \in [1, |A_i|]$. To encrypt a message $(x_1, x_2, \ldots, x_k)$, form the sum

$$c = b_{1,x_1} + b_{2,x_2} + \cdots + b_{k,x_k}$$

Public-Key Cryptography and Computational Number Theory
K. Alster, J. Urbanowicz and H. C. Williams (Eds.)

as the ciphertext, where b_{i,x_i} is the x_i^{th} element of B_i. To decrypt, compute $ch^{-1} \pmod f$ first, and then employ the easy collection $(A_1, \ldots, A_k)$ to recover x_i for all $i \in [1, k]$ in time polynomial in $\log q$.

In 1994, M. Qu and S.A. Vanstone [6] proved a theorem, which they subsequently used to show that Webb's public-key cryptosystem [10] is not secure. Their work follows earlier work of Magliveras and others [5, 2, 4, 3] on *factorization bases* for finite permutation groups. In the late 1970's S.S. Magliveras proposed a private key cryptosystem, called PGM (Permutation Group Mappings), based on certain factorization bases, called *logarithmic signatures*, for finite permutation groups. In subsequent years, factorizations in finite, preferably non-solvable groups, have been further examined in the context of cryptography by Magliveras, Memon, Tran van Trung, Horváth, Vanstone and Qu.

Since this paper deals with abelian groups, we adopt the terminology used by Qu and Vanstone because it is simpler, and because some of the notions relevant to non-abelian groups become degenerate in the case of abelian groups. Factorizations for abelian groups of the type we study here were extensively considered by L. Rédei [7, 8, 9] in the context of polynomials over finite fields.

Throughout this article $G = \mathbb{Z}_p \oplus \cdots \oplus \mathbb{Z}_p$ is the elementary abelian group of order p^s. In the sequel we view G not only as a group but also as a vector space over $\mathbb{Z}_p$.

Let A_i, $i \in [1, s]$, be ordered subsets of G with $|A_i| > 1$. The ordered collection $(A_i : i \in [1, s])$ is called a *factorization* of type $(p, \ldots, p)$ for G if $|A_i| = p$ for each $i \in [1, s]$, and each element $g \in G$ can be uniquely expressed as a sum in the form

$$g = a_1 + a_2 + \cdots + a_s, \quad a_i \in A_i.$$

A factorization $(A_i : i \in [1, s])$ would be called a *logarithmic signature* in the case of a general (not necessarily abelian) group G by the authors of [3].

A factorization $(A_i : i \in [1, s])$ of G is called *transversal* if there exists a subgroup chain $G = G_0 > G_1 > \cdots > G_{s-1} > G_s = 0$ such that A_i is a complete set of coset representatives of G_i in G_{i-1}.

A factorization $(A_i : i \in [1, s])$ of G is called a *translation* of a factorization $(B_i : i \in [1, s])$ if there exist $g_i \in G$, $i \in [1, s]$, such that $(A_i = B_i + g_i : i \in [1, s])$.

Note that the above definitions are different from those used by Magliveras et al. [3]. The authors of [3] use *exact transversal* for the kind of factorization Qu and Vanstone call *transversal*. Moreover, the authors of [3] use *transversal* for the result of a *sandwich* transformation of an exact transversal. In abelian groups a sandwich transformation degenerates into a translation. The theorem proved by M. Qu and S.A. Vanstone in [6], and used to show that Webb's public-key cryptosystem is not secure, can be stated as follows.

Theorem 1.1. *Any factorization of type $(p, \ldots, p)$ of G must be a translation of a transversal factorization of G.*

We would like to point out that if the factorization of G is not of type $(p, \ldots, p)$, then the conclusion of the theorem is generally not true. More specifically, for $s > 2$, it is possible to have $G = A + B$, with $0 \in A \cap B$, $|G| = |A| \cdot |B|$, where neither A nor B is a subgroup. [1].

In this article we prove a theorem, which is stronger than Theorem 1.1, using a method which is simpler than Qu and Vanstone's. We refer the reader to [6] for the details of breaking Webb's cryptosystem by using Qu and Vanstone's theorem.

2. Our theorem

It is desirable to know how to construct all factorizations of type $(p, \ldots, p)$ of G. This problem is solved in Theorem 2.1 below, of which Theorem 1.1 is an immediate consequence. Our proof of Theorem 2.1 is simpler than Qu and Vanstone's proof of Theorem 1.1.

Let $e_i = (0, \ldots, 0, 1, 0, \ldots, 0)$ be the ith canonical basis element of G, i.e. e_i has 1 in position i, and 0's in all other positions. Let $\mathcal{N} = (N_1, N_2, \ldots, N_s)$, where N_i is an ordered subset of G of size p whose underlying set is:

$$M_i = \{je_i + z_{i,j} : j \in \mathbb{Z}_p\}$$

with $z_{i,j} \in \langle e_{i+1}, e_{i+2}, \ldots, e_s \rangle$, and $z_{i,0} = z_{i,1} = 0$. Note that in the above definition of M_i, for each pair (i, j) a single element $z_{i,j}$ is selected. Hence, $|M_i| = p$. We have,

$$N_i + N_{i+1} + \cdots + N_s = \langle e_i, \ldots, e_s \rangle$$

and N_i is a complete set of coset representatives of $N_{i+1} + \cdots + N_s$ in $N_i + N_{i+1} + \cdots + N_s$. Therefore, $\mathcal{N} = (N_1, N_2, \ldots, N_s)$ is a transversal factorization of G. A factorization $\mathcal{N}$ of the above type is said to be of *basic form.*

The theorem below shows that any factorization of G can be obtained via factorizations $\mathcal{N}$ of basic form. In our proof of this theorem, two lemmas are needed which we now state for easy reference.

Lemma 2.1 ([6]). *Let $(A_i : i \in [1, s])$ be a factorization of type $(p, \ldots, p)$ of G with $0 \in A_i$. For each $i \in [1, s]$, select any non-zero element $a_i \in A_i$. Then $\{a_i : i \in [1, s]\}$ is a set of s generators for G.*

Lemma 2.2 ([7, 8], [6]). *Let $(A_i : i \in [1, s])$ be a factorization of type $(p, \ldots, p)$ of G with $0 \in A_i$, $i \in [1, s]$. Then at least one of the A_i is a subgroup of G.*

Here is our theorem:

Theorem 2.1. *Let $\mathcal{A} = (A_1, A_2, \ldots, A_s)$ be a factorization of type $(p, \ldots, p)$ of G. Then there exist elements $g_1, g_2, , \ldots, g_s$ of G, an automorphism δ of G, and an $\mathcal{N} = (N_1, N_2, \ldots, N_s)$ of basic form, such that*

$$A_i = \delta(N_i + g_i), \qquad i \in [1, s].$$

Proof. It suffices to show that when $0 \in A_i$ for all $i \in [1, s]$, there exist an automorphism δ of G and an $\mathcal{N}$ of basic form, such that

$$A_i = \delta(N_i), \quad i \in [1, s].$$

We proceed by induction on s. The statement is trivially true when $s = 1$. Now suppose that $s > 1$ and that the statement is true for $s - 1$.

Suppose that for each $i \in [1, s]$, a_i is a fixed non-zero element of A_i. By Lemma 2.1, $\{a_i : i \in [1, s]\}$ is a basis of G. Let ψ be the automorphism of G mapping the basis element a_i to e_i for each $i \in [1, s]$. Let $B_i = \psi(A_i)$. Then $\mathcal{B} = (B_1, \ldots, B_s)$ is a factorization of G, and $0, e_i \in B_i$, $i \in [1, s]$. By Lemma 2.2, at least one of the B_i is a subgroup of G. Let B_m be such a subgroup. Then $B_m = \langle e_m \rangle$. Let η be the automorphism of G interchanging the mth and s^{th} coordinates of every vector in G. Let $C_i = \eta(B_i)$, $i \in [1, s]$. Then $C_s = \langle e_s \rangle$. Let $\pi = \pi_{s-1}$ be the projection of G to the first $s - 1$ coordinates, $G' = \pi(G)$, and $D_i = \pi(C_i)$, $i \in [1, s]$. Then $D_s = 0$, and $G' = \langle e_1, e_2, \ldots, e_{s-1} \rangle$ is elementary abelian of order p^{s-1}. By the induction hypothesis, there exist an automorphism of G', say ξ, and $s - 1$ subsets of G', say K_1, K_2, ..., K_{s-1}, in basic form with all s^{th} coordinates equal to 0, such that $D_i = \xi(K_i)$, $i \in [1, s - 1]$. Let λ be the mapping defined by

$$\lambda(x + y) = \xi^{-1}(x) + y, \quad x \in G', y \in \langle e_s \rangle.$$

Then λ is an automorphism of G, the ordered collection $\mathcal{N} = (N_i = \lambda(C_i) : i \in [1, s])$ is of basic form, and

$$A_i = (\lambda\eta\psi)^{-1}(N_i), \quad i \in [1, s].$$

Since $\delta = (\lambda\eta\psi)^{-1}$ is an automorphism of G, we have proved the theorem when $0 \in A_i$ for all i. The general statement of the theorem now follows easily. □

Since all automorphisms of G are known and all basic form factorizations $\mathcal{N}$ are easy to construct systematically, all factorizations of G can easily be produced by Theorem 2.1. Theorem 1.1 is an immediate consequence of Theorem 2.1.

In closing we note that in the proof of Theorem 2.1, instead of using Lemma 2.1, we have used the following weaker form of Lemma 2.1.

Lemma 2.3. *Let $(A_i : i \in [1, s])$ be a factorization of type $(p, \ldots, p)$ of G and $0 \in A_i$, $i \in [1, s]$. Then there exist $a_i \in A_i$, $a_i \neq 0$, $i \in [1, s]$, such that $\{a_i : i \in [1, s]\}$ is a set of generators for G.*

Acknowledgment. We wish to thank J.W.P. Hirschfeld and T. Szőnyi for bringing to our attention some facts and bibliography on related work in the context of finite geometry. We also thank the anonymous referee for his/her valuable comments.

References

[1] Hirschfeld, J.W.P., Szőnyi, T., Private communication.

[2] Magliveras, S.S., Memon, N.D., The linear complexity profile of cryptosystem PGM. Congr. Numer. 72 (1990), 51–60.

[3] Magliveras, S.S., Memon, N.D., Algebraic properties of cryptosystem PGM. J. Cryptology 5 (1992), 167–183.

[4] Magliveras, S.S., Memon, N.D., Tam, K.C., Complexity tests of cryptosystem PGM. Congr. Numer. 79 (1990), 61–68.

[5] Magliveras, S.S., Oberg, B.A., Surkan, A.J., A new random number generator from permutation groups. Rend. del Sem. Matemat. e Fis. di Milano LIV (1984), 203–223.

[6] Qu, M., Vanstone, S.A., Factorizations of elementary abelian p-groups and their cryptographic significance. J. Cryptology 7 (1994), 201–212.

[7] Rédei, L., Zwei Lückensätze über Polynome in endlichen Primkörpern mit Anwendung auf die endlichen Abelschen Gruppen und die Gaussischen Summen. Acta Math. 79 (1947), 273–290.

[8] —, Die neue Theorie der endlichen Abelschen Gruppen und Verallgemeinerung des Hauptsatzes von Hajós. Acta Math. Acad. Sci. Hungar. 16 (1965), 329–373.

[9] —, Lückenhafte Polynome über endlichen Körpern. Birkhäuser, Basel 1970. English translation: Lacunary polynomials over finite fields. North Holland, Amsterdam 1973.

[10] Webb, W., Cryptography using complementing subsets of polynomials over finite fields. In: Finite Fields, Coding Theory and Advances in Communications and Computing (ed. by G.L. Mullen and P. Jau-Shyong Shiue), 411–415. Marcel Dekker, New York, 1993.

Optimizations for NTRU

Jeffrey Hoffstein and Joseph Silverman

Abstract. In this note we describe a variety of methods that may be used to increase the speed and efficiency of the NTRU public key cryptosystem.

1991 Mathematics Subject Classification: 94A60, 11T71.

1. An overview of NTRU

The NTRU Public Key Cryptosystem is based on ring theory and relies for its security on the difficulty of solving certain lattice problems. In this section we will briefly review the properties of NTRU that are relevant to the topics in this paper. For further details and a security analysis of NTRU, see [HPS, S1, S2].

A general formulation of the NTRU Public Key Cryptosystem uses a ring R and two (relatively prime) ideals p and q in R. A rough outline of the key creation, encryption, and decryption processes is as follows:

- **Key creation**
  Bob creates a public key h by choosing elements $f, g \in R$, computing the mod q inverse f_q^{-1} of f, and setting

  $$h \equiv f_q^{-1} * g \pmod{q}.$$

  Bob's private key is the element f. Bob also precomputes and stores the mod p inverse f_p^{-1} of f.
- **Encryption**
  In order to encrypt a plaintext message $m \in R$ using the public key h, Alice selects a random element $r \in R$ and forms the ciphertext

  $$e \equiv r * h + m \pmod{q}.$$

- **Decryption**
  In order to decrypt the ciphertext e using the private key f, Bob first computes

  $$a \equiv f * e \pmod{q}.$$

Public-Key Cryptography and Computational Number Theory
K. Alster, J. Urbanowicz and H. C. Williams (Eds.)

He chooses $a \in R$ to satisfy this congruence and to lie in a certain prespecified subset R_a of R. He next does the mod p calculation

$$f_p^{-1} * a \pmod p,$$

and the value he computes is equal to m modulo p.

Remark. In practice the elements f, g, r, and m are taken from certain large prespecified subsets R_f, R_g, R_r, and R_m of R. These sets (and R_a) are chosen large enough so that it is infeasible for an attacker to find f or g from h or to find r or m from e, while at the same time they are chosen so that the decryption process works.

The general NTRU cryptosystem described above makes no specification of the ring R or its subsets R_f, R_g, R_m, R_r, R_a. A standard implementation of NTRU uses the ring of convolution polynomials

$$R = \frac{\mathbf{Z}[X]}{(X^N - 1)}.$$

The sets R_f, R_g, R_m, R_r are sets of "small" polynomials, which means that their coefficients are chosen to be small, generally either binary $\{0, 1\}$ or trinary $\{-1, 0, 1\}$ and possibly with a specified number of nonzero coefficients. A typical choice for (p, q) is $(3, 128)$, and in order for decryption to work properly it is necessary that the coefficients of the polynomial

$$p * r * g + m * f \tag{1}$$

lie in an interval of length at most q. That is, the difference between the largest coefficient and the smallest coefficient should not exceed q. In this case we will say that the polynomial is *narrow* mod q. If p, q, and the sets R_f, R_g, R_m, R_r are chosen appropriately, then the polynomial (1) will be narrow mod q for most choices of f, g, m, r.

2. Guarding against chosen ciphertext attacks

There are various ways to construct adaptive chosen ciphertext attacks against the NTRU cryptosystem, see for example [HS2, JJ]. Padding techniques of Fujisaki and Okamoto [FO1, FO2] can be used to prevent such attacks. Further, if the NTRU plaintext is not scrambled in some way, then there may be dictionary attacks against part or all of the plaintext. An adaptation for NTRU of the Fujisaki–Okamoto method with added message scrambling works as follows (see [HS3] for details).

Let M be Alice's plaintext. She chooses a collection of random bits R and uses an all-or-nothing transformation on the concatenation $M \| R$. For example, she could split each of M and R into equal size pieces $M = M_1 \| M_2$ and $R = R_1 \| R_2$

and then use hash functions $\mathcal{H}_1$ and $\mathcal{H}_2$ to compute

$$m_1 = (M_1 \| R_1) \oplus \mathcal{H}_1(M_2 \| R_2) \qquad \text{and} \qquad m_2 = (M_2 \| R_2) \oplus \mathcal{H}_2(m_1).$$

(Here $\oplus$ denotes bitwise exclusive or.) The NTRU plaintext m is set equal to $m = m_1 \| m_2$. The NTRU "random" element r is also computed using a hash function

$$r = \mathcal{H}_3(M \| R).$$

(Of course, r is no longer random. We have moved the randomness into the plaintext as the bit string R.)

Alice computes the ciphertext $e \equiv r * h + m \pmod{q}$ is as usual. When Bob decrypts the message, he recovers the plaintext m, and then he inverts the all-or-nothing transformation to recover M and R. However, before accepting the plaintext M as valid, he recomputes $r = \mathcal{H}_3(M \| R)$ and checks that e is actually equal to $r*h+m \pmod{q}$. If this is not true, then he rejects the message. This prevents Alice from sending Bob specially constructed ciphertexts e whose decryption might allow her to deduce information about Bob's private key.

It is important to note that Alice has no control over the "random" element r. So if she sends Bob a purported ciphertext e, then the quantity $r * h$ that Bob computes will be essentially a random element modulo q. If R/qR is sufficiently large, there is little chance of choosing particular values for r and mounting an adaptive attack.

The cost of guarding against chosen ciphertext attacks is that Bob needs to reencrypt the message. (Both Bob and Alice also need to compute some hash functions, but in practice, these take very little time.) Thus Bob needs to compute the extra product $r * h \bmod q$. However, Bob's interest in computing this product is simply to compare it with $e - m$. It may be possible to do this without computing the full product. For example, if NTRU is implemented with a convolution polynomial ring $R = \mathbf{Z}[X]/(X^N - 1)$, then Bob might compute only k coefficients of $r * h$ and just compare those k coefficients to $e - m$. For an appropriate choice of k, he can be fairly confident that if the k coefficients match, then all of the coefficients will match. The probability of any particular coefficient of $r*h$ matching the corresponding coefficient of $e - m$ is approximately $1/q$. For a typical value such as $q = 128$, it would suffice to check (say) 12 coefficients to achieve a spoofing probability of under 2^{-84}.

3. Invertibility of f modulo p

In practical situations, the most time consuming part of the NTRU encryption and decryption processes is multiplication in the ring R. The NTRU encryption process involves one product $r*h \bmod q$, while decryption requires the computation of two products $f * e \bmod q$ and $f_p^{-1} * a \bmod p$. Similarly, the most time consuming part of the NTRU key creation process is the computation of the inverses f_p^{-1} and f_q^{-1}.

In order to speed both the key creation and the decryption processes, the user can choose the element f to have the form

$$f = 1 + p * f_1 \quad \text{with } f_1 \in R.$$

Notice that an f of this form has the property that $f_p^{-1} = 1$, so it is not necessary to compute the inverse modulo p, and the second multiplication in the decryption process disappears.

Remark. It is important to verify that the specified set of elements of the form $f = 1 + p * f_1$ provides the desired level of security. In practice, this means that one must assume that an attacker can use his search methods on the set of f_1's. In particular, the lattice search methods described in [HPS, S1, S2] can be easily modified to search for the pair (f_1, g), so one must analyze these alternative lattices.

Remark. In practice, the prime p is the "small" prime and the prime q is the "large" prime (e.g., $(p, q) = (2, 127)$ or $(3, 128)$). The performance gain using $f = 1 + p * f_1$ comes from eliminating an inversion modulo p in key creation and a multiplication modulo p in decryption, so this may not seem significant. However, it may happen that operations modulo p are as costly as operations modulo q, even if p is smaller than q. To take a specific example, if $p = 3$ and $q = 128$, then basic operations modulo q are no more difficult on a computer than those modulo p. [In terms of computational efficiency, the ideal situation would be to take $p = 2$ and $q = 128$, but then $e \equiv m \pmod{p}$, so the system becomes insecure. This is why it is important that the ideals p and q be relatively prime in the ring R.]

Remark. If the coefficients of the polynomial

$$a = p * g * r + m * f$$

lie in an interval of length greater than q, then the decryption process will not work properly. For an appropriate choice of parameters, this will happen very rarely. If we take f in the form $f = 1 + p * f_1$, then

$$a = m + p * (g * r + m * f_1),$$

so we see that coefficients that "wrap" (i.e., that lie outside the chosen mod q interval) will each affect a single coefficient of the decryption. This means that if a small amount of error correction is put into m, then it will be possible to correct messages that have a small amount of wrapping. Although this cannot be recommended for high security applications, since redundancy in m might open some lines of statistical attacks on long transcripts of messages, it could be useful in moderate security situations where one wants to virtually eliminate the possibility of a wrapping decryption failure.

4. Taking p to be a polynomial

If f is taken in the form $f = 1 + p * f_1$, then the polynomial

$$p * r * g + m * f$$

that is being formed during the decryption process becomes

$$p * (r * g + m * f_1) + m. \tag{2}$$

It is clear that as p becomes larger, there is more likelihood that the polynomial will be too wide. On the other hand, there are many computational advantages to taking q to be a power of 2, which would seem to force p to be at least 3.

However, as the description of NTRU makes clear, there is no particular reason that p needs to be an integer. It could instead be a polynomial, as long as the ideals generated by p and q are relatively prime in the ring R. Notice that this condition on p and q is equivalent to requiring that the elements p, q, and $X^N - 1$ generate the unit ideal in $\mathbf{Z}[X]$.

There is a second issue related to the practical matter of retrieving the plaintext m. The plaintext polynomial $m(X)$ needs to be small, say with binary or trinary coefficients; yet the decryption process only recovers $m(X)$ modulo p. In other words, we recover the image of $m(X)$ in the quotient ring

$$\frac{R}{pR} = \frac{\mathbf{Z}[X]}{(X^N - 1, p)}.$$

For example, if $m(X)$ has binary coefficients, then there are 2^N possible values for $m(X)$, and we want to ensure that the map

$$\{\text{binary polynomials } m(X)\} \longrightarrow R/pR \tag{3}$$

is injective. Even more than this, we need to know how to invert this map in order to recover m.

We also want the coefficients of p to be very small, since we need the polynomial (2) to be narrow. A natural candidate for p would be a polynomial of the form $p = X^k \pm 1$. Unfortunately, the elements

$$X^k \pm 1 \text{ and } X^N - 1 \text{ and } 128$$

are not relatively prime in the ring $\mathbf{Z}[X]$.

A second natural candidate is a polynomial of the form $p = X^k \pm X^j \pm 1$. These polynomials usually satisfy the relative primality condition, but the quotient R/pR will not be large enough and the reduction map (3) will not be injective. For example, if $p = X^2 + X + 1$ and N is a large prime, then R/pR will have approximately 1.618^N elements. (Here $1.618\ldots$ is the golden ratio $(1+\sqrt{5})/2$.)

We thus turn to the polynomial

$$p = X + 2. \tag{4}$$

(The polynomial $X - 2$ works similarly, but is slightly less efficient.) We will assume that N is odd and $N \geq 7$. We begin by verifying that the polynomial $p = X + 2$

satisfies the relative primality condition. We write

$$X^N - 1 = X^N + 2^N - 2^N - 1 = (X^N + 2^N) - 128 \cdot 2^{N-7} - 1. \tag{5}$$

Notice that $X^N + 2^N = (X+2)u(X)$ with $u(X) = X^{N-1} - 2X^{N-2} + \cdots + 2^{N-1}$, so we can rewrite (5) as

$$1 = (X+2)u(X) - (X^N - 1) - 128 \cdot 2^{N-1}.$$

This shows that the three elements $X+2$, $X^N - 1$, and 128 generate the unit ideal in $\mathbf{Z}[X]$.

Next we consider the quotient ring R/pR when $p = X + 2$. An elementary calculation shows that this ring has $2^N + 1$ elements. We will prove in Section 5 that the map (3) is injective and we will describe a fast algorithm for inverting it. This suggests that taking $p = X + 2$ and $q = 128$ provides for an efficient implementation of the NTRU public key cryptosystem. A standard security analysis of the underlying lattice problem shows that $N = 251$ provides a security level comparable to that given by RSA with 1024 bit keys.

5. Working with p equal to $X + 2$

In this section we will prove that binary plaintext messages $m(X)$ can be uniquely recovered if they are known modulo $X+2$. We begin by proving that distinct $m(X)$ give distinct residue classes. For this section, we fix the notation

$$R = \frac{\mathbf{Z}[X]}{(X^N - 1)}, \qquad R_m = \{\text{binary polynomials in } R\}, \qquad p = p(X) = X + 2.$$

Proposition.
(a) *The evaluation map*

$$R/p(X)R \longrightarrow \mathbf{Z}/(2^N+1)\mathbf{Z}, \qquad a(X) \longmapsto a(-2),$$

is an isomorphism.
(b) *The map*

$$R_m \longrightarrow R/p(X)R$$

is one-to-one.

Proof. The evaluation map $a(X) \mapsto a(-2)$ gives an isomorphism $\mathbf{Z}[X]/(X+2) \cong \mathbf{Z}$. The element $X^N - 1$ in the lefthand ring corresponds to the integer $(-2)^N - 1 = -(2^N + 1)$ in $\mathbf{Z}$. (Note we are assuming that N is odd.) Hence the quotient of $\mathbf{Z}[X]/(X+2)$ by the ideal $(X^N - 1)$ (which is the same as $R/p(X)R$) is isomorphic to the quotient of $\mathbf{Z}$ by $2^N + 1$. This proves (a).

Using the identification $R/p(X)R \cong \mathbf{Z}/(2^N + 1)\mathbf{Z}$ from (a), we need to show that if $m(X)$ and $m'(X)$ are binary polynomials satisfying

$$m(-2) \equiv m'(-2) \pmod{2^N + 1},$$

then $m(X) = m'(X)$. We first observe that if $m(X)$ is a binary polynomial, then the largest possible value of $m(-2)$ is

$$1 + (-2)^2 + (-2)^4 + \cdots + (-2)^{N-1} = \frac{2^{N+1} - 1}{3},$$

and the smallest (i.e., most negative) possible value of $m(-2)$ is

$$(-2)^1 + (-2)^3 + (-2)^5 + \cdots + (-2)^{N-2} = -\frac{2^N - 2}{3}.$$

The difference of these two limits is $2^N - 1$, so we see that the congruence class of $m(-2)$ modulo $2^N + 1$ determines exactly the value of $m(-2)$. It remains to show that different binary polynomials have different values at $X = -2$.

Suppose that $m(X)$ and $m'(X)$ are binary polynomials satisfying $m(-2) = m'(-2)$. Let $A(X) = m(X) - m'(X) \neq 0$ and suppose that $A(X) \neq 0$. Then we can write

$$A(X) = aX^k + bX^{k+1} + \cdots + cX^{N-1} \quad \text{with } a \neq 0.$$

We know that every coefficient of $A(X)$ is -1, 0, or 1, so in particular $a = \pm 1$. On the other hand, we also know that $A(-2) = 0$, so

$$0 = A(-2) \equiv a(-2)^k \pmod{2^{k+1}},$$

which implies that 2 divides a. This contradiction shows that $A(X) = 0$, so $m(X) = m'(X)$, which concludes the proof of the proposition. □

Theorem. *Let N be an odd integer and let $a(X) = a_0 + a_1X + \cdots + a_{N-1}X^{N-1} \in \mathbf{Z}[X]$ be a polynomial. The algorithm given below transforms $a(X)$ into a polynomial $b(X) \in \mathbf{Z}[X]$ with the following properties:*

- $\deg b(X) \leq N - 1$,
- *$b(X)$ is a binary polynomial, that is, its coefficients are all 0 or 1 (with one exception noted below),*
- $b(-2) \equiv a(-2) \pmod{2^N + 1}$.

The one exception occurs when

$$a(-2) \equiv \frac{2^{N+1} + 2}{3} \pmod{2^N + 1},$$

in which case one can take $b(X) = 2 + X^2 + X^4 + \cdots + X^{N-1}$. The algorithm terminates after its main loop has been executed no more than $2N + 1$ times.

Algorithm. In the following description, all indices on coefficients of a are treated modulo N. Thus $a_N = a_0$, $a_{N+1} = a_1$, etc.

```
replace each a_i with its congruence class modulo 2^N + 1.
set i = 0.
LOOP:
write a_i = u + 2v with u = 0 or 1.
set a_i = u, a_{i+1} = a_{i+1} + v, a_{i+2} = a_{i+2} + v.
set i = i + 1.
```

```
let w = min{⌊a_i/2⌋, a_{i+1}}.
set a_i = a_i − 2w and a_{i+1} = a_{i+1} − w.
if i < N then go to LOOP.
if a_i ∉ {0,1} then go to LOOP.
if a_{i+1} ∉ {0,1} and i < 2N then set i = i+1 and go to LOOP.
if i < 2N then return the polynomial a(X).
else return the polynomial 2 + X^2 + X^4 + ··· + X^{N−1}.
```

Proof. The algorithm has two main operations. In the first operation, three consecutive coefficients α, β, γ of $a(X)$ are altered by the rule

$$\textbf{Rule I}: \quad (\alpha,\beta,\gamma) \longrightarrow (u, \beta+v, \gamma+v), \qquad \text{where } \alpha = u + 2v.$$

In the second operation, two consecutive coefficients α, β are altered by the rule

$$\textbf{Rule II}: \quad (\alpha,\beta) \longrightarrow (\alpha-2w, \beta-w), \qquad \text{where } w = \min\{\lfloor \alpha/2 \rfloor, \beta\}.$$

We first observe that these rules do not change the value of $a(X)$ at $X = -2$, since for Rule I we have

$$\begin{aligned}\alpha\cdot(-2)^i + \beta\cdot(-2)^{i+1} + \gamma\cdot(-2)^{i+2} \\ &= (u+2v)\cdot(-2)^i + \beta\cdot(-2)^{i+1} + \gamma\cdot(-2)^{i+2} \\ &= u\cdot(-2)^i + (\beta+v)\cdot(-2)^{i+1} + (\gamma+v)\cdot(-2)^{i+2}\end{aligned}$$

and for Rule II we have

$$\alpha\cdot(-2)^i + \beta\cdot(-2)^{i+1} = (\alpha-2w)\cdot(-2)^i + (\beta-w)\cdot(-2)^{i+1}.$$

Also note that both rules leave all coefficients positive.

The invariance of the value of $a(-2)$ under the two substitution rules is not quite true if $i = N-2$ or $i = N-1$, since then the indices on some of the coefficients are reduced modulo N. For example, if $i = N-2$, then Rule I changes the value of $a(-2)$ as follows:

$$\begin{aligned}\alpha\cdot(-2)^{N-2} + \beta\cdot(-2)^{N-1} + \gamma\cdot(-2)^0 \\ &\to u\cdot(-2)^{N-2} + (\beta+v)\cdot(-2)^{N-1} + (\gamma+v)\cdot(-2)^0 \\ &= \alpha\cdot(-2)^{N-2} + \beta\cdot(-2)^{N-1} + \gamma\cdot(-2)^0 + v(1+2^N).\end{aligned}$$

Thus the value of $a(-2)$ changes, but its congruence class modulo 2^N+1 does not change.

Similarly, if $i = N-1$, then the following formulas show that both rules change the value of $a(-2)$ by a multiple of 2^N+1:

$$\begin{aligned}\alpha\cdot(-2)^{N-1} + \beta\cdot(-2)^0 + \gamma\cdot(-2)^1 \\ &\longrightarrow u\cdot(-2)^{N-1} + (\beta+v)\cdot(-2)^0 + (\gamma+v)\cdot(-2)^1 \\ &= \alpha\cdot(-2)^{N-1} + \beta\cdot(-2)^0 + \gamma\cdot(-2)^1 - v(1+2^N).\end{aligned}$$

$$\begin{aligned}\alpha\cdot(-2)^{N-1} + \beta\cdot(-2)^0 &\longrightarrow (\alpha-2w)\cdot(-2)^{N-1} + (\beta-w)\cdot(-2)^0 \\ &= \alpha\cdot(-2)^{N-1} + \beta\cdot(-2)^0 - w(1+2^N).\end{aligned}$$

We have now verified that repeated applications of Rules I and II do not change the value of $a(-2) \bmod 2^N + 1$, and they also leave all coefficients nonnegative. We next study the effect of the rules on the coefficients.

Consider first Rule I as applied to three consecutive coefficients (α, β, γ) of $a(X)$. If $\alpha \in \{0, 1\}$, then Rule I has no effect; and in any case after Rule I is applied, the value of α will be either 0 or 1.

Next consider Rule II as applied to two consecutive coefficients (α, β) of $a(X)$. If $\alpha \in \{0, 1\}$, then Rule II has no effect, since $w = 0$. In any case, after Rule II is applied, we have either $\alpha \in \{0, 1\}$ or else $\beta = 0$, depending on which of $\alpha/2$ and β is smaller.

Now consider what happens on each iteration through the main loop of the algorithm. We first apply Rule I to three consecutive coefficients (a_i, a_{i+1}, a_{i+2}), replacing them with coefficients $(\epsilon_i, a'_{i+1}, a'_{i+2})$ satisfying $\epsilon_i \in \{0, 1\}$, and then we apply Rule II to the pair of consecutive coefficients (a'_{i+1}, a'_{i+2}). There are two possible outcomes. We end up with either

$$(\epsilon_i, a''_{i+1}, 0) \qquad \text{or} \qquad (\epsilon_i, \epsilon_{i+1}, a''_{i+2}),$$

with $\epsilon_i, \epsilon_{i+1} \in \{0, 1\}$. In particular, the ith coefficient of $a(X)$ becomes 0 or 1, and at most the next two coefficients are altered. Thus after the main loop has been iterated N times, every coefficient of $a(X)$ will have been changed to 0 or 1, but the first two coefficients a_0 and a_1 may have been subsequently changed to some other value when we applied the two rules to a_{N-2} and a_{N-1}. However, after the rules have been applied to a_{N-1}, at most one of a_0 and a_1 can be nonbinary.

So after N iterations of the main loop, we are in the situation where at most one a_i is nonbinary. We may also assume that the nonbinary coefficient satisfies $0 \le a_i < 2^N + 1$, since it is always permissible to reduce the coefficients of $a(X)$ modulo $2^N + 1$. (In practice, the nonbinary value will be much smaller than $2^N + 1$ if the original coefficients of $a(X)$ were not too large.) Further, each additional pass through the loop will leave at most one coefficient of $a(X)$ nonbinary.

Let (α, β, γ) be consecutive coefficients of $a(X)$ with α the nonbinary coefficient. As noted above, after applying the Rule I we get $(\epsilon, \beta', \gamma')$, and when we then apply Rule II, we get either $(\epsilon, \beta'', 0)$ or $(\epsilon, \epsilon', \gamma'')$, where $\epsilon, \epsilon' \in \{0, 1\}$. We are going to estimate the size of β'' and γ'' and show that they are considerably smaller than α.

The application of Rule I says $\beta' = \beta + v$ and $\gamma' = \gamma + v$, where $\alpha = \epsilon + 2v$. Note that $v \le \alpha/2$. Now suppose that Rule II yields $(\epsilon, \beta'', 0)$. This occurs if $\gamma' \le \beta'/2$ and we have $\beta'' = \beta' - 2\gamma'$. Then

$$\beta'' = \beta' - 2\gamma' \le \beta' = \beta + v \le \beta + \alpha/2 \le 1 + \alpha/2, \quad \text{since } \beta \in \{0, 1\}.$$

Similarly, if Rule II yields $(\epsilon, \epsilon', \gamma'')$, then $\beta'/2 \le \gamma'$, we have $\gamma'' = \gamma' - \beta'$, and

$$\gamma'' = \gamma' - \beta' \le \gamma' = \gamma + v \le \gamma + \alpha/2 \le 1 + \alpha/2, \quad \text{since } \gamma \in \{0, 1\}.$$

To recapitulate, we have shown that after N iterations of the main loop, there is at most one nonbinary coefficient in $a(X)$, call it A. Further, each subsequent iteration of the main loop yields a polynomial with at most one nonbinary coefficient, call it B, satisfying

$$0 \le B \le A/2 + 1.$$

We know that the initial nonbinary coefficient is at most 2^N, so it takes at most $N+1$ iterations to ensure that the sole nonbinary coefficient is at most 2. (Notice if $A = 2$, the bound for B is 2.)

Finally, we need to consider what happens if we obtain a polynomial with one coefficient equal to 2 and the other coefficients all binary. If we at the effect of Rule I on this coefficient and the two subsequent coefficients, we find four possibilities:

$$(2,0,0) \to (0,1,1), \quad (2,0,1) \to (0,1,2), \quad (2,1,0) \to (0,0,0), \quad (2,1,1) \to (0,0,1).$$

Thus $a(X)$ becomes binary unless three consecutive coefficients equal $(2,0,1)$, in which case they are permuted to $(0,1,2)$. Thus repeated application of Rule I will eventually eliminate the 2, unless the coefficients of $a(X)$ following the 2 look like $0,1,0,1,0,\ldots,0,1$, in which case after at most N iterations $a(X)$ will become the polynomial $2 + X^2 + X^4 + \cdots + X^{N-1}$.

This completes the proof of the theorem except for the assertion that the value of the exceptional polynomial at $X = -2$ satisfies

$$2 + 4 + 4^2 + \cdots + 4^{(N-1)/2} \equiv \frac{2^{N+1} + 2}{3} \pmod{2^N + 1}.$$

We leave the proof of this congruence, valid for all odd integers N, as an exercise for the reader. □

Remark. If the coefficients of $a(X)$ are positive and small compared to 2^N, then a more careful analysis of the size of the coefficients shows that after the first N iterations of the algorithm, the sole nonbinary coefficient A satisfies

$$A \le 2 \max_{0 \le i < N} a_i.$$

(The 2 can certainly be improved. We leave the proof of this formula to the reader.) Hence except in the exceptional case, the main loop of the algorithm is actually executed no more than

$$N + 1 + \log_2 \max_i a_i$$

times.

6. Low Hamming weight products

In this section we briefly describe a method for speeding up the encryption and decryption processes through the use of products of low Hamming weight polynomials. For further details and a discussion of similar constructions for Diffie–Hellman over $\mathrm{GF}(2^n)$ and on Koblitz elliptic curves, see [HS1].

The most time consuming part of NTRU encryption is computation of the product $r(X) * h(X) \bmod q$, and the most time consuming part of NTRU decryption is computation of the product $f(X) * e(X) \bmod q$. The polynomials $h(X)$ and $e(X)$ have coefficients that are more or less randomly distributed modulo q, while one normally take $r(X)$ and $f(X)$ to have binary (i.e., 0 or 1) or trinary (i.e., -1, 0, or 1) coefficients.

For concreteness, suppose that $r(X)$ is a binary polynomial with d ones. Then computation of the product $r(X) * h(X) \bmod q$ requires approximately dN operations, where one operation is an addition and a remainder modulo q. We further require that the set of $r(X)$ polynomials be sufficiently large so that an attacker cannot find $r(X)$ by an exhaustive search, and there is also a lattice norm requirement on the number of nonzero coefficients in $r(X)$. The search space for $r(X)$ has essentially $\binom{N-1}{d-1}$ elements, since the rotation $X^i r(X)$ is really equivalent to $r(X)$. For information about the lattice norm requirement, see [HPS, S1, S2].

It is possible to significantly reduce the computational requirements by taking $r(X)$ to be a product of polynomials with fewer ones. Thus suppose that we write $r(X) = r_1(X)r_2(X)$, where r_1 and r_2 are binary polynomials with d_1 and d_2 ones respectively. Then $r(X)$ will have approximately $d_1 d_2$ ones. (In practice, it will have a few twos and an occasional three mixed in with the ones, but that will not affect matters very much.) Notice that the computation of the product

$$r(X) * h(X) = r_1(X) * \big(r_2(X) * h(X)\big)$$

requires only $(d_1 + d_2)N$ operations, so the computational complexity is proportional to the sum of d_1 and d_2. On the other hand, the search space for the pair of polynomials (r_1, r_2) has size approximately $\binom{N-1}{d_1-1}\binom{N-1}{d_2-1}$, so is proportional to the product of the r_1 search space and the r_2 search space. (In practice, there may be a meet-in-the-middle search that reduces the size of the search space, see [HS1] for details.) Further, the number of nonzero coefficients in $r_1(X)r_2(X)$ is essentially the product $d_1 d_2$. Thus one might say that using a product $r = r_1 r_2$ requires computation proportional to the sum $d_1 + d_2$ while giving security proportional to the product $d_1 d_2$. In rough terms, this explains why one obtains significant performance gains without changing the level of security.

Similar remarks apply to the private key polynomial $f(X)$; or, if $f(X)$ is taken in the form $f(X) = 1 + p * f'(X)$, to the polynomial $f'(X)$. For example, one might take $f(X)$ to have the form

$$f(X) = 1 + p * (f_1(X) * f_2(X) + f_3(X)),$$

where f_1, f_2, f_3 are quite sparse binary polynomials. Further, to help reduce decryption failures caused by coefficients in the polynomial (1) spreading further than q, one might further choose f_1, f_2, f_3 so that the polynomial $f_1 * f_2 + f_3$ is purely binary, that is, has only 0 and 1 coefficients. In principle this will cut down somewhat on the size of the search space, but in practice there does not appear to be any way to determine if a triple (f_1, f_2, f_3) is in the desired space without actually computing at least part of the quantity $f_1 f_2 + f_3$. Thus the effective search space remains the collection of all triples (f_1, f_2, f_3) having the specified number of nonzero coefficients.

Acknowledgements. The authors would like to thank Jill Pipher for her assistance, Eliane Jaulmes and Antoine Joux for their work on adaptive attacks against NTRU, E. Fujisaki and T. Okamoto for correspondence related to their work on protection against general adaptive attacks, and Rafał Nessel for his suggestions and corrections to the original version of this paper.

References

[FO1] Fujisaki, E., Okamoto, T., Secure integration of asymmetric and symmetric encryption schemes. In: Advances in Cryptology – CRYPTO '99 (ed. by M. Wiener; Lecture Notes in Comput. Sci. 1666), 537–554. Springer, Berlin 1999.

[FO2] —, —, How to Enhance the Security of Public-Key Encryption at Minimum Cost. IEICE Transactions on Fundamentals of Electronics, Communications and Computer Sciences, Vol. E83-A, No. 1, Special Issue on Cryptography and Information Security (January 2000).

[HPS] Hoffstein, J., Pipher, J., Silverman, J.H., NTRU: a ring-based public key cryptosystem. In: Algorithmic Number Theory – ANTS III (ed. by J.P. Buhler; Lecture Notes in Comput. Sci. 1423), 267–288. Springer, Berlin 1998.

[HS1] Hoffstein, J., Silverman, J.H., Small Hamming Weight Products in Cryptography. Preprint, September 2000.

[HS2] —, —, Reaction Attacks Against the NTRU Public Key Cryptosystem. NTRU Technical Report #015, August 1999, <www.ntru.com>.

[HS3] —, —, Protecting NTRU Against Chosen Ciphertext and Reaction Attacks. NTRU Technical Report #016, June 2000, <www.ntru.com>.

[JJ] Jaulmes, E., Joux, A., A chosen-ciphertext attack against NTRU. In: Advances in Cryptology – CRYPTO 2000 (ed. by M. Bellare; Lecture Notes in Comput. Sci. 1880), 20–35. Springer, Berlin 2000.

[S1] Silverman, J.H., Estimated Breaking Times for NTRU Lattices. NTRU Technical Report #012, March 1999, <www.ntru.com>.

[S2] —, Dimension-Reduced Lattices, Zero-Forced Lattices, and NTRU Public Key Cryptosystem. NTRU Technical Report #013, March 1999, <www.ntru.com>.

The efficiency and security of a real quadratic field based key exchange protocol

Michael J. Jacobson, Jr., Renate Scheidler and Hugh C. Williams

Abstract. Most cryptographic key exchange protocols make use of the presumed difficulty of solving the discrete logarithm problem (DLP) in a certain finite group as the basis of their security. Recently, real quadratic number fields have been proposed for use in the development of such protocols. Breaking such schemes is known to be at least as difficult a problem as integer factorization; furthermore, these are the first discrete logarithm based systems to utilize a structure which is not a group, specifically the collection of reduced ideals which belong to the principal class of the number field. For this structure the DLP is essentially that of determining a generator of a given principal ideal.

Unfortunately, there are a few implementation-related disadvantages to these schemes, such as the need for high precision floating point arithmetic and an ambiguity problem that requires a short, second round of communication. In this paper we describe work that has led to the resolution of some of these difficulties. Furthermore, we discuss the security of the system, concentrating on the most recent techniques for solving the DLP in a real quadratic number field.

1. Introduction and motivation

In 1976, W. Diffie and M. Hellman introduced their by now legendary protocol for communicating a secret cryptographic key across an insecure channel. In their original scheme, two communication partners first agree on a large prime p and a nonzero element $g \in \mathbb{F}_p$; p and g are public. The two parties generate respective secret exponents a and b, and subsequently exchange g^a and g^b. From their own secret exponent and the information received from the other communicant, each party can now generate the secret common key g^{ab}. The only known attack on this scheme is for an eavesdropper to solve an instance of the *discrete logarithm problem* (DLP) in $\mathbb{F}_p$: given $g, g^a \in \mathbb{F}_p$, find a.

It was soon recognized that the Diffie–Hellman protocol could be extended to any sufficiently large group with fast arithmetic and a difficult discrete logarithm problem. Many such groups were suggested, including arbitrary finite fields, matrices over a finite field [15], the unit group of $\mathbb{Z}/n\mathbb{Z}$ with n a product of two primes [13], the group of points on an elliptic curve [14, 8], the Jacobian of a hyperelliptic curve [9], and the class group of an imaginary quadratic number field [3].

Public-Key Cryptography and Computational Number Theory
K. Alster, J. Urbanowicz and H. C. Williams (Eds.)

More careful inspection of the scheme reveals that only a semi-group structure is required, and two non-group versions were proposed, using as underling structure the infrastructure of a real quadratic number field and function field, respectively [17, 18].

In order to conduct key exchange in the set $\mathcal{R}$ of reduced principal ideals of a real quadratic number field K, one requires an operation under which $\mathcal{R}$ is closed and associative. The most natural way to do this is to assign to two reduced principal ideals the reduced principal ideal "closest" to their (generally nonreduced) product; that is, the unique reduced ideal whose generator is closest to the product of the two respective generators. In this model, a reduced principal ideal $\mathfrak{r}$ is publicized beforehand. Alice and Bob generate respective secret exponents a and b and compute the reduced principal ideal closest to $\mathfrak{r}^a$ and $\mathfrak{r}^b$, respectively. The common key the unique reduced principal ideal closest to $\mathfrak{r}^{ab}$.

Unfortunately, ideal generators are generally too large to allow for efficient arithmetic. Moreover, in practice, they are only given up to a limited precision, so it may be impossible to determine the unique reduced principal ideal closest to a given ideal. The arithmetic of [17] opted for computing two possible candidates – adjacent reduced principal ideals – for the closest ideal. To make sure that both parties obtain the same two candidates for the final key ideal, generators had to be given to p correct bits where $2^p > 3072\sqrt{D}B^2$ and B is an upper bound on the exponents a and b. The problem of ambiguity was eliminated in [4], maintaining essentially the same precision p for computing the key ideal. In this paper, we give a more elegant version of the protocol of [17] which achieves five-fold ambiguity with considerably less precision, namely $2^p > 46B^2 \log_2 B$; note that this bound is independent of the radicand D of the field. In practice, the two communicants always obtain the same ideal after their exchange, but for the paranoid user, we describe how communicating five more bits guarantees uniqueness.

Our paper is organized as follows. In the next section, we review the basics of real quadratic number fields and their ideals. Section 3 introduces the so called (f, p) representations used to represent ideals and the algorithms underlying our protocol. In Section 4, we present the actual key exchange protocol. In the remaining sections we discuss implementation and security issues.

2. Real quadratic number fields

In this section we will review some basic properties of real quadratic number fields from a computational perspective. This material can be found in greater detail in [21] and [16].

Let D be a positive squarefree integer and let $K = \mathbb{Q}(\sqrt{D})$ be the real quadratic number field of discriminant $\Delta = (4/\sigma^2)D$ where $\sigma = 2$ if $D \equiv 1 \pmod 4$ and $\sigma = 1$ otherwise. For an element $\alpha = a + b\sqrt{D} \in K$ with $a, b \in \mathbb{Q}$, the conjugate of α is $\overline{\alpha} = a - b\sqrt{D}$. The maximal order of K is $\mathcal{O} = \mathbb{Z}[\omega]$ where $\omega = (\sigma - 1 + \sqrt{D})/\sigma$.

Every nonzero ideal $\mathfrak{a}$ in $\mathcal{O}$ is a $\mathbb{Z}$-module of rank 2 with a $\mathbb{Z}$-basis of the form $\{SQ/\sigma, S(P + \sqrt{D})/\sigma\}$ with $S, Q, P \in \mathbb{Z}$, $Q > 0$, $\sigma \mid Q$ and $\sigma Q \mid D - P^2$. If $\mathfrak{a}$ is primitive, then $S = 1$; in this case, write $\mathfrak{a} = (Q, P)$ for brevity. A primitive ideal $\mathfrak{a} = (Q, P)$ is reduced if Q is a minimum in $\mathfrak{a}$; that is, no nonzero $\alpha \in \mathfrak{a}$ satisfies $|\alpha| < Q$ and $|\overline{\alpha}| < Q$. If $\mathfrak{b}_1 = (Q_0, P_0)$ is any primitive ideal, then the recursion

$$q_{j-1} = \left\lfloor \frac{P_{j-1} + \sqrt{D}}{Q_{j-1}} \right\rfloor, \quad P_j = q_{j-1}Q_{j-1} - P_{j-1}, \quad Q_j = \frac{D - P_j^2}{Q_{j-1}}, \tag{2.1}$$

$j \in \mathbb{N}$, defines a sequence of pairwise equivalent ideals $\mathfrak{b}_j = (Q_{j-1}, P_{j-1})$ $(j \in \mathbb{N})$. This, of course, is the continued fraction expansion of $(P + \sqrt{D})/Q$, where the q_j values are the *partial quotients* and the $(P_j + \sqrt{D})/Q_j$ values are the *complete quotients*. Here, $\mathfrak{b}_j = (\Psi_j)\mathfrak{b}_1$ where

$$\Psi_j = \prod_{i=1}^{j-1} \psi_i \text{ and } \psi_i = \frac{P_i + \sqrt{D}}{Q_{i-1}} \quad (1 \le i \le j-1). \tag{2.2}$$

for $j \in \mathbb{N}$. It is easy to see that $\Psi_{j+2} = q_j\Psi_{j+1} + \Psi_j$ and $\Psi_j\overline{\Psi_j} = (-1)^{j-1}Q_{j-1}/Q_0$ for $j \in \mathbb{N}$. If $\mathfrak{b}_1$ is reduced, then $\mathfrak{b}_j$ is reduced for all $j \in \mathbb{N}$, in which case

$$1 + \frac{1}{\sqrt{\Delta}} < \psi_j < \sqrt{\Delta} \text{ and } \psi_{j+1}\psi_j > 2 \quad (j \in \mathbb{N}). \tag{2.3}$$

If $\mathfrak{b}_1$ is nonreduced, then $\mathfrak{b}_j$ is reduced as soon as $Q_{j-1} > 0$ and $P_{j-1} < \sqrt{D}$; this happens after no more than $O(\log(Q_0/\sqrt{D}))$ iterations of (2.1). Furthermore, if $\mathfrak{b}_j$ is the first reduced ideal in the sequence $(\mathfrak{b}_i)_{i\in\mathbb{N}}$, then $\Psi_j \le 1$ (see Corollary 4.6 of [16]) and $|\overline{\Psi_i}| < 1$, $0 < P_i < \sqrt{D}$, and $\sigma \le Q_i < 2\sqrt{D}$ for $i \ge j$. The process of obtaining $\mathfrak{b}_{j+1}$ from $\mathfrak{b}_j$ is called a *forward step*. Also, by the symmetry property of the continued fraction of ω, we see that ψ_i is a complete quotient in that continued fraction.

Suppose $\mathfrak{b}_{j+1} = (Q_j, P_j)$ is reduced for some $j \in \mathbb{N}$. Then one can perform a *backward step* to obtain $\mathfrak{b}_j = (Q_{j-1}, P_{j-1})$ from the recursion

$$Q_{j-1} = \frac{D - P_j^2}{Q_j}, \quad q_{j-1} = \left\lfloor \frac{P_j + \sqrt{D}}{Q_{j-1}} \right\rfloor, \quad P_{j-1} = q_{j-1}Q_{j-1} - P_j. \tag{2.4}$$

If we set $\mathfrak{b}_1^* = (Q_0^*, P_0^*) = \mathfrak{b}_j$, so $Q_0^* = Q_{j-1}$ and $P_0^* = P_{j-1}$, then (2.4) generates a sequence of reduced ideals $\mathfrak{b}_i^* = (Q_{i-1}^*, P_{i-1}^*)$ $(i \in \mathbb{N})$. Here, we have $\mathfrak{b}_i^* = (\Psi_i^*)\mathfrak{b}_1^*$ where $\Psi_i^* = \Psi_{j-i+1}$ and in fact $\mathfrak{b}_i^* = \mathfrak{b}_{j-i+1}$. Analogous to (2.2), we have

$$\Psi_i^* = \prod_{j=1}^{i-1} \psi_j^* \text{ where } \psi_j^* = \frac{\sqrt{D} - P_{j-1}^*}{Q_{j-1}^*} = \frac{1}{\psi_{i-j}} \quad (1 \le j \le i-1) \tag{2.5}$$

as well as $|\Psi_i^*\overline{\Psi_i^*}| = Q_{i-1}^*/Q_0$ and $\Psi_{i+2}^* = \Psi_i^* - q_i^*\Psi_{i+1}^*$ for $1 \le i \le j-2$.

Every principal ideal $\mathfrak{a}$ has a unique generator θ such that $1 \le \theta < \epsilon$ where ϵ is the unique fundamental unit of K with $\epsilon > 1$. For $i \in \mathbb{N}$, let $\mathfrak{b}_i = (\Psi_i)$ with $\mathfrak{b}_1 = \mathcal{O}$. If $\mathfrak{a}$ is nonreduced, then there exists a unique index $j \in \mathbb{N}$ with $\Psi_j < \theta < \Psi_{j+1}$. We define $\rho(\mathfrak{a}) = \mathfrak{b}_{j+1}$ and $\rho^{-1}(\mathfrak{a}) = \mathfrak{b}_j$. Write $\rho^0(\mathfrak{a}) = \mathfrak{a}$ and $\rho^n(\mathfrak{a}) = \rho(\rho^{n-1}(\mathfrak{a})) = \rho^{-1}(\rho^{n+1}(\mathfrak{a}))$ for $n \in \mathbb{Z}$.

3. (f, p) representations and their arithmetic

In order to perform arithmetic with principal ideals, we represent a (generally unknown) primitive ideal $\mathfrak{a}$ by a known ideal $\mathfrak{b} = (\theta)\mathfrak{a}$ (where (θ) is the fractional principal ideal generated by $\theta \in K$) and an approximation of θ to a given precision and accuracy. More precisely:

Definition 3.1. Let $p \in \mathbb{N}$ and $f \in \mathbb{R}$ with $1 \le f < 2^p$. An (f, p) *representation* of a primitive ideal $\mathfrak{a}$ is a pair $(\mathfrak{b}, d)$ where

1. $\mathfrak{b}$ is an ideal equivalent to $\mathfrak{a}$,
2. $d, p \in \mathbb{Z}$ with $d > 0$ and $p \ge 0$,
3. if $\theta \in K$ with $\mathfrak{b} = (\theta)\mathfrak{a}$, then $\left| \dfrac{2^p\theta}{d} - 1 \right| < \dfrac{f}{2^p}$.

An (f, p) representation $(\mathfrak{b}, d)$ of $\mathfrak{a}$ is *reduced* if $\mathfrak{b}$ is a reduced ideal.

Informally speaking, $d2^{-p}$ is an approximation of the relative generator θ of $\mathfrak{b}$ with respect to $\mathfrak{a}$ to accuracy $f2^{-p}$, and p is the precision of the approximation. Note that for any primitive ideal $\mathfrak{a}$ and any $p \in \mathbb{N}$, $(\mathfrak{a}, 2^p + 1)$ is always a $(1, p)$ representation, and hence an (f, p) representation for any $f \in [1, 2^p)$, of $\mathfrak{a}$.

We now investigate the arithmetic of (f, p) representations.

Algorithm MULT

Input: (f_i, p) representations $(\mathfrak{b}_i, d_i)$ of primitive ideals $\mathfrak{a}_i$ where $\mathfrak{b}_i = (Q_i, P_i)$ for $i = 1, 2$.

Output: An (f, p) representation $((S)\mathfrak{b}, d)$ of $\mathfrak{a}_1\mathfrak{a}_2$ where $\mathfrak{b} = (Q, P)$ is primitive and $f = 1 + f_1 + f_2 + f_1 f_2 2^{-p}$.

Algorithm:

1. Compute $G = \gcd\left(\dfrac{Q_1}{\sigma}, \dfrac{Q_2}{\sigma}\right)$ and solve $\dfrac{Q_1}{\sigma} X \equiv G \left(\bmod \dfrac{Q_2}{\sigma}\right)$ for $X \in \mathbb{Z}$.
2. Compute $S = \gcd\left(\dfrac{P_1 + P_2}{\sigma}, G\right)$ and solve $Y \dfrac{P_1 + P_2}{\sigma} + ZG = S$ for $Y, Z \in \mathbb{Z}$.
3. Set $U \equiv XY(P_2 - P_1) + Z\,\dfrac{D - P_1^2}{Q_1} \left(\bmod \dfrac{Q_2}{S}\right)$.
4. Set $Q = \dfrac{Q_1 Q_2}{\sigma S^2}$, $P \equiv P_1 + U \dfrac{Q_1}{\sigma S} \pmod{Q}$, $d = \left\lceil \dfrac{d_1 d_2}{2^p} \right\rceil$.

Write $((S)\mathfrak{b}, d) = \mathrm{MULT}(\mathfrak{b}_1, d_1, \mathfrak{b}_2, d_2)$ for the output of Algorithm MULT.

Lemma 3.2. *If $d_1, d_2 > 2^p$, then $d > 2^p$, and Algorithm* MULT *produces the correct output using $O(\log D)$ integer operations.*

Proof. The expressions for Q and P are simply given by the well-known ideal multiplication formulas, and the complexity result is clear. Let $\mathfrak{b}_i = (\theta_i)\mathfrak{a}_i$ for $i = 1, 2$. Then $\mathfrak{b} = (\theta_1\theta_2)\mathfrak{a}_1\mathfrak{a}_2$. Set $f_0 = f_1 + f_2 + f_1 f_2 2^{-p} = f - 1$. Then

$$1 - \frac{f_0}{2^p} < \frac{2^{2p}\theta_1\theta_2}{d_1 d_2} < 1 + \frac{f_0}{2^p}\,.$$

It is clear that $d > 2^p$ and $2^p\theta_1\theta_2/d < 1 + f2^{-p}$. If $\eta = d - d_1d_2 2^{-p}$, then

$$2^p\theta_1\theta_2 > (d-\eta)\Big(1 - \frac{f_0}{2^p}\Big) = d\Big(1 - \frac{f_0 + 2^p\eta d^{-1}}{2^p}\Big) + \eta\frac{f_0}{2^p} > d\Big(1 - \frac{f_0+1}{2^p}\Big). \quad \square$$

For the following algorithm, we assume that $s \in \mathbb{N}$ is sufficiently large so that

$$\frac{d_1}{2^s} < \frac{1}{10}.$$

Algorithm REDUCE

Input: An (f,p) representation $((S)\mathfrak{b}_1, d_1)$ of some ideal $\mathfrak{a}$ where $\mathfrak{b}_1 = (Q_0, P_0)$ is primitive.

Output: A reduced $(f+2,p)$ representation $(\mathfrak{b}, d)$ of $\mathfrak{a}$ where $\mathfrak{b} = (Q,P)$.

Algorithm:

1. Set $T_{-2} = -P_0 2^s + \lfloor 2^s\sqrt{D}\rfloor$, $T_{-1} = 2^s Q_0$, $M = \lfloor SQ_0 2^{p+s}/d_1\rfloor$, $j = 0$.
2. While $Q_{j-1} < 0$ or $P_{j-1} > \sqrt{D}$ do
   (a) increase j by 1;
   (b) set $q_{j-1} = \left\lfloor \dfrac{P_{j-1} + \sqrt{D}}{Q_{j-1}} \right\rfloor$, $P_j = q_{j-1}Q_{j-1} - P_{j-1}$, $Q_j = \dfrac{D - P_j^2}{Q_{j-1}}$;
   (c) set $T_{j-1} = q_{j-1}T_{j-2} + T_{j-3}$.
3. If $T_{j-2} \le M$, then
   (a) Repeat
      i. increase j by 1;
      ii. set $q_{j-1} = \left\lfloor \dfrac{P_{j-1} + \sqrt{D}}{Q_{j-1}} \right\rfloor$, $P_j = q_{j-1}Q_{j-1} - P_{j-1}$, $Q_j = \dfrac{D - P_j^2}{Q_{j-1}}$;
      iii. set $T_{j-1} = q_{j-1}T_{j-2} + T_{j-3}$
      until $T_{j-2} > M$.
   (b) Set $\mathfrak{b} = (Q_{j-1}, P_{j-1})$, $d = \left\lceil \dfrac{d_1 T_{j-2}}{SQ_0 2^s} \right\rceil$.

   else

   (c) Set $Q_0^* = Q_{j-1}$, $P_0^* = P_{j-1}$, $T_{-1}^* = T_{j-2}$, $T_0^* = T_{j-3}$, $j = 0$.
   (d) Repeat
      i. increase j by 1;
      ii. set $Q_j^* = \dfrac{D - (P_{j-1}^*)^2}{Q_{j-1}^*}$, $q_j^* = \left\lfloor \dfrac{P_{j-1}^* + \sqrt{D}}{Q_j^*} \right\rfloor$, $P_j^* = q_j^* Q_j^* - P_{j-1}^*$;
      iii. set $T_j^* = T_{j-2}^* - q_j^* T_{j-1}^*$
      until $T_{j-1}^* \le M$.
   (e) Set $\mathfrak{b} = (Q_{j-1}^*, P_{j-1}^*)$, $d = \left\lceil \dfrac{d_1 T_{j-2}^*}{SQ_0 2^s} \right\rceil$.

Write $(\mathfrak{b}, d) = \text{REDUCE}((S)\mathfrak{b}_1, d_1)$ for the output of Algorithm REDUCE.

With the notation of the algorithm, we let

$$\mathfrak{b}_j = (Q_{j-1}, P_{j-1}),\ d_j = \left\lceil \frac{d_1 T_{j-2}}{SQ_0 2^s} \right\rceil,\ \mathfrak{b}_j^* = (Q_{j-1}^*, P_{j-1}^*),\ d_j^* = \left\lceil \frac{d_1 T_{j-2}^*}{SQ_0 2^s} \right\rceil$$

for $j \in \mathbb{N}$. We first show that all the loops in the above algorithm terminate.

Lemma 3.3. *Let s be as described above and assume that $d_1 > 2^p$. Then the loops in steps* 2, 3(a), *and* 3(d) *in Algorithm* REDUCE *terminate.*

Proof. The while loop in step 2 terminates after $O(\log(Q_0/D))$ iterations, once the first reduced ideal $\mathfrak{b}_l$ ($l \in \mathbb{N}$) is encountered. Then $\Psi_l \leq 1$. We consider two cases.

Case 1. $T_{l-2} \leq M$. Then it suffices to show the existence of an index $k > l$ with $T_{k-2} > M$.

For $i \in \mathbb{N}$, we have $\mathfrak{b}_i = (\Psi_i)\mathfrak{b}_1$ where Ψ_i is given by (2.2). Write $\Psi_i = (G_{i-2} + B_{i-2}\sqrt{D})/Q_0$ with $G_{i-2}, B_{i-2} \in \mathbb{Z}$ and $B_{i-2} \geq 0$. Set

$$\hat{\Psi}_i = \frac{G_{i-2} + B_{i-2}2^{-s}\lfloor 2^s\sqrt{D}\rfloor}{Q_0}.$$

Then

$$0 < \Psi_i - \hat{\Psi}_i < \frac{B_{i-2}}{Q_0 2^s}$$

and $T_{i-2} = 2^s Q_0 \hat{\Psi}_i$ for $i \in \mathbb{N}$. It is easy to see that $T_{i-2} > M$ if and only if $\hat{\Psi}_i > S2^p/d_1$.

Since $\Psi_1 = 1$ and the sequence $(\Psi_i)_{i \geq l}$ is increasing, there exists an index $k > l$ for which $\Psi_{k-1} \leq 3S/2 < \Psi_k$. We have $\Psi_k = \psi_{k-1}\Psi_{k-1}$ with $\psi_{k-1} = (P_{k-1} + \sqrt{D})/Q_{k-2} < 2\sqrt{D}$, hence $\Psi_k < 3S\sqrt{D}$. Together with $|\overline{\Psi}_k| < 1$, obtain

$$0 < \frac{B_{k-2}}{Q_0 2^s} \leq \frac{\Psi_k + |\overline{\Psi}_k|}{2^{s+1}\sqrt{D}} < \frac{3S\sqrt{D} + 1}{2^{s+1}\sqrt{D}} < S \cdot 2^{1-s} < \frac{S}{2} \tag{3.1}$$

and hence $\hat{\Psi}_k > \Psi_k - B_{k-2}/Q_0 2^s > S > S2^p/d_1$. Then $T_{k-2} > 2^{p+s}Q_0 S/d_1$ and hence $T_{k-2} > M$.

Case 2. $T_{l-2} > M$. Then $T_{-1}^* > M$ and we need to establish the existence of an index $k \in \mathbb{N}$ with $T_k^* \leq M$. We have $\mathfrak{b}_i^* = (\Psi_i^*)\mathfrak{b}_1^*$ where the Ψ_i^* are given by (2.5), and as before, we define an approximation $\hat{\Psi}_i^*$ to Ψ_i^* ($i \in \mathbb{N}$) for which

$$0 < \operatorname{sgn}(B_{i-2})(\Psi_i^* - \hat{\Psi}_i^*) < \frac{|B_{i-2}^*|}{Q_0 2^s}$$

and $T_{i-2}^* = 2^s Q_0 \hat{\Psi}_i^*$ for $i \in \mathbb{N}$. Once again, $T_{i-2}^* > M$ if and only if $\hat{\Psi}_i^* > S2^p/d_1$.

Now $B_{-1}^* > 0$ implies that $\Psi_1^* > \hat{\Psi}_1^* > S2^p/d_1$. Therefore, there must exist an index $k \in \mathbb{N}$ with $\Psi_k^* > S2^p/d_1 \geq \Psi_{k+1}^*$. If $T_{k-1}^* \leq M$, there is nothing to prove, so assume $T_{k-1}^* > M$. Then $\hat{\Psi}_{k+1}^* > S2^p/d_1 > \Psi_{k+1}^*$ and hence $B_{k-1}^* < 0$. The identity $\Psi^*_{k+2} = \psi^*_{k+1}\Psi^*_{k+1}$ implies

$$B^*_k = \frac{G^*_{k-1} - B^*_{k-1}P^*_k}{Q^*_k} = \frac{Q_0\Psi^*_{k+1} - B^*_{k-1}(P^*_k + \sqrt{D})}{Q^*_k} > 0.$$

Therefore $\hat{\Psi}^*_{k+2} < \Psi^*_{k+2} < \Psi^*_{k+1} \leq S2^p/d_1$ and hence $T^*_k \leq M$. □

Lemma 3.4. *Let $((S)\mathfrak{b}_1, d_1)$ be the input of Algorithm* REDUCE *and let s be as described above. Assume that $2^p < d_1 < 9S^2Q_02^{p-2}$.*

1. *Let $j \in \mathbb{N}$ be defined by the last iteration of the loop in step* 3(a) *of Algorithm* REDUCE, *so $\mathfrak{b}_j$ is reduced and $T_{j-3} \leq M < T_{j-2}$. Then $d_{j-1} \leq 2^p < d_j$, and if $P_{j-1} \neq \lfloor\sqrt{D}\rfloor$, then $d_j \leq 3Q_{j-1}2^{p-1}$.*
2. *Let $j \in N$ be defined by the last iteration of the loop in step* 3(d) *of Algorithm* REDUCE, *so $\mathfrak{b}^*_j$ is reduced and $T^*_{j-1} \leq M < T^*_{j-2}$. Then $d^*_{j+1} \leq 2^p < d^*_j$, and if $P^*_{j-1} \neq \lfloor\sqrt{D}\rfloor$, then $d^*_j \leq 3Q^*_{j-1}2^{p-1}$.*

Proof. We use the same notation as in the proof of the previous lemma. It is easy to see that $T_{i-2} > M$ if and only if $d_i > 2^p$, so $d_{j-1} \leq 2^p < d_j$; similarly, $T^*_{i-2} > M$ if and only if $d^*_i > 2^p$, so $d_{j+1} \leq 2^p < d_j$. Furthermore, $d_12^{-s} < 1/4$ and $2^p + 1/2 < 3 \cdot 2^{p-1}$.

Once again, let $\mathfrak{b}_l$ be first reduced ideal computed in the algorithm and suppose first that $T_{l-2} \leq M$. Let k be as in case 1 of the proof of Lemma 3.3. Then $j \leq k$ and $B_{k-2}/Q_0 < 2S$ by (3.1). Since B_i strictly increases with i, it follows that

$$\frac{d_1}{S}\Psi_{j-1} < \frac{d_1}{S}\left(\hat{\Psi}_{j-1} + \frac{B_{j-3}}{Q_02^s}\right) < d_{j-1} + \frac{d_1}{2^s}\frac{B_{k-2}}{SQ_0} < 2^p + 1/2 < 3 \cdot 2^{p-1}.$$

If $P_{j-1} \neq \lfloor\sqrt{D}\rfloor$, then $\psi_{j-1} = Q_{j-1}/(\sqrt{D} - P_{j-1}) < Q_{j-1}$, hence

$$\frac{d_1}{S}\hat{\Psi}_j \leq \frac{d_1}{S}\Psi_j = \frac{d_1}{S}\psi_{j-1}\Psi_{j-1} < 3Q_{j-1}2^{p-1},$$

and we have $d_j \leq 3Q_{j-1}2^{p-1}$.

Now suppose that $T_{l-2} > M$ and let k be as in case 2 of the proof of Lemma 3.3. Then $j \leq k+1$. Let $i \leq k$, so $\Psi^*_i > S2^p/d_1$. Then

$$|\overline{\Psi^*_i}| = \frac{Q^*_{i-1}}{\Psi^*_iQ_0} < \frac{2\sqrt{D}d_1}{SQ_02^p} < \frac{9}{2}S\sqrt{D},$$

$$|\overline{\Psi^*_{k+1}}| = |\overline{\psi^*_k}||\overline{\Psi^*_k}| < \frac{P^*_{k-1} + \sqrt{D}}{Q^*_{k-1}}\frac{d_1Q^*_{k-1}}{SQ_02^p} < \frac{2\sqrt{D}d_1}{SQ_02^p} < \frac{9}{2}S\sqrt{D}.$$

Together with $\Psi^*_i, \Psi^*_{k+1} < 1$, this implies for all $i \leq k+1$,

$$\frac{d_1|B^*_{i-2}|}{SQ_02^s} \leq \frac{d_1}{S2^s}\frac{|\Psi^*_i| + |\overline{\Psi^*_i}|}{2\sqrt{D}} < \frac{d_1}{2^s}\left(\frac{1}{2S\sqrt{D}} + \frac{9}{4}\right) < \frac{1}{4}.$$

If $j \le k$, then

$$\begin{aligned}\frac{d_1}{S}\hat{\Psi}_j^* &< \frac{d_1}{S}\left(\frac{1}{\psi_j^*}\left(\hat{\Psi}_{j+1}^* + \frac{|B_{j-1}^*|}{Q_0 2^s}\right) + \frac{|B_{j-2}^*|}{Q_0 2^s}\right) \\ &< \frac{1}{\psi_j^*}\left(d_{j+1}^* + \frac{1}{4} + \frac{1}{4}\psi_j^*\right) < \frac{3 \cdot 2^{p-1}}{\psi_j^*},\end{aligned}$$

where we use $\psi_j^* < 1$ for the last inequality. Since $\psi_j^* = (\sqrt{D} - P_{j-1}^*)/Q_{j-1}^* > 1/Q_{j-1}^*$ if $P_{j-1}^* \neq \lfloor\sqrt{D}\rfloor$, we obtain $d_j^* \le 3Q_{j-1}^* 2^{p-1}$.

If $j = k+1$, then $d_1\Psi_j^*/S < 2^p$, so $d_1\hat{\Psi}_j^*/S < 2^p + 1/4 < 3 \cdot 2^{p-1}$ and hence $d_j^* < 3 \cdot 2^{p-1}$. □

Corollary 3.5. *Assume the conditions and notation of Lemma* 3.4.

1. $d_i - 2 < d_1\Psi_i/S < d_i + 1$ *for* $i \in \{j, j-1\}$.
2. $d_i^* - 2 < d_1\Psi_i^*/S < d_i^* + 1$ *for* $i \in \{j, j+1\}$.

Proof. If k is defined as in case 1 of the proof of Lemma 3.3, we have by (3.1), $0 < d_1(\Psi_i - \hat{\Psi}_i)/S < d_1 2^{1-s} < 1$ for all $i \le k$. If k is as in case 2 of the same proof, then again $d_1|\Psi_i^* - \hat{\Psi}_i^*|/S < 1/4$ for all $i \le k+1$. Also, if $j = k+1$, then $d_1\Psi_j^*/S > 2^p - 1/4 > 2^p - 1$, so $|\overline{\Psi_j^*}| < d_1 Q_{j-1}^*/SQ_0(2^p - 1)$ and

$$|\overline{\Psi_{j+1}^*}| = |\overline{\psi_j^*}||\overline{\Psi_j^*}| < \frac{2\sqrt{D}d_1}{SQ_0(2^p-1)} < \frac{18\sqrt{D}S2^{p-2}}{2^p-1}.$$

Together with $\Psi_{j+1}^* < 1$, we obtain

$$\begin{aligned}\frac{d_1}{S}|\Psi_{j+1}^* - \hat{\Psi}_{j+1}^*| &< \frac{d_1}{S}\frac{B_{j-1}^*}{Q_0 2^s} \le \frac{d_1}{S2^s}\frac{\Psi_{j+1}^* + |\overline{\Psi_{j+1}^*}|}{2\sqrt{D}} < \frac{d_1}{2^s}\left(\frac{1}{2S\sqrt{D}} + \frac{9 \cdot 2^{p-2}}{2^p-1}\right) \\ &= \frac{d_1}{2^s}\left(\frac{1}{2S\sqrt{D}} + \frac{9}{4}\right) + \frac{d_1}{2^s}\frac{9/4}{2^p-1} < \frac{1}{2},\end{aligned}$$

where we use the fact that $d_1 2^{-s} < 1/10$ for the last inequality. Now is easy to see that for any two positive real numbers α and $\hat{\alpha}$, the inequality $|\alpha - \hat{\alpha}| < 1$ implies $\lceil\hat{\alpha}\rceil - 2 < \alpha < \lceil\hat{\alpha}\rceil + 1$. □

Theorem 3.6. *Assuming the conditions of Lemma* 3.4 *and the notation of Algorithm* REDUCE, *the following hold.*

1. *$(\mathfrak{b}, d)$ is a reduced $(f+2, p)$ representation of $\mathfrak{a}$.*
2. *The loops in step 3 in the algorithm are executed at most $O(\log(SD))$ and $O(\log D)$ times, respectively.*
3. *$T_i, T_i^* = O(SQ_0\sqrt{D}2^s)$ throughout the algorithm.*

Proof. Let $(S)\mathfrak{b}_1 = (\theta)\mathfrak{a}$, $\mathfrak{b} = (\Psi)\mathfrak{b}_1$, $d = \lceil d_1\hat{\Psi}/S\rceil$.

1. From $f < 2^p$ and $d_1(1 - f2^{-p}) < 2^p\theta < d_1(1 + f2^{-p})$, we obtain from the previous corollary,

$$\begin{aligned}\frac{2^p\Psi\theta}{S} &< \frac{\Psi}{S}d_1\left(1 + \frac{f}{2^p}\right) \le (d+1)\left(1 + \frac{f}{2^p}\right)\\ &< d\left(1 + \frac{f}{2^p} + \frac{2}{d}\right) < d\left(1 + \frac{f+2}{2^p}\right)\end{aligned}$$

and

$$\begin{aligned}\frac{2^p\Psi\theta}{S} &> \frac{\Psi}{S}d_1\left(1 - \frac{f}{2^p}\right) \ge (d-2)\left(1 - \frac{f}{2^p}\right)\\ &> d\left(1 - \frac{f}{2^p} - \frac{2}{d}\right) > d\left(1 - \frac{f+2}{2^p}\right).\end{aligned}$$

2. Suppose step 3(a) is executed j times. Then by (2.3) $\Psi_j > 2^{\lfloor (j-1)/2 \rfloor}$, and by Corollary 3.5 and Lemma 3.4,

$$\Psi_j < \frac{S}{d_1}(d_j + 1) < \frac{S}{2^p}(3Q_{j-1}2^{p-1} + 1) = O(S\sqrt{D})$$

because $\mathfrak{b}_j$ is reduced.

Now let j be the number of iterations of step 3 (d). Then again by (2.3) $\Psi_j^* < 2^{-\lfloor (j-1)/2 \rfloor}\Psi_1^*$ where $d_1\Psi_1^*/S = O(\sqrt{D}2^p)$. It follows that

$$2^p - 2 < d_j^* - 2 \le \frac{d_1\Psi_j^*}{S} < 2^{-\lfloor (j-1)/2 \rfloor}O(\sqrt{D}2^p),$$

so $2^{\lfloor (j-1)/2 \rfloor} = O(\sqrt{D})$.

3. Let j be as in part 1 of Lemma 3.4. Since $d_1 > 2^p$ and $d_j \le 3Q_{j-1}2^{p-1} < 3\sqrt{D}2^p$, we see that $\hat{\Psi}_j \le Sd_j/d_1 < 3\sqrt{D}S$. Using the bound $B_{j-2} < 2Q_0S$ from (3.1), we obtain for $1 \le i \le j$:

$$T_{i-2} = 2^sQ_0\hat{\Psi}_i \le 2^sQ_0\Psi_i \le 2^sQ_0\Psi_j < 2^sQ_0\hat{\Psi}_j + B_{j-2} = O(2^s\sqrt{D}SQ_0).$$

Now let j be as in part 2 of Lemma 3.4. Since $\Psi_i^* \le \Psi_1^* = O(S\sqrt{D})$ and $|\overline{\Psi_i^*}| < 1$ for $1 \le i \le j$, we obtain

$$\begin{aligned}T_{i-2}^* &= 2^sQ_0\hat{\Psi}_i^* \le 2^sQ_0\Psi_i^* + |B_{i-2}^*|\\ &\le 2^sQ_0\Psi_i^* + \frac{Q_0}{2\sqrt{D}}(\Psi_i^* + |\overline{\Psi_i^*}|) = O(2^s\sqrt{D}Q_0S).\end{aligned}$$ □

We combine the previous two algorithms as follows:

Algorithm MR

Input: Reduced (f_i, p) representations $(\mathfrak{b}_i, d_i)$ of primitive ideals $\mathfrak{a}_i$ $(i = 1, 2)$.

Output: A reduced (f, p) representation $(\mathfrak{b}, d)$ of $\mathfrak{a}_1\mathfrak{a}_2$ where $f = 3 + f_1 + f_2 + f_1f_22^{-p}$.

Algorithm:

1. $((S)\mathfrak{c}, e) = \text{MULT}(\mathfrak{b}_1, d_1, \mathfrak{b}_2, d_2)$.
2. $(\mathfrak{b}, d) = \text{REDUCE}((S)\mathfrak{c}, e)$.

Write $(\mathfrak{b}, d) = \text{MR}(\mathfrak{b}_1, d_1, \mathfrak{b}_2, d_2)$.

Theorem 3.7. *Let $(\mathfrak{b}_1, d_1), (\mathfrak{b}_2, d_2)$ be the inputs and $(\mathfrak{b}, d)$ the output of Algorithm* MR. *Write $\mathfrak{b}_1 = (Q_1, P_1)$, $\mathfrak{b}_2 = (Q_2, P_2)$, $\mathfrak{b} = (Q, P)$, and suppose that $2^p < d_i < 3Q_i2^p$ for $i = 1, 2$. Then $(\mathfrak{b}, d)$ is a reduced (f, p) representation of $\mathfrak{a}_1\mathfrak{a}_2$ with $2^p < d < 3Q2^p$. Also, Algorithm* MR *requires $O(\log D)$ integer operations to compute $(\mathfrak{b}, d)$, and the largest value computed in the algorithm is bounded by $O(D^{3/2}2^p)$.*

Proof. We have $(S)\mathfrak{c} = \mathfrak{b}_1\mathfrak{b}_2$. Let $\mathfrak{c} = (\tilde{Q}, \tilde{P})$. Then $((S)\mathfrak{c}, e)$ is a (g, p) representation of $\mathfrak{a}_1\mathfrak{a}_2$ where $g = 1 + f_1 + f_2 + f_1f_22^{-p}$ and $2^p < e \leq d_1d_22^{-p} < 9S^2\tilde{Q}2^{p-2}$. Also, $(\mathfrak{b}, d)$ is an (f, p) representation of $\mathfrak{a}_1\mathfrak{a}_2$ with $2^p < d < 3Q2^{p-1}$ and $f = g + 2$. The complexity result is clear since $S \leq \min\{Q_1, Q_2\} < 2\sqrt{D}$, and the largest value computed is $O(\tilde{Q}S\sqrt{D}2^p) = O(D^{3/2}2^p)$. □

We point out that the upper bound $3Q2^{p-1}$ on d required $P < \lfloor\sqrt{D}\rfloor$. In the very unlikely case where equality occurs, we can only achieve $d < 3aQ_{j-1}2^{p-1}$ where $a = (\sqrt{D} - \lfloor\sqrt{D}\rfloor)^{-1} > 1$. However, letting $\mathfrak{b} = (\Psi)\mathfrak{c}$, one can still achieve $e|\Psi - \hat{\Psi}|/S < 1$ at the expense of higher precision s' in Algorithm REDUCE. Here, s' must be chosen so that

$$\frac{e}{2^{s'}} < \frac{1}{10}.$$

Let T_j ($j \in \mathbb{N}$) be the sequence of values computed in steps 2 and 3 of Algorithm REDUCE with respect to precision s, and let T'_j be the analogous sequence if precision s' is used. Then one can recover T'_{j-1} and T'_j from T_{j-1} and T_j using the formulas

$$(2^{2s}D - \delta^2)T'_{j-1} = (2^{s+s'}D - \delta\delta')T_{j-1} + (2^s\delta' - s^{s'}\delta)(Q_jT_j - P_{j+1}T_{j-1}),$$
$$(2^{2s}D - \delta^2)T'_j = (2^{s+s'}D - \delta\delta')T_j + (2^s\delta' - s^{s'}\delta)(P_{j+1}T_j + Q_{j+1}T_{j-1}),$$

where $\delta = \lfloor\sqrt{D}2^s\rfloor$ and $\delta' = \lfloor\sqrt{D}2^{s'}\rfloor$. Analogous formulas can be derived for T^*_{j-1} and T^*_j. In any case, it is always possible to achieve $a < 2$, for example, if we choose D to be of the form $D = A^2 + R$ with $A, R \in \mathbb{N}$ and $A + 1 \leq R \leq 2A$.

We conclude this section with a method for exponentiation with (f, p) representations. This algorithm is an adaptation of the standard binary exponentiation technique.

Algorithm EXP

Input: $n \in \mathbb{N}$ and a reduced (f_0, p) representation $(\mathfrak{b}_0, d_0)$ of some ideal $\mathfrak{a}$.

Output: A reduced (f, p) representation $(\mathfrak{b}, d)$ of $\mathfrak{a}^n$ for suitable $f \in [1, 2^p)$.

Algorithm:

1. Compute the binary representation of n, say $n = \sum_{i=0}^{k} b_i 2^{k-i}$ ($b_0 = 1$, $b_i \in \{0,1\}$ for $1 \le i \le k$, $k = \lfloor \log_2 n \rfloor$).
2. Set $(\mathfrak{c}, e) = (\mathfrak{b}_0, d_0)$.
3. For $i = 1$ to k do
   (a) $(\mathfrak{b}, d) = \mathrm{MR}(\mathfrak{c}, e, \mathfrak{c}, e)$;
   (b) if $b_i = 1$ then replace $(\mathfrak{b}, d)$ by $\mathrm{MR}(\mathfrak{b}, d, \mathfrak{b}_0, d_0)$;
   (c) $(\mathfrak{c}, e) = (\mathfrak{b}, d)$.

Write $(\mathfrak{b}, d) = \mathrm{EXP}(\mathfrak{b}_0, d_0, n)$.

Before we analyze Algorithm EXP, we require an auxiliary lemma.

Lemma 3.8. *Let $p, k \in \mathbb{N}$ and $h \in \mathbb{R}$ with $p \ge 8$ and $h \ge \max\{16, k\}$. Define a sequence $(a_i)_{i \ge 0}$ recursively via*

$$a_0 \in \mathbb{R}_+, \quad a_i = 6 + \left(\left(1 + \frac{1}{h}\right)^2 + \frac{3}{2^p}\right) a_0 + \left(2 + \frac{1}{h}\right) a_{i-1} \quad (i \in \mathbb{N}).$$

Then $a_k < (3.43 a_0 + 9.9) 2^k$.

Proof. For brevity, set $g = 2 + h^{-1}$. Then it is easy to verify that the closed form for $(a_i)_{i \ge 0}$ is

$$a_i = \left(g^{i+1} - g + 1 + \frac{g^i - 1}{g - 1} \cdot \frac{3}{2^p}\right) a_0 + 6\,\frac{g^i - 1}{g - 1} \quad (i \in \mathbb{N}).$$

Now $(g^i - 1)/(g - 1) < g^i$ and the multiple of a_0 in the above equality is bounded above by $g^{i+1} + 3 \cdot 2^{-8} g^i$. Using $h \ge 16$ and $p \ge 8$, we obtain

$$\begin{aligned} a_i &< g^i\left(\left(2 + \frac{1}{16} + \frac{3}{256}\right) a_0 + 6\right) < g^i (2.08 a_0 + 6) \\ &= 2^i \left(1 + \frac{1}{2h}\right)^i (2.08 a_0 + 6) < 2^i \exp\left(\frac{i}{2h}\right) (2.08 a_0 + 6). \end{aligned}$$

Since $h \ge k$, we have $a_k \le 2^k \exp(0.5)(2.08 a_0 + 6) < (3.43 a_0 + 9.9) 2^k$. □

Theorem 3.9. *Let $(\mathfrak{b}_0, d_0)$ and n be the inputs and $(\mathfrak{b}, d)$ the output of Algorithm* EXP. *Write $\mathfrak{b}_0 = (\tilde{Q}, \tilde{P})$, $\mathfrak{b} = (Q, P)$, and suppose that $d_0 < 9 S^2 \tilde{Q} 2^{p-2}$.*

1. *$\mathfrak{b}$ is a reduced (f, p) representation of $\mathfrak{b}_0$ with $2^p < d < 3Q2^{p-1}$.*
2. *Suppose $p \ge 8$ and let $h \in \mathbb{R}_+$ with $h \ge \max\{16, \log_2 n\}$. Set $e = 3.43 f_0 + 9.9$. If $hen < 2^p$, then $f < en$ and hence $hf < 2^p$.*

Also, Algorithm EXP *requires $O(\log n \log D)$ integer operations.*

Proof. After the i-th iteration of step 3 ($1 \le i \le k$), let $\mathfrak{b} = \mathfrak{b}_i$ and $d = d_i$. If we set $s_0 = b_0 = 1$ and $s_i = 2 s_{i-1} + b_i$ for $1 \le i \le k$, then $(\mathfrak{b}_i, d_i)$ is an (f_i, p)

representation of $\mathfrak{a}^{s_i}$ where

$$f_i = 3 + f_0 + \left(3 + 2f_{i-1} + \frac{f_{i-1}^2}{2^p}\right) + f_0\left(3 + 2f_{i-1} + \frac{f_{i-1}^2}{2^p}\right)2^{-p}$$

for $1 \le i \le k$. Set $f = f_k$. Since $s_k = n$, Algorithm EXP produces a reduced (f,p) representation of $\mathfrak{a}^n$ with $2^p < d < 3Q2^{p-1}$.

Let $(a_i)_{i\ge 0}$ be as in Lemma 3.8 with $a_0 = f_0$. Since $h \ge k$, we have $a_k < e2^k \le en$, so $ha_i < 2^p$ for $0 \le i \le k$. Then $hf_0 < 2^p$ and inductively, $f_i < a_i$, so $f_i < en$ and $hf_i < 2^p$; in particular, $f < en$ and $hf < 2^p$. □

4. The protocol

Our goal is to establish a protocol by which both parties are able to compute a unique reduced principal ideal $\mathfrak{k}$. This ideal will be an (f,p) representation of $\mathfrak{r}^{ab}$ for suitable f where $\mathfrak{r}$ is a public reduced starting ideal and a and b are the respective secret exponents. Since the two communicants follow different exponentiation procedures for computing a candidate for $\mathfrak{k}$, they may not arrive at the same ideal. However, we will see that their respective candidates lie within only five reduction steps (2.1) of each other. By exchanging five more bits of information, both parties will be able to agree on a unique key ideal.

Lemma 4.1. *Let $\mathfrak{r}$ be a reduced principal ideal and let $p, a, b, B \in \mathbb{Z}$ with $B \ge 36$, $0 < a, b \le B$, and $2^p \ge 46B^2 \max\{16, \log_2 B\}$. If $(\mathfrak{k}, d) = \mathrm{EXP}(\mathrm{EXP}(\mathfrak{r}, 2^p+1, a), b)$, then $\mathfrak{k}$ is an (f,p) representation of $\mathfrak{r}^{ab}$ with $16f < 2^p$.*

Proof. Clearly, $(\mathfrak{r}, 2^p + 1)$ is a reduced $(1,p)$ representation of $\mathfrak{r}$. Let h equal $\max\{16, \log_2 B\}$, then $h \ge \max\{16, \log_2 a, \log_2 b\}$. Let $(\mathfrak{a}, e) = \mathrm{EXP}(\mathfrak{r}, 2^p + 1, a)$. Since $h(3.43 \cdot 1 + 9.9)B < 2^p$, by Theorem 3.9, $(\mathfrak{a}, e)$ is a reduced (g,p) representation of $\mathfrak{r}^a$ with $g < 13.33a \le 13.33B$ and $hg < 2^p$. Similarly, $h(3.43g + 9.9)b < h(45.72B^2 + 9.9B) < 46B^2 \max\{16, \log_2 B\} < 2^p$. Since $h \ge 16$, $(\mathfrak{k}, d)$ is a reduced (f,p) representation of $\mathfrak{r}^{ab}$ with $16f < 2^p$. □

Theorem 4.2. *Let $(\mathfrak{b}_1, d_1)$ and $(\mathfrak{b}_2, d_2)$ be two (f,p) representations of some ideal $\mathfrak{a}$ with $p \ge 3$ and $7f \le 2^p$. Let $\mathfrak{k} = (\kappa)\mathfrak{b}_1$, $\rho^{-1}(\mathfrak{k}) = (\kappa_-)\mathfrak{b}_1$, $\mathfrak{m} = (\mu)\mathfrak{b}_2$, $\rho^{-1}(\mathfrak{m}) = (\mu_-)\mathfrak{b}_2$ where $\mathfrak{k}$ and $\mathfrak{m}$ are reduced ideals with $d_1\kappa, d_2\mu > 2^p - 2$ and $d_1\kappa_-, d_2\mu_- < 2^p + 1$. Then $\mathfrak{k} \in \{\rho^{-2}(\mathfrak{m}), \rho^{-1}(\mathfrak{m}), \mathfrak{m}, \rho(\mathfrak{m}), \rho^2(\mathfrak{m})\}$.*

Proof. Suppose $\mathfrak{k} = \rho^i(\mathfrak{m})$ with $|i| \ge 3$ and assume without loss of generality that $i > 0$ (the case $i < 0$ can be proved by reversing the roles of $\mathfrak{k}$ and $\mathfrak{m}$). Then $\rho^{-1}(\mathfrak{k}) = (\alpha)\mathfrak{m}$ where $\alpha > 2$ by (2.3). Let $\mathfrak{b}_i = (\theta_i)\mathfrak{a}$ for $i = 1, 2$ and assume that $\theta_1, \theta_2, \kappa, \kappa_-, \mu, \mu_- > 0$. Then $\kappa_-\theta_1 > 2\mu\theta_2$. Now

$$\kappa_-\theta_1 < \kappa_-\frac{d_1}{2^p}\left(1 + \frac{f}{2^p}\right) < \frac{2^p + 1}{2^p}\left(1 + \frac{f}{2^p}\right),$$

$$\mu\theta_2 > \mu\frac{d_2}{2^p}\Big(1-\frac{f}{2^p}\Big) > \frac{2^p-2}{2^p}\Big(1-\frac{f}{2^p}\Big),$$

so $(2^p+1)(1+f2^{-p}) > 2(2^p-2)(1-f2^{-p})$, implying

$$\frac{f}{2^p} > \frac{1}{3}\cdot\frac{2^p-5}{2^p-1} = \frac{1}{3}\Big(1-\frac{4}{2^p-1}\Big) > \frac{1}{3}\cdot\frac{3}{7} = \frac{1}{7},$$

a contradiction. □

Corollary 4.3. *Let* $\mathfrak{r}$, p, B, a, b *be as in Lemma* 4.1 *and let*

$$(\mathfrak{k}, k) = \mathrm{EXP}(\mathrm{EXP}(\mathfrak{r}, 2^p+1, a), b) \quad \text{and} \quad (\mathfrak{m}, m) = \mathrm{EXP}(\mathrm{EXP}(\mathfrak{r}, 2^p+1, b), a).$$

Then $\mathfrak{k} \in \{\rho^{-2}(\mathfrak{m}), \rho^{-1}(\mathfrak{m}), \mathfrak{m}, \rho(\mathfrak{m}), \rho^2(\mathfrak{m})\}$.

Proof. By Lemma 4.1, $(\mathfrak{k}, k)$ and $(\mathfrak{m}, m)$ are (f,p) representations of $\mathfrak{r}^{ab}$ with $16f < 2^p$.

Let $(\mathfrak{a}, e) = \mathrm{EXP}(\mathfrak{r}, 2^p+1, a)$, and let $(\mathfrak{b}_1, d_1)$ be the quantity computed by Algorithm EXP on input $(\mathfrak{a}, e, b)$ before the very last reduction step. By Theorem 3.9 and Lemma 4.1, $(\mathfrak{b}_1, d_1)$ is a (g,p) representation of $\mathfrak{r}^{ab}$ with $16g < 16f < 2^p$. Once $(\mathfrak{b}_1, d_1)$ is found, Algorithm EXP generates $(\mathfrak{k}, k)$ and $(\rho^{-1}(\mathfrak{k}), k_-)$. If $\mathfrak{k} = (\kappa)\mathfrak{b}_1$ and $\rho^{-1}(\mathfrak{k}) = (\kappa_-)\mathfrak{b}_1$, then by Lemma 3.4, $k_- \le 2^p < k$, and by Corollary 3.5, $k_- - 2 < d_1\kappa_- < k_- + 1$ and $k - 2 < d_1\kappa < k+1$. It follows that $d_1\kappa > 2^p - 2$ and $d_1\kappa_- < 2^p + 1$.

Similarly, let $(\mathfrak{b}, e') = \mathrm{EXP}(\mathfrak{r}, 2^p+1, b)$, and let $(\mathfrak{b}_2, d_2)$ be the quantity generated by algorithm EXP on input $(\mathfrak{b}, e', a)$ before the very last reduction step when computing $(\mathfrak{m}, m) = \mathrm{EXP}(\mathfrak{b}, e', a)$. Then $(\mathfrak{b}_2, d_2)$ is also a (g,p) representation of $\mathfrak{r}^{ab}$ with $16g < 2^p$, and if $\mathfrak{m} = (\mu)\mathfrak{b}_2$ and $\rho^{-1}(\mathfrak{m}) = (\mu_-)\mathfrak{b}_2$, then $d_2\mu > 2^p - 2$ and $d_2\mu_- < 2^p + 1$. Therefore, the conditions of Theorem 4.2 are satisfied and the corollary follows. □

Lemma 4.4. *Assume the notation and conditions of Theorem* 4.2 *and suppose that* $p \ge 9$ *and* $16f < 2^p$.

1. *If* $d_1\kappa_- < 7\cdot 2^{p-3} + 1$, *then* $\mathfrak{k} \in \{\rho^{-2}(\mathfrak{m}), \rho^{-1}(\mathfrak{m}), \mathfrak{m}\}$.
2. *If* $d_1\kappa > 5\cdot 2^{p-2} - 1$, *then* $\mathfrak{k} \in \{\rho^2(\mathfrak{m}), \rho(\mathfrak{m}), \mathfrak{m}\}$. *If* $d_2\mu > 5\cdot 2^{p-2} - 1$, *then* $\mathfrak{k} \in \{\rho^{-2}(\mathfrak{m}), \rho^{-1}(\mathfrak{m}), \mathfrak{m}\}$.
3. *If* $d_1\kappa < 5\cdot 2^{p-2} + 1$, *then* $\mathfrak{k} \ne \rho^2(\mathfrak{m})$.
4. *If* $d_1\kappa_- > 7\cdot 2^{p-3} - 1$, *then* $\mathfrak{k} \ne \rho^{-2}(\mathfrak{m})$.
5. *If* $d_1\kappa_- > 7\cdot 2^{p-3} - 1$ *and* $d_2\mu < 5\cdot 2^{p-2} + 1$, *then* $\mathfrak{k} \ne \rho^{-1}(\mathfrak{m})$.
6. *Suppose* $d_1\kappa < 5\cdot 2^{p-2} + 1$. *If* $d_2\mu < 7\cdot 2^{p-2} + 1$, *then* $\mathfrak{k} \ne \rho^{-2}(\mathfrak{m})$, *and if* $d_2\mu > 7\cdot 2^{p-2} - 1$, *then* $\mathfrak{k} \ne \mathfrak{m}$.
   Suppose $d_2\mu < 5\cdot 2^{p-2} + 1$. *If* $d_1\kappa < 7\cdot 2^{p-2} + 1$, *then* $\mathfrak{k} \ne \rho^2(\mathfrak{m})$, *and if* $d_1\kappa > 7\cdot 2^{p-2} - 1$, *then* $\mathfrak{k} \ne \mathfrak{m}$.

Proof. We only prove parts 1 and 5; the other parts follow similarly. Note that the second statements in parts 2 and 6 are immediate corollaries of the respective first statements of these parts. By Theorem 4.2, $\mathfrak{k} \in \{\rho^{-2}(\mathfrak{m}), \rho^{-1}(\mathfrak{m}), \mathfrak{m}, \rho(\mathfrak{m}), \rho^2(\mathfrak{m})\}$.

For part 1, we have

$$\kappa_-\theta_1 < \frac{7 \cdot 2^{p-3}+1}{2^p}\Big(1+\frac{f}{2^p}\Big) < \Big(\frac{7}{8}+\frac{1}{2^9}\Big)\Big(1+\frac{1}{16}\Big) = \frac{7633}{2^{13}}.$$

Suppose $\mathfrak{k} \in \{\rho(\mathfrak{m}), \rho^2(\mathfrak{m})\}$, then we reason as in the proof of Theorem 4.2 that

$$\kappa_-\theta_1 \geq \mu\theta_2 > \Big(1-\frac{1}{2^{p-1}}\Big)\Big(1-\frac{f}{2^p}\Big) > \Big(1-\frac{1}{2^8}\Big)\Big(1-\frac{1}{16}\Big) = \frac{7650}{2^{13}},$$

a contradiction.

Assume the conditions of part 5. Then

$$\mu\theta_2 < \frac{5 \cdot 2^{p-2}+1}{2^p}\Big(1+\frac{f}{2^p}\Big) < \Big(\frac{5}{4}+\frac{1}{2^9}\Big)\Big(1+\frac{1}{16}\Big) = \frac{10897}{2^{13}}$$

and

$$\kappa_-\theta_1 > \frac{7 \cdot 2^{p-3}-1}{2^p}\Big(1-\frac{f}{2^p}\Big) > \Big(\frac{7}{8}-\frac{1}{2^9}\Big)\Big(1-\frac{1}{16}\Big) = \frac{6705}{2^{13}}.$$

If $\mathfrak{k} = \rho^{-1}(\mathfrak{m})$, then $\mathfrak{m} = \rho^2(\mathfrak{k}_-)$, so $\mu\theta_2 > 2\kappa_-\theta_1$ by (2.3), contradicting the above inequalities. □

Lemma 4.5. *Let $D \in \mathbb{N}$ be squarefree with $D \equiv 3 \pmod 4$ and let $\mathfrak{k}$ be a reduced ideal, $\mathfrak{k} = (Q, P)$ and $\rho(\mathfrak{k}) = (Q_+, P_+)$. Then $Q \not\equiv Q_+ \pmod 4$.*

Proof. If $Q \equiv Q_+ \pmod 4$, then $QQ_+ \equiv 0$ or $1 \pmod 4$. By (2.1), $QQ_+ = D - P_+^2$; this is $3 \pmod 4$ if P_+ is even and $2 \pmod 4$ if P_+ is odd. □

Throughout our protocol, $\mathfrak{k}, \mathfrak{m}, \kappa, \kappa_-, \mu, \mu_-, k, k_-, m, m_-$ are as in Corollary 4.3 and its proof.

Protocol:

Alice and Bob publicly agree on

- a large squarefree positive integer $D \equiv 3 \pmod 4$;
- a reduced principal ideal $\mathfrak{r}$ in $\mathbb{Z}(\sqrt{D})$,
- a bound $B \in \mathbb{N}$ on the exponents.

Both precompute $p = \lceil \log_2(46B^2 \max\{16, \log_2 B\}) \rceil$.

Alice

- secretly generates $a \in \mathbb{N}$, $a \leq B$;
- computes $(\mathfrak{a}, d_a) = \mathrm{EXP}(\mathfrak{r}, 2^p+1, a)$, here $\mathfrak{a} = (Q_a, P_a)$;
- sends (Q_a, P_a, d_a) to Bob.

Bob

- secretly generates $b \in \mathbb{N}$, $b \leq B$;
- computes $(\mathfrak{b}, d_b) = \mathrm{EXP}(\mathfrak{r}, 2^p+1, b)$, here $\mathfrak{b} = (Q_b, P_b)$;

- sends (Q_b, P_b, d_b) to Alice.

Alice

- computes $(\mathfrak{k}, k) = \mathrm{EXP}(\mathfrak{b}, d_b, a)$, here $\mathfrak{k} = (Q, P)$; she also knows $(\rho^{-1}(\mathfrak{k}), k_-)$;
- sets $q \equiv Q \,(\mathrm{mod}\, 4)$, $0 \le q \le 3$;
- sets
$$b_1 = \begin{cases} 0 \text{ if } k_- \le 7 \cdot 2^{p-3}, \\ 1 \text{ if } k_- > 7 \cdot 2^{p-3}, \end{cases} \quad b_2 = \begin{cases} 0 \text{ if } k \le 5 \cdot 2^{p-2}, \\ 1 \text{ if } k > 5 \cdot 2^{p-2}, \end{cases} \quad b_3 = \begin{cases} 0 \text{ if } k \le 7 \cdot 2^{p-2}, \\ 1 \text{ if } k > 7 \cdot 2^{p-2}, \end{cases}$$
- sends (b_1, b_2, b_3, q) to Bob.

Bob

- computes $(\mathfrak{m}, m) = \mathrm{EXP}(\mathfrak{a}, d_a, b)$.
- determines an ideal $\mathfrak{l} = (\tilde{Q}, \tilde{P})$ according to the following rules:

```
if b_2 = 0 then
    if b_1 = 0 then
        if m > 7·2^{p-2} then set l = ρ^{-2}(m), else set l = ρ^{-1}(m)
    else
        if m > 5·2^{p-2} then set l = ρ^{-1}(m), else set l = m
else (in this case b_2 = 1)
    if m > 5·2^{p-2} then
        set l = m
    else
        if b_3 = 0 then set l = m, else set l = ρ(m).
```

- sets $\mathfrak{k} = \mathfrak{l}$ if $\tilde{Q} \equiv Q \,(\mathrm{mod}\, 4)$ and $\mathfrak{k} = \rho(\mathfrak{l})$ if $\tilde{Q} \not\equiv Q \,(\mathrm{mod}\, 4)$.

We see that after both parties have completed their respective double exponentiations, Alice simply communicates five more bits to Bob. Upon receipts of these bits, Bob is able to determine the ideal $\mathfrak{k} = (Q, P)$ computed by Alice. The common key is a substring of P or Q of suitable size, where the last two bits of Q should not be used in the key. In the precomputation, $\mathfrak{r}$ can be obtained by applying a few iterations of (2.1) to the ideal $\mathcal{O} = (1, 0)$. Similarly, Bob can easily compute $\mathfrak{l}$ and $\mathfrak{k}$ using (2.1) and (2.4).

It remains to be shown that the ideal computed by Bob is in fact the same ideal $\mathfrak{k}$ that Alice generates.

Theorem 4.6. *After the protocol is executed, both communicants have computed the same ideal* $\mathfrak{k}$.

Proof. By Corollary 4.3, $\mathfrak{m} \in \{\rho^{-2}(\mathfrak{k}), \rho^{-1}(\mathfrak{k}), \mathfrak{k}, \rho(\mathfrak{k}), \rho^2(\mathfrak{k})\}$. If we show that if $\mathfrak{l} \in \{\mathfrak{k}, \rho^{-1}(\mathfrak{k})\}$, then by Lemma 4.5, $\mathfrak{k} = \mathfrak{l}$ if $\tilde{Q} \equiv Q \,(\mathrm{mod}\, 4)$ and $\mathfrak{k} = \rho(\mathfrak{l})$ otherwise. We note that by Corollary 3.5, $k - 2 < d_1\kappa < k + 1$, $k_- - 2 < d_1\kappa_- < k_- + 1$, and $m - 2 < d_2\mu < m + 1$.

Suppose first that $b_2 = 0$, so $k \le 5 \cdot 2^{p-2}$ and hence $d_1\kappa < 5 \cdot 2^{p-2} + 1$.

Case $b_1 = 0$. Then $d_1\kappa_- < 7 \cdot 2^{p-3} + 1$, so by part 1 of Lemma 4.4, $\mathfrak{k} \in \{\rho^{-2}(\mathfrak{m}), \rho^{-1}(\mathfrak{m}), \mathfrak{m}\}$. If $m > 7 \cdot 2^{p-2}$, then $d_2\mu > 7 \cdot 2^{p-2} - 1$, so by part 6 of the same lemma, $\mathfrak{k} \neq \mathfrak{m}$. Hence, $\mathfrak{k} \in \{\rho^{-2}(\mathfrak{m}), \rho^{-1}(\mathfrak{m})\}$ and $\mathfrak{l} = \rho^{-2}(\mathfrak{m}) \in \{\mathfrak{k}, \rho^{-1}(\mathfrak{k})\}$.

Similarly, if $m \le 7 \cdot 2^{p-2}$, then $d_2\mu < 7 \cdot 2^{p-2} + 1$, so again by part 6 of the same lemma $\mathfrak{k} \ne \rho^{-2}(\mathfrak{m})$. Hence, $\mathfrak{k} \in \{\rho^{-1}(\mathfrak{m}), \mathfrak{m}\}$ and $\mathfrak{l} = \rho^{-1}(\mathfrak{m}) \in \{\mathfrak{k}, \rho^{-1}(\mathfrak{k})\}$.

Case $b_1 = 1$. Then $d_1\kappa < 5 \cdot 2^{p-2} + 1$ and $d_1\kappa_- > 7 \cdot 2^{p-3} - 1$, so by parts 3 and 4 of Lemma 4.4, $\mathfrak{k} \in \{\rho^{-1}(\mathfrak{m}), \mathfrak{m}, \rho(\mathfrak{m})\}$. If $m > 5 \cdot 2^{p-2}$, then by part 2 of the lemma, $\mathfrak{k} \in \{\rho^{-1}(\mathfrak{m}), \mathfrak{m}\}$, and if $m \le 5 \cdot 2^{p-2}$, then by part 5, $\mathfrak{k} \in \{\mathfrak{m}, \rho(\mathfrak{m})\}$. In either case, $\mathfrak{l} \in \{\mathfrak{k}, \rho^{-1}(\mathfrak{k})\}$.

Now suppose that $b_2 = 1$. Then part 2 of Lemma 4.4 yields $\mathfrak{k} \in \{\rho^2(\mathfrak{m}), \rho(\mathfrak{m}), \mathfrak{m}\}$. If $m > 5 \cdot 2^{p-2}$, then again by part 2, $\mathfrak{k} = \mathfrak{m} = \mathfrak{l}$. If $m \le 5 \cdot 2^{p-2}$, then part 6 shows that $\mathfrak{k} \in \{\rho(\mathfrak{m}), \mathfrak{m}\}$ if $b_3 = 0$ and $\mathfrak{k} \in \{\rho^2(\mathfrak{m}), \rho(\mathfrak{m})\}$ if $b_3 = 1$. In either case, $\mathfrak{l} \in \{\mathfrak{k}, \rho^{-1}(\mathfrak{k})\}$. □

5. Implementation

The algorithms described above were implemented using the C++ computer algebra library LiDIA [10]. All computations in this section were performed using the GNU g++ compiler version 2.91.66 on a Pentium III 800 MHz computer running Linux.

A number of optimizations have been incorporated into our implementation. In practice, algorithm MULT can be improved by testing for two commonly-occurring cases: squaring an ideal ($\gcd(Q_1/\sigma, Q_2/\sigma) = Q_1/\sigma$), and the case $\gcd(Q_1/\sigma, Q_2/\sigma)$ equals 1. Descriptions of the corresponding algorithms can be found in [2] and [5].

We now need to discuss how the value of s was selected. In Theorem 3.9 we need $d_0 < 9S^2\tilde{Q}2^{p-2}$, where $S^2\tilde{Q} \le Q_1Q_2 < (2\sqrt{D})(2\sqrt{D}) = 4D$. Thus, if $d_0 < 9D2^p$, we have $d_0 < 9S^2\tilde{Q}2^{p-2}$. We also require $d_1(= d_0)$ to satisfy $d_1/2^s < 1/10$ in Lemma 3.4. Thus, if $2^s > 90D2^p$, the EXP algorithm must work. However, in practice this is a much larger value of s than is actually needed.

Suppose we select some C and K such that

$$2^s > CK2^p/\sqrt{D}$$

and d_1 in Lemma 3.4 satisfies $d_1 < K2^p$. Although we have been at pains to point out how backward steps can be performed in our algorithms, it turns out that for large values of D we require very few backward steps. The reason for this is that the first reduced ideal $\mathfrak{b}_l$ that we find is such that $\Psi_l \le 1$ (see Lemma 3.3). Indeed, it is often much smaller than 1. Since, as we shall see below, K can be taken to be small, we almost always have $d_1\Psi_l/S < 2^p$ and, as a consequence, we expect $d_1\hat{\Psi}_l/S < 2^p$ or $T_{l-2} \le M$. If we now refer to the proof of Corollary 3.5, we may assume that

$$\frac{d_1\Psi_j}{S} - \frac{d_1\hat{\Psi}_j}{S} = \frac{\eta d_1 B_{j-2}}{Q_0 2^s S},$$

where $0 < \eta < 1$ and $j \le k$. Since $B_{j-2} \le B_{k-2}$, we get

$$\frac{B_{j-2}}{Q_0 2^s} \le \frac{\Psi_k + |\overline{\Psi_k}|}{2^{s+1}\sqrt{D}} \le \frac{\psi_{k-1}\Psi_{k-1} + 1}{2^{s+1}\sqrt{D}} < \frac{3\psi_{k-1}S/2 + 1}{2^{s+1}\sqrt{D}} < \frac{(5/4)\psi_{k-1}S}{2^s\sqrt{D}}.$$

It follows that

$$\frac{\eta d_1 B_{j-2}}{Q_0 2^s S} < \frac{(5/4)d_1}{2^s\sqrt{D}}\psi_{k-1};$$

thus, if $\psi_{k-1} < 4C/5$, then $d_1\Psi_j/S - d_1\hat{\Psi}_j/S < 1$. Also, since $B_{j-3} < B_{j-2}$, we get

$$\frac{d_1\hat{\Psi}_j}{S} < \frac{d_1\psi_{j-1}\Psi_{j-1}}{S} < \psi_{j-1}\left(\frac{d_1\hat{\Psi}_j}{S}+1\right);$$

thus,

$$d_j < (2^p+1)\psi_{j-1}+1 < \psi_{j-1}(2^p+2).$$

Therefore, after executing step 1 of Algorithm MR we would have

$$e < \gamma_1\gamma_2\frac{(2^p+2)^2}{2^p}+1 < \frac{5}{4}\gamma_1\gamma_2 2^p.$$

Here γ_1 (>1) and γ_2 (>1) are complete quotients in the continued fraction expansion of ω. If $K > (5/4)\gamma_1\gamma_2$, then $e < K2^p$.

By the Gauß–Kuzmin law, we expect that if γ is any complete quotient in the continued fraction of any real number, then the approximate probability that $\gamma > b$ is $\log_2(1+1/b)$. Putting $C = \sqrt{5K/4}$, we see that the probability that $\gamma_1 > 4C/5$, $\gamma_2 > 4C/5$, or $\psi_{k-1} > 4C/5$ is very likely no more than $3\log_2(1+5/(4C)) < 15/(4C\log 2)$. If we put $C = (5/4)^{1/3}D^{1/12}$, we see that if D is large (>512 bits, say) we get $C > 2^{43}$, which means that it is very likely that we have $\psi_{k-1} < 4C/5$ and $e < K2^p$. Thus, if

$$2^s > \frac{2^p}{D^{1/4}},$$

$\hat{\Psi}_i$ should be a sufficiently good approximation of Ψ_i for the algorithms to work. Indeed, we found that values of s such that $2^s > D^{3/4}$ for $B=\sqrt{D}$ and $2^s > D^{1/4}$ for $B = D^{1/4}$ are perfectly adequate as long as D is large.

The reduction algorithm REDUCE primarily consists of expanding the continued fraction of $(P+\sqrt{D})/Q$, and hence we have used the more efficient Tenner's algorithm [21] for both the "forward" and "backward" reduction stages. In addition, we have tried two separate strategies to further speed the algorithm. In each iteration, $q_{j-1}=1$ can be recognized easily by testing whether

$$Q_{j-1} \le P_{j-1}+\sqrt{D} < 2Q_{j-1}.$$

By handling this case specially, we avoid all multiplications and divisions for the computation of P_j and Q_j for that iteration. Since the Gauß–Kuzmin law predicts that $q_{j-1}=1$ about 41% of the time, this should, and does, yield a fairly significant speed-up.

Our second method to speed REDUCE was to replace the simple continued fraction algorithm by the nearest integer continued fraction. The advantage of this algorithm is that the number of iterations required to find a reduced ideal is only about 70% of that using the simple continued fraction. However, the smaller number of iterations comes at the price of having $q_{j-1} \neq 1$ for all j, so we cannot combine our two methods. We found that using the nearest integer continued fraction for the forward steps and using the simple continued fraction while testing for partial quotients of 1 was slightly faster than other combinations of the two approaches, so we used it in our implementation. In the forward reduction steps, we compute T_{j-1} using the formula $T_{j-1} = q_{j-1}T_{j-2} - T_{j-3}$.

In order to test the efficiency of our protocol, we have computed numerous examples using discriminants of 512, 768, and 1024 bits. For each size of radicand, we have executed the key exchange protocol 100 times for each of 20 randomly chosen prime radicands of the given size. We have carried these computations out using exponent bounds $B = \sqrt{D}$ (with $2^s < D^{3/4}$) and $B = D^{1/4}$ (with $2^s < D^{1/4}$). In all cases, we take $\mathfrak{r} = \rho^5((1))$. Table 5.1 contains the average CPU time per communication partner for a single execution of the protocol, for each combination of the parameter sizes.

Table 5.1. Average CPU time (in seconds) per key exchange per partner

$\log_2 D$	$B = D^{1/4}$, $2^s < D^{1/4}$	$B = \sqrt{D}$, $2^s < D^{3/4}$
512	0.48	1.20
768	1.33	2.90
1024	2.74	6.12

After profiling our implementation, we found that by far the most time-consuming part of the algorithm was ideal reduction — 97% compared to 3% for ideal multiplication. Clearly, any attempts to further optimize the protocol should begin there. One possible improvement would be to incorporate the NUCOMP algorithm of Shanks [19], which essentially combines the multiplication and reduction stages. In addition to keeping the sizes of the intermediate operands close to $\sqrt{D}$ (as opposed to operands close to D in Algorithm MULT), many of the reduction steps are replaced by slightly less expensive simple continued fraction steps.

One disadvantage of using the nearest integer continued fraction for reduction is that the backward steps in REDUCE are sometimes necessary, since some reduced ideals may be skipped in this type of continued fraction expansion. However, on average only 1 backward step was required per call to reduce, and the overall time was not greatly affected.

It should be noted that in all cases both partners do in fact end up with the same ideal, even though the precision used is less than what is theoretically necessary in the worst case. Thus, in practice, the second round of the protocol is never executed.

6. Security

6.1. Principal ideal testing

As shown in [17], our scheme can be broken by solving the principal ideal problem: given a reduced principal ideal $\mathfrak{a}$, find a generator μ, or $\log|\mu|$, such that $(\mu) = \mathfrak{a}$. It is also mentioned in [17] that solving the principal ideal problem is at least as difficult as factoring.

The best algorithms available for solving the principal ideal problem employ index-calculus techniques and are of subexponential complexity. Abel [1] was able show that under the Generalized Riemann Hypothesis (GRH) this problem has complexity

$$\exp\Big((1.44 + o(1))\sqrt{\log D \log\log D}\Big).$$

Improvements by Vollmer [20] suggest that the constant 1.44 can likely be reduced to $\sqrt{2} \approx 1.41$. To date, no algorithm of complexity $\exp\big((\log D)^{1/3}(\log\log D)^{2/3}\big)$ is known to exist.

The most efficient algorithm in practice is due to Jacobson [6], and is based on principles used in the self-initializing quadratic sieve factoring algorithm. Using this algorithm, it is possible to solve the principal ideal problem for values of $D \approx 10^{66}$ in a little over a day on a 298 MhZ UltraSPARC-II [6].

Notice that for all of these algorithms it is easy to verify unconditionally the correctness of a solution. The assumption of the GRH is needed to prove the complexity and the termination of the algorithms.

Currently, it appears that computing $R = \log\epsilon$, the regulator of K, is somewhat easier than solving the principal ideal problem in practice [5]. Computing the regulator is nevertheless a special case of solving the principal ideal problem ($\mathfrak{a} = (1)$), and hence security estimates based on the state of the art of regulator computation will yield conservative parameter selection guidelines.

Using a recent distributed implementation of the self-initializing quadratic sieve based algorithm in [5], we have successfully computed the regulators of a 90 decimal digit discriminant and a 101 decimal digit discriminant, both of which are significantly larger than those presented in [5]. Using a cluster of 16 550 MHz Pentium III computers, we found that for the 90-digit

$$\begin{aligned} D = {} & 215224698103728400410483771240601671668634200915018506046263 \\ & 918977716591590126558308631804 \end{aligned}$$

we get (under the GRH)

$$\begin{aligned} R &= 1314117837933813360543450767405060115166686144.03321787 \\ h &= 1, \end{aligned}$$

where h denotes the ideal class number of K. This computation took a total of 10 days (approximately 5.2 months of CPU time). For the 101-digit

$$\begin{aligned} D = {} & 13022194102190350410319085329793205127319464132884776163361 5 \\ & 78366571379092583560263087397184669099836 \end{aligned}$$

we get (under the GRH)

$$\begin{aligned} R &= 31780254623174755539291764915494863617276316347826 0.945231457 \\ h &= 1, \end{aligned}$$

and the entire computation on the cluster took 87 days (approximately 3.8 years of CPU time). More details about these computations will appear in a forthcoming paper.

6.2. The choice of D

As demonstrated in [5, Chapter 7], the performance of the self-initializing quadratic sieve is highly sensitive to the quadratic character of D. In fact, the two discriminants listed above were constructed so that they would be especially amenable to this algorithm by forcing $(D/p) = 1$ for as many small primes p as possible. Conversely, selecting D such that $(D/p) = -1$ for as many small primes as possible will cause the regulator computation and the principal ideal problem to be much harder than average using this algorithm.

Table 6.1 contains run-times, taken from [5] and [7], for a number of various sized discriminants. The wide range of run-times between 60 and 63 digits and between 72 and 73 digits is due to the varying quadratic characters of the discriminants. For example, the 73-digit discriminant was constructed so that $(D/p) = 1$ for all primes $p \leq 233$, and less than 14 hours of CPU time were required to compute its regulator. On the other hand, the 72-digit discriminant was constructed so that $(D/p) = -1$ for all primes $p \leq 337$, and approximately 18 days of CPU time were required to compute its regulator.

Table 6.1. Regulator computation CPU times for various sized discriminants

no. of digits in Δ	CPU time
54	6.26 m
58	17.79 m
60 — 63	49.56 m — 2.17 d
67	5.40 h
72 — 73	13.57 h — 18 d
80	2.05 d

The radicand D must also be selected such that there are sufficiently many reduced principal ideals to foil exhaustive search or baby-step giant-step attacks. Selecting D such that $(D/p) = -1$ for many small primes does have a minimizing effect on the product hR, but a result of Littlewood [11] states that under the GRH we still have $hR \gg \sqrt{D}/\log\log\Delta$. Thus, if we also force D to be prime, h will be odd and the Cohen-Lenstra heuristics predict that $h = 1$ about 75% of the time, so we will have $R \approx \sqrt{D}$.

Fortunately, constructing such values of D can be done quite expeditiously using special purpose sieve devices for solving systems of simultaneous linear congruences [7]. We used the Manitoba Scalable Sieve Unit [12] to generate three special values of D in this manner. The first 526-bit

$$\begin{aligned} D_1 = {} & 21400905888438118758623964453344480776164900535917750565452 9\\ & 10904563090086200913324692607413981496938229384779456844137 7\\ & 91200636936944043856483330455658015624 7 \end{aligned}$$

was constructed so that $(D_1/p) = -1$ for all odd $p < 587$. The 778-bit

$$\begin{aligned} D_2 = {} & 7966695318289831703557735835343750025821809025449055523789 50\\ & 1447644933464728330475857535786547318911020800150337524057 05\\ & 4856533597043164165292453381445659636409971003185269148427 97\\ & 7528267619057493105054635628562285717837308221817449 43 \end{aligned}$$

has $(D_2/p) = -1$ for all odd $p < 727$, and the 1040-bit

$$\begin{aligned} D_3 = {} & 6284227670864449154436031638981048861226040073651914078565 68\\ & 8783248553165317642043628603623973553739703743196297141749 71\\ & 7092662848835497798900090185427290638450635063676329584116 25\\ & 9373733832564372134693724476495383297910815905930482687261 82\\ & 7973673919580354014126545636872858815939039488990305202869 97\\ & 00578991536 87 \end{aligned}$$

has $(D_3/p) = -1$ for all odd $p < 937$. All three D values are prime.

We have compared the speed of our protocol using these specially constructed D values with that using randomly selected prime D of the same sizes. For each of 20 random D values of the specified size and the special D value of that size, the key exchange protocol was performed 100 times. The average CPU time per key exchange for each communication partner are given in Table 6.2. As in the previous section we have used our LiDIA implementation on a Pentium III 800 MHz computer running Linux. In all cases, we have used $B = D^{1/4}$, $2^s < D^{1/4}$, and $\mathfrak{r} = \rho^5((1))$.

Table 6.2. Protocol execution times (CPU seconds) for special D values

$\log_2 D$	Random	Special
526	0.49	0.49
778	1.29	1.31
1040	3.55	2.97

Even though the key exchange protocol does not execute any slower for these special radicands, using the state-of-the-art algorithm to solve the principal ideal problem will be significantly harder than for random radicands.

Acknowledgement. The authors gratefully acknowledge Detlef Hühnlein and Sachar Paulus for making an early version of their paper [4] available to us. We derived our idea of (f, p) representations from their manuscript.

Addendum[1)]

Although we have specified in our paper that all the operations take place in the principal ideal class of a real quadratic field, our technique will still work if we start with an arbitrary reduced ideal $\mathfrak{r}$ instead of a principal reduced ideal. This is because our routines will still find ideals which are close to and equivalent to $\mathfrak{r}^{ab}$, and both communication partners will still find the same key ideal $\mathfrak{k}$.

In order to see that this is the case, all that we need do at the end of Section 2 is let $\mathfrak{r}$ be any arbitrary reduced ideal which we will consider fixed in the sequel. Then every ideal $\mathfrak{a}$ equivalent to $\mathfrak{r}$ has a unique relative generator $\theta \in K$ such that $\mathfrak{a} = (\theta)\mathfrak{r}$, $1 \leq \theta < \varepsilon$. For $i \in \mathbb{N}$, let $\mathfrak{b}_i = (\Psi_i)\mathfrak{b}_1$, where $\mathfrak{b}_1 = \mathfrak{r}$. If $\mathfrak{a}$ is non-reduced, then there exists a unique index $j \in \mathbb{N}$ with $\Psi_j < \theta < \Psi_{j+1}$. Then, as before, we define $\rho(\mathfrak{a}) = \mathfrak{b}_{j+1}$, $\rho^{-1}(\mathfrak{a}) = \mathfrak{b}_j$, $\rho^0(\mathfrak{a}) = \mathfrak{a}$, $\rho^n(\mathfrak{a}) = \rho(\rho^{n-1}(\mathfrak{a})) = \rho^{-1}(\rho^{n+1}(\mathfrak{a}))$ for $n \in \mathbb{Z}$.

To determine any reduced ideal, simply select a prime $p < \sqrt{\Delta}/2$ and let $\mathfrak{s} = (Q, P)$, where $Q = p$ and $Q \mid D^2 - P$. By results given in [21], for example, any ideal $\mathfrak{r}$ found by performing the continued fraction algorithm on $\mathfrak{s}$ will be reduced and equivalent to $\mathfrak{s}$.

There is some advantage to this beyond the increased generality in that we need no longer be concerned that h is small, although that is not in practice likely to be a problem. Even if h is quite large, there is no way a cryptanalyst can make use of the information that the number of reduced ideals in any ideal class is small because he does not know to which class $\mathfrak{r}^{ab}$ belongs. In other words we are making use here of the full infrastructure — the set of all reduced ideals — not just the set of reduced principal ideals. Hence the structure we are using must contain $O(hR) \gg (D^{1/2}/\log\log D)$ ideals under the ERH (see [11]), regardless of the size of h.

There is, of course, the possibility that when h is large, the ideal $\mathfrak{r}$ selected by the communication partners might belong to a small cyclic subgroup of the class group of K. In that case, it would be easy for a cryptanalyst to break the scheme by exhaustive search. However, by adapting the reasoning used in [3] to the real case, we see that the chance of selecting such an $\mathfrak{r}$ when h is large is very remote. We also point out that in the case where h is very large (R is small) there is no

1) Received by the editors on April 12, 2001.

need to make use of (f, p) representations. This is because the number of reduced ideals in any ideal class is $O(R)$. Thus, the communication partners need only agree beforehand on selecting that key ideal $\mathfrak{k} = (P, Q)$ in the class of $\mathfrak{r}^{ab}$ such that P is largest and Q is least, for example. Examples of values of D for which R is small are those of Richaud–Degert type ($D = M^2 + m$, where $m \mid 4M$), but there are many others.

References

[1] Abel, C.S., Ein Algorithmus zur Berechnung der Klassenzahl und des Regulators reellquadratischer Ordnungen. Ph.D. Thesis, Universität des Saarlandes, Saarbrücken, Germany, 1994.

[2] Buchmann, J., Düllmann, S., Williams, H.C., On the complexity and efficiency of a new key exchange system. In: Advances in Cryptology – EUROCRYPT '89 (ed. by J.-J. Quisquater and J. Vandewalle; Lecture Notes in Comput. Sci. 434), 597–616. Springer, Berlin 1990.

[3] Buchmann, J.A., Williams, H.C., A key-exchange system based on imaginary quadratic fields. J. Cryptology 1 (1988), 107–118.

[4] Hühnlein, D., Paulus, S., On the implementation of cryptosystems based on real quadratic number fields. In: Seventh Annual Workshop on Selected Areas in Cryptography – SAC 2000 (Lecture Notes in Comput. Sci.). Springer, Berlin (to appear).

[5] Jacobson, M.J., Jr., Subexponential Class Group Computation in Quadratic Orders. Ph.D. Thesis, Technische Universität Darmstadt, Darmstadt, Germany, 1999.

[6] —, Computing discrete logarithms in quadratic orders. J. Cryptology 13 (2000), 473–492.

[7] Jacobson, M.J., Jr., Williams, H.C., New quadratic polynomials with high densities of prime values. Math. Comp. (submitted).

[8] Koblitz, N., Elliptic curve cryptosystems. Math. Comp. 48 (1987), 203–209.

[9] —, Hyperelliptic cryptosystems. J. Cryptology 1 (1988), 94–97.

[10] The LiDIA Group, LiDIA: a C++ library for computational number theory. Software, Technische Universität Darmstadt, Germany, 1997. See `http://www.informatik.tu-darmstadt.de/TI/LiDIA`.

[11] Littlewood, J.E., On the class-number of the corpus $P(\sqrt{-k})$. Proc. London Math. Soc. (2) 27 (1928), 358–372.

[12] Lukes, R.F., Patterson, C.D., Williams, H.C., Numerical sieving devices: their history and some applications. Nieuw Arch. Wisk. (4) 13 (1995), 113–139.

[13] McCurley, K.S., A key distribution scheme based on factoring. J. Cryptology 1 (1988), 95–105.

[14] Miller, V. Use of elliptic curves in cryptography. In: Advances in Cryptology – CRYPTO '85 (ed. by H.C. Williams; Lecture Notes in Comput. Sci. 218), 417–426. Springer, Berlin 1986.

[15] Odoni, R.W.K., Varadharajan, V., Sanders, P.W., Public-key distribution in matrix rings. Electronics Letters 20 (1984), 386–387.

[16] van der Poorten, A.J., te Riele, H.J.J., Williams, H.C., Computer verification of the Ankeny-Artin-Chowla conjecture for all primes less than 100 000 000 000. Math. Comp. 70 (2001), 1311–1328.

[17] Scheidler, R., Buchmann, J.A., Williams, H.C., A key-exchange protocol using real quadratic fields. J. Cryptology 7 (1994), 171–199.

[18] Scheidler, R., Stein, A., Williams, H.C., Key-exchange in real quadratic congruence function fields. Des. Codes Cryptogr. 7 (1996), 153–174.

[19] Shanks, D., On Gauss and composition I, II. In: Number Theory and Applications (Banff, 1988) (ed. by Richard A. Mollin), 163–179. Kluwer, Dordrecht 1989.

[20] Vollmer, U., Asymptotically fast discrete logarithms in quadratic number fields. In: Algorithmic Number Theory – ANTS IV (ed. by W. Bosma; Lecture Notes in Comput. Sci. 1838), 581–594. Springer, Berlin 2000.

[21] Williams, H.C., Wunderlich, M.C., On the parallel generation of the residues for the continued fraction factoring algorithm. Math. Comp. 48 (1987), 405–423.

Extending the binary gcd algorithms

Przemysław Kubiak

Abstract. Based on a certain observation regarding divisibility by 2^k in representation of multiple of gcd, we develop a procedure which speeds up the extended binary gcd algorithms. We test usage of two such algorithms in inverting modulo a prime—for tests we get random primes having 160–1030 bits in length (the most interesting from cryptographical point of view).

Furthermore, based on the same ideas, we show how to extend the accelerated integer gcd algorithm presented by Weber [10]. We also estimate execution time of this extended version.

1. Introduction

First, recall that a gcd algorithm solely computes the greatest common divisor d of two integers A and B, while an extended gcd algorithm also computes the Bezout coefficients, i.e., two integers u and v such that $uA + vB = d$. The extended gcd algorithms are very useful when inverting A modulo B (which is only possible when A and B are coprime). In this case, the inverse of A mod B is given by integer u.

There are a few variants of extended gcd algorithm, e.g. Lehmer–Euclid's algorithm (see [5] or [7]) or binary gcd algorithm, as well as its plus-minus version ([1]). It is known that extended Lehmer–Euclid's method is fast, but extended binary algorithms are simpler—unlike Lehmer–Euclid's, do not perform division on large numbers at all.

We study here three extended binary gcd methods. Their basic gcd algorithms all deal with the tailing bits of some multiples a and b of looked for gcd. Suppose that a and b are odd, with $b > a$. Then, the main idea is to find some small r, s for building a short linear combination of a and b of the form $t = bs - ar$ that has many zero tailing bits. If t has k tailing bits equal to 0, one uses a shift $t := t/2^k$, and we replace b by t. Then, the length of the new b is shorter than the length of the old one. If $r = 1$ or $s = 1$, then $\gcd(a, b) = \gcd(a, t)$; otherwise, the short linear combination can introduce some (small) spurious factors for computing the gcd.

For $r, s = 1$ we get the binary gcd algorithm, for $r \in \{-1, 1\}$, $s = 1$ we get its plus-minus version. When $s = 1$ and $r \geq 1$ we obtain the dmod operation, and for $r, s \geq 1$—the 'combined shift'. The two latest are elements of accelerated gcd algorithm.

Public-Key Cryptography and Computational Number Theory
K. Alster, J. Urbanowicz and H. C. Williams (Eds.)

In Section 2 we first describe the basic extended binary gcd method, and then this algorithm is speeded up using the following idea.

Let A and B be positive integers and B be odd. Furthermore, let D (given e.g. as a linear combination of a and b) be a multiple of $\gcd(A, B)$, then

$$uA + vB = D$$

for some $u, v \in \mathbf{Z}$. If, in addition, D is divisible by 2^k, then there exists an integer y (less than 2^k, but not less than 0) such that *both* $u + yB$, $v - yA$ are also divisible by 2^k.

According to the author's knowledge, this observation has never been developed in the existing literature of the subject.

In Section 3 we discuss the extended plus-minus binary gcd method.

Observations made in Sections 2, 3 are used (in Section 4) to extend the accelerated integer gcd algorithm. The accelerated gcd algorithm is (according to [10]) very fast, and there was a hope that its extended version will be fast too—we show that the extended version will be about three times slower (so, relatively very fast).

As we shall see near the end of this article, the results of tests point out that even small precomputed arrays can be very useful in extending the binary gcd algorithms, especially on processors with relatively slow multiplication.

The author is very grateful to the anonymous referee for his/her valuable comments.

2. The extended binary gcd algorithm

2.1. The basic extended algorithm

Let $\mathrm{Val}_2(a)$ denote the dyadic valuation of the integer a, i.e. the greatest integer n such that $2^n | a$. To describe this integer we will also use the expression $2^n \,\|\, a$, defined as follows:

$$2^n \,\|\, a \Longleftrightarrow 2^n \mid a \,\wedge\, 2^{n+1} \nmid a.$$

We start with recalling the binary gcd (BGCD) algorithm.

BGCD algorithm

Input: $A, B \in \mathbf{N}$.

Output: $d = \gcd(A, B)$.

Algorithm:

preparation phase:

1. Let $n = \min(\mathrm{Val}_2(A), \mathrm{Val}_2(B))$, assign $a := A/2^n$, $b := B/2^n$.
2. If $2 \mid a$ then $a := a/2^{\mathrm{Val}_2(a)}$
3. else if $2 \mid b$ then $b := b/2^{\mathrm{Val}_2(b)}$.
4. If $a > b$ exchange a and b.

a and b are odd, with $0 < a \leq b$,
actual algorithm:

5. `While` $a \neq b$ `do:`
6.   `While` $a < b$ `do:`
7. k := $\mathrm{Val}_2(b - a)$;
8. b := $(b - a)/2^k$.
9.   Exchange a and b.
10. d := $2^n a$ (or $2^n b$).

This algorithm was first analysed by Brent [2] in 1976 who proved (under heuristic hypotheses) that the average number of executions of line 8 is of the form $A_2 \ln N$ for random odd input pairs A, B with $0 < A \leq B \leq N$. Then Vallée [9] in 1998 really proves this fact and analyses some other parameters: She shows that the average number of executions of line 9 is of the form $A_1 \ln N$, and the average value of the sum of all the integers k of line 7 is of the form $A_3 \ln N$, with $A_3 = 2A_2$, so that the average value of integer k is 2. Brent made extensive computations about the constant $A_2 \sim 1.018501$, and Vallée presents some results of simulations that suggest that $A_1 \sim 0.555$, $A_3 \sim 2.025$. The summary of results of [9] can be also found in [3].

The BGCD algorithm can be extended in such a way that we get numbers $u, v \in \mathbf{Z}$ satisfying

$$d = uA + vB, \tag{2.1}$$

cf. for example [7] or [5]. To make the paper self-contained we describe such an extension.

Obviously, the same u, v satisfy

$$d' = uA' + vB',$$

where $A' = A/2^n$, $B' = B/2^n$, $d' = d/2^n$ for $2^n \parallel \gcd(A, B)$. Moreover d', and at least one of the numbers A', B' are odd. Thus, without loss of generality, we can assume that $2 \nmid \gcd(A, B)$. For convenience we take B being odd through the rest of the paper.

It will also be convenient to express a and b from BGCD algorithm, as a linear combination of A, B:

$$u_a A + v_a B = a,$$
$$u_b A + v_b B = b.$$

The above equations are called here "basic equations".

Integers A and B are fixed for whole process, while a and b are being changed during algorithm's processing time (but both are kept odd after preparation phase).

The whole idea of the extended binary gcd (EBGCD) algorithm is to change the values of u_a, u_b, v_a, v_b in accordance with changes of a and b (to keep the basic equations right). At the beginning of the algorithm we perform the following, starting substitutions (B is odd, hence $n = 0$ in line 1):

$$a := A, \quad b := B$$

and

$$\begin{aligned}(u_a, v_a) &:= (1,0) \quad \text{or} \quad (1-B, A),\\ (u_b, v_b) &:= (0,1) \quad \text{or} \quad (B, 1-A).\end{aligned}$$

Suppose that A is odd, so both starting a and b are odd. Let $b > a$ (see lines 4, 9 of BGCD algorithm).

According to line 8 we assign

$$\begin{aligned}b &:= (b-a)/2^k,\\ u_b &:= u_b - u_a,\\ v_b &:= v_b - v_a.\end{aligned} \tag{2.2}$$

Now we must divide by 2^k the left side of the changed basic equation. We do this in the following way:

$$\begin{aligned}&\texttt{for } (i = 1; i \le k; i{+}{+})\\ &\quad \texttt{if } (2 \mid u_b) \texttt{ then}\\ &\qquad u_b := u_b/2\\ &\qquad v_b := v_b/2\\ &\quad \texttt{else}\\ &\qquad u_b := (u_b + B)/2\\ &\qquad v_b := (v_b - A)/2\end{aligned} \tag{2.3}$$

Let us consider the i-th iteration of the above loop. Suppose that $2 \mid u_b$. The integer $(b-a)/2^{i-1}$ (the operation $(b-a)/2^k$ from (2.2) can be viewed as division by 2 of the difference $b-a$, performed k times) is still divisible by $2^{k-(i-1)}$, hence $v_b B$ is also divisible by 2. Because of odd parity of B we have $2 \mid v_b$.

On the other hand, if $2 \nmid u_b$ then $2 \mid (u_b + B)$. Moreover

$$\begin{aligned}u_b A + v_b B &= (b-a)/2^{i-1},\\ (u_b + B)A + (v_b - A)B &= (b-a)/2^{i-1}.\end{aligned}$$

and $2^{k-(i-1)} \mid (b-a)/2^{i-1}$, thus $2 \mid (v_b - A)$.

After execution of (2.2), (2.3) the equality $u_b A + v_b B = b$ is still satisfied, and b is odd.

Next, we carry on the whole procedure as long as $a \ne b$.

So, we perform the binary gcd algorithm on the right sides of the basic equations, and at the same time we are manipulating variables (u_b, v_b) and (u_a, v_a). When $a = b$, (u_a, v_a) (and obviously (u_b, v_b)) are the numbers we are looking for.

Before formulae (2.2) we made the assumption that A is odd. When the input A is even, then the starting a is also even. In such a case, after substitution $a := a/2^{\mathrm{Val}_2(a)}$ (see line 2 of the BGCD), we perform on variables u_a, v_a operations analogous to (2.3), and the equation $u_a A + v_a B = a$ is satisfied with odd a.

During execution of the EBGCD algorithm we have $0 < a, b \le \max(A, B)$. Moreover, it can be easily guaranteed that $|u_b| < B$, $|v_b| < A$ after execution of (2.2), (2.3). In consequence, after exchanging variables forced by line 9, the pair (u_a, v_a) satisfies the same inequalities. So, values of the variables are bounded—in this version of the algorithm the upper bounds for $|u_b|$ and $|v_b|$ can be easily ensured by keeping $0 < u_b < B$.

Note that, since B is odd, all decisions concerning u_b in (2.3) are independent of v_b. Thus we can omit calculations of v_b, v_a, and return on the output u and d only. Nevertheless, if we have to return v as well, it can be effectively found just before the exit as follows:

$$|v| = \frac{uA - d}{B} \quad \text{for } u > 0,\ v < 0,$$

where division $(uA - d) : B$ is an "exact division" (see [6]).

2.2. Speeding up the extended binary gcd algorithm

Let us denote by t_3, t_1, t_2 the differences $b-a$, u_b-u_a, v_b-v_a from (2.2). The right side of the equality

$$t_1 A + t_2 B = t_3$$

can be divided by 2^k in one step (i.e., the substitution $b := (b-a)/2^k$ can be done quickly), whereas to divide the left side we perform k iterations of the loop (2.3). Nevertheless, for $k \le 32$ we can execute the whole loop (2.3) in one stage (instead of k iterations). Note that $k > 32$ is supposed to be extremely rare: once a 2^{32} executions of line 7 on average. Let $2^k \,|\, t_3$ ($k \le 32$), then

$$t_1 A + t_2 B \equiv 0 \pmod{2^k},$$
$$(t_1 + yB)A + (t_2 - yA)B \equiv 0 \pmod{2^k}$$

for any integer y. Let y be such that

$$t_1 + yB \equiv 0 \pmod{2^k}. \tag{2.4}$$

Then

$$(t_2 - yA)B \equiv 0 \pmod{2^k},$$

and since $\gcd(B, 2) = 1$, we have

$$t_2 - yA \equiv 0 \pmod{2^k}.$$

How to find y? From (2.4) we have $yB \equiv -t_1 \pmod{2^k}$, and

$$y \equiv t_1 \cdot (-B^{-1}) \pmod{2^k}. \tag{2.5}$$

We take nonnegative y, less than 2^k (as we remember $k \le 32$, hence multiplying long variables by y is fast on present-day processors). The value $(-B^{-1}) \pmod{2^{32}}$

can be found once, at the very beginning of the algorithm (very effective inverting modulo a power of two is described in [6]).

As we can see, the pseudocode (2.3) can be replaced by the following instructions:

$$\begin{aligned} y &:= t_1 \cdot (-B^{-1}) \pmod{2^k} \\ t_1 &:= (t_1 + yB)/2^k \\ t_2 &:= (t_2 - yA)/2^k \end{aligned} \tag{2.6}$$

where the values of t_1 and t_2 are, after (2.2), in u_b and v_b, respectively.

We emphasise that (2.6) are most often executed for k not greater than 2 (about $\frac{3}{4}$ events). Thus it is worth to do some precomputations: $2 \cdot B$ and $3 \cdot B$ can be obtained at the very beginning (i.e., just after a call) of the algorithm, and then might be stored in the array `B[ ]` (`B[`y`]`$= y \cdot B$ for $y = 1, 2, \ldots,$ `LMT`-1, where `LMT` denotes an upper bound of the array's size, and here `LMT` $= 4$). Instructions (2.6) are therefore transformed to (we omit computation of t_2 and, as a consequence, of v_b and v_a):

```
y := t1 · (−B^−1) (mod 2^k)
if (y > 0) then
    if (y < LMT) then
        t1 := t1 + B[y]                                   (2.7)
    else
        t1 := t1 + yB
t1 := t1/2^k
```

If the EBGCD algorithm is called repeatedly for the same odd B, it is worth to build the array `B[ ]` already before the first call, with `LMT` equal 256 for instance. Also $(-B^{-1}) \pmod{2^{32}}$ we compute before the first call. Then we do not have to multiply (long by short number) for $y <$ `LMT`. Note that the lowest possible value for `LMT` is 2, because `B[1]`$=B$ is always at our disposal.

In the case of $k > 32$ the instructions (2.6) (or (2.7)) may be performed in at most $q + 1$ stages, where $k = q \cdot 32 + r$ for some $0 \le r < 32$.

Suppose that $0 < u_a, u_b < B$. To find y (see (2.5)) we do not have to know the whole t_1, but only the last word of the difference:

$$y \equiv (u_b - u_a) \cdot (-B^{-1}) \pmod{2^k}. \tag{2.8}$$

Therefore for $y > 0$ we do not worry about the sign of $u_b - u_a$, instead of substitution $t_1 := t_1 + yB$ in (2.6) we perform:

$$\begin{aligned} t_1 &:= u_b + yB, \\ t_1 &:= t_1 - u_a. \end{aligned}$$

Note that $u_b \ge 0$, and $u_b + yB$ has execution time only a bit greater than instruction $y \cdot B$ (addition might already being performed while multiplication is being performed—we do not have to write the result of $y \cdot B$ to memory).

We make comparisons only if $y = 0$, and then for $u_a > u_b$ we have

$$\begin{aligned} t_1 &:= u_a - u_b \quad (\text{in } t_1 \text{ value } |t_1|), \\ t_1 &:= t_1/2^k, \\ t_1 &:= B - t_1. \end{aligned}$$

And for $u_b > u_a$:

$$\begin{aligned} t_1 &:= u_b - u_a, \\ t_1 &:= t_1/2^k. \end{aligned}$$

As a result we get $0 < t_1 < B$.

2.3. Extending the outer loop

Let us return to BGCD. This algorithm has two nested loops. The external loop corresponds to an exchange. Between two exchanges, there is a sequence of internal loops. Each internal loop consists in subtractions and shifts, and can be written as:

$$b = a + 2^{k_1} c_1, \quad c_1 = a + 2^{k_2} c_2, \quad c_2 = a + 2^{k_3} c_3, \quad \dots, \quad c_{\ell-1} = a + 2^{k_\ell} c_\ell,$$

with $0 < c_\ell \leq a$. Then, the global result of the above sequence of internal loops can be expressed in one substitution:

$$b := (b - xa)/2^m, \tag{2.9}$$

where

$$x = 1 + 2^{k_1} + 2^{k_1+k_2} + 2^{k_1+k_2+k_3} + \cdots + 2^{k_1+k_2+k_3+\cdots+k_{\ell-1}},$$

while the exponent m is equal to $m = k_1 + k_2 + k_3 + \cdots + k_{\ell-1} + k_\ell$ (hence, we get x being odd, and less than 2^m). Obviously, after substitution (2.9) we have $b = c_\ell$, thus $b \leq a$, and line 9 of BGCD is executed.

Instruction (2.9) is done gradually, in the internal loop, but u_b does not have to be actualized gradually as well. We can wait till the end of the internal loop, when collecting x is completed, and then change u_b according to (2.9). From the basic equations we get

$$\begin{aligned} u_b &:= (u_b - xu_a + yB)/2^m, \\ v_b &:= (v_b - xv_a - yA)/2^m, \end{aligned} \tag{2.10}$$

with $y \equiv (u_b - xu_a) \cdot (-B^{-1}) \pmod{2^m}$.

Suppose that $|u_a|, |u_b| < B$, then $|v_a|, |v_b| \leq A$ (or even less than A if only $\gcd(A, B) < \min(A, B)$). We shall to keep $|u_b| < B$. It is immediate that

$$\begin{aligned} u_b - xu_a &= \operatorname{sgn}_1(u_b) \cdot \bigl(|u_b| - \operatorname{sgn}_1(u_a) \cdot \operatorname{sgn}_1(u_b) \cdot x \cdot |u_a|\bigr), \\ v_b - xv_a &= -\operatorname{sgn}_1(u_b) \cdot \bigl(|v_b| - \operatorname{sgn}_1(u_a) \cdot \operatorname{sgn}_1(u_b) \cdot x \cdot |v_b|\bigr), \end{aligned}$$

where

$$\operatorname{sgn}_1(c) = \begin{cases} 1 & \text{for } c \geq 0, \\ -1 & \text{for } c < 0. \end{cases}$$

There are two cases now:

- $\operatorname{sgn}_1(u_a) \cdot \operatorname{sgn}_1(u_b) = 1$,
- $\operatorname{sgn}_1(u_a) \cdot \operatorname{sgn}_1(u_b) = -1$.

In the first case we take $y \equiv (|u_b| - x \cdot |u_a|) \cdot (-B^{-1}) \pmod{2^m}$, where only the last m bits of the $|u_b| - x \cdot |u_a|$ are needed. From (2.10), multiplying A and B by $\operatorname{sgn}_1(u_b) \cdot y$ instead of y, we obtain the following substitutions:

$$\begin{aligned} u_b &:= \operatorname{sgn}_1(u_b) \cdot (|u_b| + yB - x \cdot |u_a|)/2^m, \\ v_b &:= -\operatorname{sgn}_1(u_b) \cdot (|v_b| + yA - x \cdot |v_a|)/2^m. \end{aligned}$$

Let $y \equiv (|u_b| + x \cdot |u_a|) \cdot B^{-1} \pmod{2^m}$ in the second case, then instead of (2.10) we perform:

$$\begin{aligned} u_b &:= \operatorname{sgn}_1(u_b) \cdot (|u_b| + x \cdot |u_a| - yB)/2^m, \\ v_b &:= -\operatorname{sgn}_1(u_b) \cdot (|v_b| + x \cdot |v_a| - yA)/2^m. \end{aligned}$$

In both cases inequality $|u_b| < B$ holds.

Positive integers are usually stored in arrays of 32-bit computer words. To characterize variables from $\mathbf{Z}$ one can associate sign's bit with the array. Since, according to (2.9), u_b will be changed (or both u_b, v_b, if we keep track also on v_b), hence we put $\operatorname{sgn}_1(u_b)$ on purpose before the parentheses above: we want to know if the sign of u_b is also changed.

Obviously, we have $m \geq k$ (k from line 7 of BGCD), in many cases $m > k$. Thus, on processors with fast multiplication, computing the new coefficients (u, v) (or just u) only after each exchange (in the external loop of BGCD) is surely more efficient than computing them after each operation $b := (b - a)/2^k$ in the internal loop. However, for other processors the method described in Subsection 2.2 may be faster, as it significantly reduces the number of multiplications $y \cdot B$, and does not perform multiplications $x \cdot |u_a|$.

The average asymptotic value of m can be expressed by A_3/A_1. The results of simulations presented in [9] suggest that

$$\frac{2.025 - 0.005}{0.555 + 0.005} \leq \frac{A_3}{A_1} \leq \frac{2.025 + 0.005}{0.555 - 0.005},$$

i.e.

$$3.607 < \text{the average asymptotic value of } m < 3.690.$$

So, it is more than 2, but still less than 32.

Hence, instead of (2.3), we will perform operations analogous to extended dmod operation (see Section 4). The only difference are x and m (in dmod operation they are obtained immediately, in BGCD they are collected in the internal loop).

3. The extended plus-minus binary gcd algorithm

3.1. The inner loop

In preparation phase the plus-minus binary gcd (PMBGCD) algorithm proceeds exactly like BGCD, thus only the main part of the algorithm is given:

PMBGCD algorithm

a and b are odd, with $0 < a \le b$

5. `While` $a \neq b$ `do:`
6.   `While` $a < b$ `do:`
7.   `if` $a \equiv b \pmod 4$ `then` $t := b - a$;
8.   `else` $t := b + a$;
9. $k := \mathrm{Val}_2(t)$;
10. $b := t/2^k$.
11.   Exchange a and b.
12. $d := 2^n a$ (or $2^n b$).

In line 8 we certainly have $a \equiv -b \pmod 4$ (a and b, as odd numbers, can only be equal to 1 or 3 mod 4). Note that $k \ge 2$, and even for $t = b + a$ we have $b := (b+a)/2^k < (b+b)/2^k = b/2^{k-1}$. Thus the value of b decreases.

Another version, without comparing a and b, but allowing $a, b < 0$, is presented in [4]—extension of this algorithm is given in [1]. A description of plus-minus binary gcd is also given in [6].

Avoiding comparisons is very desirable from systolic algorithms' point of view (see [6]). On the other hand, to compare long numbers usually suffices to check only their most significant computer words—thus this operation does not take too much and is not very inconvenient on systems with single processor (on such systems we are less interested in systolic parallelization). Moreover, due to comparison the greater number is always shorten in the above procedure.

Like in Subsection 2.2, let us extend the above algorithm (an extended method we denote as EPMBGCD). From basic equations, the operations performed on a and b in lines 7–8 are equivalent to:

$$\begin{aligned} &a \equiv b \pmod 4: \\ &\qquad \underbrace{(u_b - u_a)}_{t_1} A + \underbrace{(v_b - v_a)}_{t_2} B = \underbrace{(b - a)}_{t_3}, \\ &a \not\equiv b \pmod 4: \\ &\qquad \underbrace{(u_b + u_a)}_{t_1} A + \underbrace{(v_b + v_a)}_{t_2} B = \underbrace{(b + a)}_{t_3}. \end{aligned} \tag{3.1}$$

We shall to keep $|u| < B$ (only just before the exit a substitution $u := u + B$ will be done, if necessary).

Define the function sgn(opr) as follows:

$$\operatorname{sgn}(\mathrm{opr}) = \begin{cases} 1 & \text{for } \mathrm{opr} = `+', \\ -1 & \text{for } \mathrm{opr} = `-'. \end{cases}$$

Then operations performed on u_a, u_b in (3.1) can be written as:

$$u_b \,\mathrm{opr}\, u_a = \operatorname{sgn}_1(u_b) \cdot \big(|u_b| + \operatorname{sgn}(\mathrm{opr}) \cdot \operatorname{sgn}_1(u_a) \cdot \operatorname{sgn}_1(u_b) \cdot |u_a|\big),$$

where sgn(opr) depends on satisfying the congruence $a \equiv b \pmod 4$. Note that u_b will be changed (see line 10). Finally, we have two cases:

- $\operatorname{sgn}(\mathrm{opr}) \cdot \operatorname{sgn}_1(u_a) \cdot \operatorname{sgn}_1(u_b) = -1$ and we have subtraction: $|u_b| - |u_a|$,
- $\operatorname{sgn}(\mathrm{opr}) \cdot \operatorname{sgn}_1(u_a) \cdot \operatorname{sgn}_1(u_b) = 1$ and we have addition: $|u_b| + |u_a|$.

Adding to and subtracting from the left side of the equality $t_1A + t_2B = t_3$ a number $\operatorname{sgn}_1(u_b) \cdot yAB$, we have in case of subtraction

$$|u_b| + yB - |u_a|,$$

and we take the least nonnegative $y \equiv (|u_b| - |u_a|) \cdot (-B^{-1}) \pmod{2^k}$. Similarly to Subsection 2.2, the array `B[ ]` can be used. If $y \geq$ `LMT`, but $(-y) \pmod{2^k} <$ `LMT`, then the element `B[`$(-y) \pmod{2^k}$`]`, denoted for short as `B[`$-y$`]`, can be exploited:

$$-\big(|u_a| + \mathtt{B[}-y\mathtt{]} - |u_b|\big).$$

The above expression is divisible by 2^k, and the sign of u_b is changed.

In case of addition we have

$$|u_b| + |u_a| + yB.$$

Since $|u_a|, |u_b| < B$, thus $|u_b| + |u_a| < 2B$. Here we do not postpone addition, y can be found after its execution:

$$\begin{aligned} t_1 &:= |u_a| + |u_b|, \\ y &:= t_1 \cdot (-B^{-1}) \pmod{2^k}. \end{aligned}$$

Two cases should be distinguished:

- $y < 2^k - 1$, i.e., $y \leq 2^k - 2$, then
  $t_1 := t_1 + yB \quad (t_1 < 2^kB)$,
  $t_1 := t_1/2^k \qquad (t_1 < B)$,
- not $(y < 2^k - 1)$, and because $y < 2^k$ we have $y = 2^k - 1$, i.e., $y \equiv -1 \pmod{2^k}$
  $t_1 := t_1 - B$ (the sign of t_1 will be changed if $t_1 < B$),
  $t_1 := t_1/2^k$.

Again, the array `B[ ]` (by taking `B[`y`]` or `B[`$-y$`]`) can be used.

Both in subtraction and addition, if we make use of `B[`$-y$`]` and we have `LMT` $= 2^s$ for some $s \geq 1$, then for $k \leq s+1$ the only 'chance' to perform multiplication is to have $k = s+1$ and $y =$ `LMT` simultaneously. On the other hand, the probability that $2^k \,\|\, t_3$ for some $k \leq s+1$, equals (in EPMBGCD algorithm) $1 - \frac{1}{2^s}$ (see appendix A for more exact estimation of probability that, in a single iteration, a multiplication is avoided).

3.2. The outer loop

Let us return to PMBGCD. This algorithm has two nested loops. The external loop corresponds to an exchange. Between two exchanges, there is a sequence of internal loops. Each internal loop consists in subtractions (or additions) and shifts and can be written as

$$b = \epsilon_1 a + 2^{k_1} c_1, \quad c_1 = \epsilon_2 a + 2^{k_2} c_2, \quad c_2 = \epsilon_3 a + 2^{k_3} c_3, \quad \dots, \quad c_{\ell-1} = \epsilon_\ell a + 2^{k_\ell} c_\ell,$$

with $0 < c_\ell \le a$ and $\epsilon_j \in \{-1, 1\}$ for $1 \le j \le \ell$. Then, the global result of the above sequence of internal loops can be expressed in the substitution (2.9), for

$$x = \epsilon_1 + \epsilon_2 2^{k_1} + \epsilon_3 2^{k_1+k_2} + \epsilon_4 2^{k_1+k_2+k_3} + \cdots + \epsilon_\ell 2^{k_1+k_2+k_3+\cdots+k_{\ell-1}},$$

and $m = k_1 + k_2 + k_3 + \cdots + k_{\ell-1} + k_\ell$ (we know that $k_j \ge 2$).

The integer x can be written as $x = x' - x''$, where x' contains powers of 2 corresponding to all $\epsilon_j = 1$ for $1 \le j \le \ell$, and x'' contains powers of 2 corresponding to all $\epsilon_j = -1$ for $1 \le j \le \ell$. If $x'' > x'$ then subtraction was the last operation on a and b.

The assignment (2.9) leads to the following:

$$\begin{aligned} u_b - x u_a &= \operatorname{sgn}_1(u_b) \cdot (|u_b| - \operatorname{sgn}_1(u_a) \cdot \operatorname{sgn}_1(u_b) \cdot \operatorname{sgn}_1(x' - x'') \cdot |x' - x''| \cdot |u_a|), \\ v_b - x v_a &= -\operatorname{sgn}_1(u_b) \cdot (|v_b| - \operatorname{sgn}_1(u_a) \cdot \operatorname{sgn}_1(u_b) \cdot \operatorname{sgn}_1(x' - x'') \cdot |x' - x''| \cdot |v_b|). \end{aligned}$$

And again, we have two cases:

- $\operatorname{sgn}_1(u_a) \cdot \operatorname{sgn}_1(u_b) \cdot \operatorname{sgn}_1(x' - x'') = 1$,
- $\operatorname{sgn}_1(u_a) \cdot \operatorname{sgn}_1(u_b) \cdot \operatorname{sgn}_1(x' - x'') = -1$.

Each of them can be easily served by choosing appropriate $0 \le y < 2^m$.

Tests pointed out that $m < 8$ in very most cases, hence usage of the table `B[ ]` with `LMT` $= 2^8$ allows us to avoid very most of multiplications $y \cdot B$. Unfortunately, multiplications $|x' - x''| \cdot |u_a|$ have to be performed.

We emphasise that elements `B[`$-y$`]` can be also used in this version of EPM-BGCD method:

Because shifts in the internal loop of PMBGCD algorithm are not smaller than 2, thus $|x' - x''| \le 2^{m-1} - 1$. For y such that $y \ge$ `LMT` and $(-y) \pmod{2^m} <$ `LMT` we have $(-y) \pmod{2^m} < 2^{m-1}$. Therefore, in the case

- $\operatorname{sgn}_1(u_a) \cdot \operatorname{sgn}_1(u_b) \cdot \operatorname{sgn}_1(x' - x'') = -1$

if the addition $(|u_b| + |x' - x''| \cdot |u_a|) + (-y) \pmod{2^m} \cdot B$ instead of subtraction $(|u_b| + |x' - x''| \cdot |u_a|) - yB$ is performed, then we obtain a number less than

$$(B - 1) + (2^{m-1} - 1) \cdot (B - 1) + 2^{m-1} \cdot B = 2^{m-1} \cdot (B - 1) + 2^{m-1} \cdot B < 2^m \cdot B.$$

Thus the result divided by 2^m will be less than B.

Similarly, in the case

- $\operatorname{sgn}_1(u_a) \cdot \operatorname{sgn}_1(u_b) \cdot \operatorname{sgn}_1(x' - x'') = 1$

we have $|u_b| - |x' - x''| \cdot |u_a| - (-y) \pmod{2^m} \cdot B > -2^m B$.

4. The extended accelerated gcd algorithm

The detailed description of the accelerated gcd (AGCD) algorithm is given in [10]. Similar fast method of finding gcd gives Jebelean [6]. The accelerated gcd algorithm seems to be very effective for input data from 256 to $32 \cdot 1024$ bits in length (see [10]). For variables of length 4096 bits it works over 5.5 times faster than the binary gcd. Hence there is a big temptation to extend such fast method.

We define $l(a) = \lfloor \log_2 a \rfloor + 1$, i.e., $l(a)$ denotes the number of bits of positive integer a. Then the accelerated gcd algorithm proceeds as follows.

AGCD algorithm

Input: A, B—odd positive integers, such that $l(B) \geq l(A)$.

Output: $d = \gcd(A, B)$.

Algorithm:

```
a := A, b := B;
While a ≠ 0 do
    if l(b) − l(a) > w then
        dmod(b,a);
    else
        (r, s) := ReducedRatMod(b, a);
        b := |ar − bs|/2^64;
    b := b/2^(Val_2(b));
    Exchange a and b;
d' := gcd(dmod(A, b), b);
d := gcd(dmod(B, d'), d'),
```

where $\operatorname{dmod}(b, a)$ operation is a sequence of following instructions:

```
While l(b) ≥ l(a) + 32 do
    if b ≢ 0 (mod 2^32) then
        x := ba^(−1) (mod 2^32);
        b := |b − xa|;
    b := b/2^32;
k := l(b) − l(a);
if b ≢ 0 (mod 2^(k+1)) then
    x := ba^(−1) (mod 2^(k+1));
    b := |b − xa|;
b := b/2^(k+1);
```

The procedure `ReducedRatMod` is called in line 6 of the AGCD algorithm. This auxiliary procedure returns integers r, s such that $ar \equiv bs \pmod{2^{64}}$ and $0 < r, |s| < 2^{32}$, and proceeds as follows:

Input: b, a—odd positive integers.

$c := ba^{-1} \pmod{2^{64}}$;
$(r_1, s_1) := (2^{64}, 0)$;
$(r_2, s_2) := (c, 1)$;
`While` $r_2 \geq 2^{32}$ `do`
    $(r_1, s_1) := (r_1, s_1) - \lfloor r_1/r_2 \rfloor \cdot (r_2, s_2)$;
    Exchange the pairs (r_1, s_1) and (r_2, s_2);
Return (r_2, s_2).

The proof of correctness of the above procedure can be found in [10].

Let us return to the main algorithm. The operation $b := |ar - bs|/2^{64}$ (performed in line 7) is called here "combined shift".

The number w, used in line 3, informs the algorithm how close $l(b)$ and $l(a)$ must be in order for a combined shift to be effective—Weber[10] chose $w = 10$ for tests.

The dmod operation preserves the gcd, but the combined shift may introduce spurious factors. They are eliminated in lines 10–11, where instructions equivalent to computing $\gcd(A, B, b)$ are executed. Results of tests, presented in [10], suggest that b entering line 10 (i.e., gcd with spurious factors) is less than 2^{64} on average (even for $A, B < 2^{64 \cdot 32}$).

As we can see, the AGCD method uses three main operations:

- dmod,
- $b := |ar - bs|/2^{64}$,
- $b := b/2^m$.

After extending each of these operations, we get extended accelerated gcd (EAGCD) algorithm which finds (u, v, d) satisfying equation (2.1).

It suffices when algorithm returns (u, d), v can be found by exact division.

4.1. dmod

Let $l(b) > l(a)$, $\gcd(a, 2) = 1$. This operation can be viewed as sequential repeating the two following instructions

- $b := |b - xa|$ where $x < 2^m \leq 2^{32}$, $x \equiv ba^{-1} \pmod{2^m}$,
- $b := b/2^m$,

In all executions of the above pair, maybe apart the last one, we have $m = 32$.

Let us denote by d the term xa. Then $|b - xa| = \operatorname{sgn}_1(b - d) \cdot (b - xa)$, and

$$\operatorname{sgn}_1(b - d) \cdot (u_b - xu_a) = \operatorname{sgn}_1(b - d) \cdot \operatorname{sgn}_1(u_b) \cdot (|u_b| - \operatorname{sgn}_1(u_a) \cdot \operatorname{sgn}_1(u_b) \cdot x|u_a|).$$

There are two cases:

1. $\mathrm{sgn}_1(u_a) \cdot \mathrm{sgn}_1(u_b) = 1$,
   then we perform the following instructions:
   (a) $c_2 := x \cdot |u_a|$,
   (b) $y := (|u_b| - c_2) \cdot (-B)^{-1} \pmod{2^m}$,
   (c) $c_1 := |u_b| + yB$,
   (d) $u_b := \underbrace{\mathrm{sgn}_1(b-d) \cdot \mathrm{sgn}_1(u_b) \cdot \mathrm{sgn}_1(c_1 - c_2)}_{\text{sign's bit}} \cdot |c_1 - c_2|$,
   (e) assign $u_b := u_b/2^m$.
2. $\mathrm{sgn}_1(u_a) \cdot \mathrm{sgn}_1(u_b) = -1$,
   (a) $c_1 := |u_b| + x|u_a|$,
   (b) $y := c_1 \cdot B^{-1} \pmod{2^m}$,
   (c) $c_2 := y \cdot B$,
   (d) follow the instructions: 1(d) and 1(e).

As always, in 1(b) we calculate only the last 32-bit computer word of the difference $|u_b| - c_2$. In both cases calculating c_1 is quite fast (without writing to memory the result of multiplication). Also in both cases we have $c_1 < 2^m B$, $c_2 < 2^m B$.

Point 1(d) can be put into words as follows: compare c_1 and c_2, subtract smaller from greater, assign difference to $|u_b|$ changing, if necessary, the sign of u_b. Note that after this $|u_b| < 2^m B$, and, as a consequence, after 1(e) $|u_b| < B$.

As we could see, the extended dmod operation is analogous to EPMBGCD algorithm computing u in the external loop.

To recapitulate cases 1 and 2, we have:

- two multiplications long by short number (c_1,c_2), including one multiplication with addition (c_1),
- one long comparison,
- one long subtraction,
- one m-bit shift.

$B^{-1} \pmod{2^{32}}$ (needed to get y) is calculated only once—at the very beginning of the algorithm, whereas $a^{-1} \pmod{2^{32}}$ (needed to get x) is calculated after each call of dmod operation (a is being changed in the main loop of AGCD, see line 9).

Considering only the number of multiplications we get

$$\text{cost of the extended dmod (including dmod)} = 3 \cdot \text{cost of dmod}$$

Considering all needed operations we get that cost of the extended dmod equals roughly $3 \cdot$ cost of dmod.

According to [10], extending the binary gcd algorithm by only dmod operation (without $b := |ar - bs|/2^{64}$) speeds up the process of finding gcd significantly.

4.2. $b := |ar - bs|/2^{64}$

We have $0 < r, |s| < 2^{32}$, a and b odd and greater than zero.

$b := |ar - bs|/2^{64}$ consists of a few instructions:

- $d_1 := a \cdot r$,
- $d_2 := b \cdot |s|$,
- if $s > 0$ then compare d_1 and d_2, and subtract smaller from greater, else add d_1 to d_2.
- shift right the result by 64 bits,

It is easy to see that

$$|ar - bs| = -\min\big(\mathrm{sgn}_1(s), \mathrm{sgn}_1(d_1 - d_2)\big) \cdot \big(b \cdot |s| - \mathrm{sgn}_1(s) \cdot ar\big),$$

hence, the corresponding operation on u_b and u_a can be described as

$$-\min\big(\mathrm{sgn}_1(s), \mathrm{sgn}_1(d_1 - d_2)\big) \cdot \mathrm{sgn}_1(u_b) \cdot \big(|u_b| \cdot |s| - \mathrm{sgn}_1(u_a) \cdot \mathrm{sgn}_1(u_b) \cdot \mathrm{sgn}_1(s) \cdot |u_a| r\big).$$

Thus, to perform a combined shift on u_b and u_a, we consider two cases:

1. $\mathrm{sgn}_1(u_a) \cdot \mathrm{sgn}_1(u_b) \cdot \mathrm{sgn}_1(s) = 1$,
   (a) $c_2 := |u_a| \cdot r$,
   (b) $c_3 := |u_b| \cdot |s|$,
   (c) $y_1 := (c_3 - c_2) \cdot (-B)^{-1} \pmod{2^{32}}$,
   (d) $c_1 := c_3 + y_1 B$,
   (e) $u_b := \underbrace{-\min(\mathrm{sgn}_1(s), \mathrm{sgn}_1(d_1 - d_2)) \cdot \mathrm{sgn}_1(u_b) \cdot \mathrm{sgn}_1(c_1 - c_2)}_{\text{sign's bit}} \cdot |c_1 - c_2|$,
   (f) $u_b := u_b/2^{32}$,
   (g) $y_2 := |u_b| \cdot (-B)^{-1} \pmod{2^{32}}$,
   (h) i. if $y_2 < 2^{32} - 1$ then
   $u_b := \mathrm{sgn}_1(u_b) \cdot (|u_b| + y_2 B)$
   ii. else ($y_2 = 2^{32} - 1$):
   $u_b := \mathrm{sgn}_1(u_b) \cdot \mathrm{sgn}_1(|u_b| - B) \cdot \big||u_b| - B\big|$,
   (i) assign $u_b := u_b/2^{32}$.
2. $\mathrm{sgn}_1(u_a) \cdot \mathrm{sgn}_1(u_b) \cdot \mathrm{sgn}_1(s) = -1$,
   (a) $c_3 := |u_b| \cdot |s|$,
   (b) $c_1 := c_3 + |u_a| \cdot r$,
   (c) $y_1 := c_1 \cdot B^{-1} \pmod{2^{32}}$,
   (d) $c_2 := y_1 B$,
   (e) follow the instructions: 1(e)–1(f).

In both cases calculating c_1 is quite fast; also in both cases we have $c_2 < 2^{32}B$, $c_3 < 2^{32}B$ and $c_1 < 2 \cdot 2^{32}B$. After 1(e) $|u_b| < 2 \cdot 2^{32}B$, and after 1(f) $|u_b| < 2B$.

Note that the sign of u_b could be changed by instruction 1(e) and the sign of new u_b could be changed by 1(h)ii.

To recapitulate cases 1 and 2, we have:

- four multiplications long by short number (c_1,c_2,c_3, and $|u_b|+y_2B$ in 1(h)ii—we pass the case $y_2 = 2^{32}-1$ over), including two multiplications with addition (c_1 and $|u_b| + y_2B$),
- two 32-bit shifts (costs of a single 32-bit and a single 64-bit shift are obviously the same),
- one long comparison,
- one long subtraction.

Finding r and s should be added to the cost of the combined shift.
Considering only the number of multiplications we get:

cost of the extended combined shift (including combined shift)
$= 3 \cdot$ cost of the combined shift.

Considering all needed operations we get that cost of the extended combined shift equals roughly $3 \cdot$ cost of the combined shift.

4.3. $b := b/2^m$

We extend this operation (shift) by simple adding to u_b some multiple of B:

- $u_b := u_b + \mathrm{sgn}_1(u_b) \cdot yB = \mathrm{sgn}_1(u_b) \cdot (|u_b| + yB)$,
  where $y \equiv |u_b| \cdot (-B^{-1}) \pmod{2^m}$,
- $u_b := u_b/2^m$.

The cost of this operation (for $m \leq 32$): one multiplication with addition and one m-bit shift.

The case $m > 32$ (is supposed to be rare) should be served in a few steps.

We finally get that the cost of the extended shift (including shift) is somewhat more than $3\cdot$ shift.

From Subsections 4.1, 4.2 and 4.3 we get:

Execution time of the AGCD algorithm equals about one third of running time of the EAGCD algorithm.

Therefore, for sufficiently long numbers the extended accelerated gcd method will be faster than the binary gcd algorithm, which gives only gcd.

If the purpose of extending the accelerated gcd method is finding inverses modulo a prime B, then to remove spurious factor from a result it suffices to assign

$$d := \gcd(\mathrm{dmod}(B, b), b),$$

instead of executing lines 10–11 of AGCD algorithm.

We know that $d = 1$, but executing the above instruction is necessary, because operations on u_a, u_b are associated with it.

5. Tests

In order to determine how observations presented in previous sections make the algorithms faster, we have implemented finding inverses modulo a prime. When $\gcd(A, B) = 1$, the algorithms become a bit simpler, the greatest power of two dividing both A and B is 2^0, and the preparing shifts $a := A/2^n$, $b := B/2^n$ will not be done. Assume that B is odd and we search for $A^{-1} \pmod B$, hence we will not calculate v from

$$uA + vB = 1$$

—as we could see in previous sections, finding u_a, u_b is (provided B is odd) completely independent from v_a, v_b.

Moreover, if we know that $\gcd(A, B) = 1$, then we do not have to wait with stopping the algorithm till $a = b$, it suffices when *one* of the numbers a, b equals 1. What is more, the results presented in [9] show that the average behaviour of the BGCD method is asymptotically independent of the assumption $\gcd(A, B) = 1$ (this agrees intuitively with the following: for randomly chosen A, B, the $\gcd(A, B)$ is almost always very small—see [7]).

32-bit and k-bit shifts (for $k < 32$) are realized in different ways:

32-bit—by rewriting whole computer 32-bit words, k-bit—by using (in Intel's processors) `shrd` instruction. Therefore, after checking whether $k \leq 32$, it must be checked whether $k \leq 31$ or $k = 32$. To simplify our implementation, we will only check whether $k \leq 31$ or $k > 31$—tests suggest (see Table 2) that we pay for it very seldom.

We checked execution times for algorithms:

(1) *Extended binary gcd*—without multiplying by y, with shifting t_3 and t_1 only by one bit in one iteration. This version is denoted by **ord_inv**.

(2) *Extended binary gcd with shifting*—still without multiplying by y. u (as t_1) is computed in the inner loop. In one iteration t_1 (and t_3) is shifted right as much as possible (but not more than by 31 bits): after each adding B to t_1 we check how many tailing zeros appear in the result. In this way we get required shift of t_1 in the minimal number of iterations. We denote this version by **shr_inv**.

(3) *Extended plus-minus binary gcd with shifting*—without multiplying by y as well. u is computed in the inner loop. Remarks about shifting—like in (2). Denoted by **±shr_inv**.

(4) *Extended binary gcd.* u is computed in the inner loop. $B^{-1} \pmod{2^{32}}$ and the table `B[ ]` are determined once, before the first call of the algorithm. We denote it by **inv**.

(5) *Extended plus-minus binary gcd*, similar to (4), denoted by **±inv**.

(6) *Fast extended binary gcd.* u is computed in the external loop. $B^{-1} \pmod{2^{32}}$ and the table `B[ ]` are determined once, like in (4). We denote it by **fst_inv**.

(7) *Fast extended plus-minus binary gcd*, similar to (6), denoted by **±fst_inv**.

In implementations of (4)–(7) we use table `B[ ]` with `LMT` = 256, elements `B[`$-y$`]` are not used. The time of computing $B^{-1} \pmod{2^{32}}$ and determining the table `B[ ]` was not included in execution time of the algorithms.

Tests are made for prime numbers in 24 lengths (from 160 to 1030 bits). For each length we randomly chose (using LiDIA package [8]) 200 primes B. Next, for every B we randomly chose 200 integers A such that $1 < A < B$, and for pairs (A, B) a function finding $A^{-1} \pmod B$ was called. Thus a given function was executed for given length $200 \cdot 200$ times.

Tests were performed under Windows, on Pentium III 500MHz processor. In Table 1 the total execution times for 40000 function calls (described above) are presented. Times are given in seconds. Results are truncated, not rounded.

n	ord_inv	shr_inv	±shr_inv	inv	±inv	fst_inv	±fst_inv	Rf	$Rf\pm$
160	4.56	4.48	4.64	3.49	3.18	3.14	3.14	29.78	32.34
180	5.38	5.22	5.40	4.04	3.63	3.55	3.55	31.93	34.27
210	6.89	6.53	6.79	5.14	4.58	4.42	4.41	32.17	35.08
230	8.16	7.64	7.96	6.02	5.33	5.12	5.08	33.03	36.19
256	9.55	8.87	9.18	7.05	6.35	6.09	6.06	31.33	33.97
310	12.79	11.66	11.99	9.22	8.05	7.65	7.55	34.34	37.00
330	14.45	13.06	13.39	10.33	8.95	8.47	8.33	35.15	37.75
410	20.35	18.06	18.40	14.34	12.31	11.65	11.44	35.51	37.78
430	22.41	19.80	20.18	15.70	13.43	12.58	12.31	36.45	38.96
510	29.58	25.80	26.13	20.71	17.79	16.68	16.35	35.34	37.39
512	30.36	26.54	26.83	21.16	18.21	16.96	16.60	36.09	38.11
530	32.09	27.89	28.16	22.13	18.72	17.38	16.92	37.67	39.90
610	42.55	36.70	37.15	28.97	24.42	22.44	21.88	38.86	41.11
630	43.41	37.26	37.44	29.57	24.83	22.95	22.27	38.38	40.52
710	54.65	46.46	46.63	36.91	30.76	28.25	27.36	39.17	41.31
730	56.41	47.84	47.92	38.09	31.75	29.31	28.39	38.73	40.73
768	62.57	52.92	53.00	42.34	35.58	32.63	31.66	38.33	40.26
810	68.97	58.11	58.17	46.26	38.35	35.07	33.87	39.64	41.77
830	71.07	59.76	59.76	48.07	40.13	36.91	35.77	38.23	40.14
910	84.98	71.15	71.00	56.76	46.78	42.62	41.12	40.09	42.07
930	90.36	75.84	76.00	60.23	49.75	45.15	43.59	40.46	42.64
1010	102.78	85.56	85.23	68.28	56.07	50.95	48.95	40.44	42.56
1024	105.87	88.14	87.85	70.54	58.46	53.11	51.23	39.74	41.68
1030	107.59	89.70	89.21	71.32	58.45	52.95	51.03	40.96	42.80

Table 1. Total execution times

Each function (1)–(7) has been implemented equally carefully. Low-level functions (written in assembler) has not been optimized for any processor—we only tried to avoid dependencies between current and next instructions, but loops have not been expanded to make more than one iteration in one stage.

We will not link the results of (4)–(7) to results of (1) ((1) is included only to indicate efficiency of (2)). For convenience, execution time of the function f is denoted by $t(f)$. As we can see, $t(\text{shr\_inv})$ and $t(\pm\text{shr\_inv})$ are close, surprisingly the ±shr_inv is even a bit slower in most cases. The ±inv is faster than the inv one. The ratio

$$\frac{t(\text{shr\_inv}) - t(\text{inv})}{t(\text{shr\_inv})} \cdot 100\%$$

is equal at least 19.55% (it varies from 19.55% to 22.58%), and the ratio

$$\frac{t(\pm\text{shr\_inv}) - t(\pm\text{inv})}{t(\pm\text{shr\_inv})} \cdot 100\%$$

is greater than 30% (it varies from 30.82% up to 34.53%). The values of ratios:

$$Rf = \frac{t(\text{shr\_inv}) - t(\text{fst\_inv})}{t(\text{shr\_inv})} \cdot 100\%,$$

$$Rf\pm = \frac{t(\pm\text{shr\_inv}) - t(\pm\text{fst\_inv})}{t(\pm\text{shr\_inv})} \cdot 100\%,$$

are placed in Table 1.

In Table 2 we present the average number of m-shifts in a single call of a given function (as we remember, the average value of m lie somewhere between 3.607 and 3.690). Results are truncated.

For EBGCD method it turned out that the average number of m-shifts for $m \leq 2$, divided by the average number of all shifts is equal about 0.39. For $m \leq 8$ we have about 0.95. The same values for EPMBGCD are equal about 0.25 and 0.92, respectively.

C3+C4 in the table means the average number of all m-shifts (a value in column 3 plus a value in column 4) in EBGCD. In other words: C3+C4 denotes the average number of exchanges. C8+C9 means the same in EPMBGCD method. Note that $0.38 \approx 0.555 \cdot \ln 2$, where 0.555 is the estimation of the asymptotic value of A_1 (see [9]). As we can see, in PMBGCD exchanges occur more often than in BGCD.

M in the table denotes the total number of 31-bit shifts, which occurred during 40000 calls of a given method. The m-shifts by more than 31-bits are, in our implementation, divided into at most 31-bit pieces. Thus the biggest possible shift is a 31-shift.

Table 2 shows that some precomputations allow us to reduce the number of multiplications $y \cdot B$ significantly. In Table 3 experimentally estimated probabilities of performing multiplication $y \cdot B$ in a single iteration of EPMBGCD algorithm's given loop are placed (these probabilities are denoted by Pr(not(`B[ ]`))). The estimations are obtained under the assumption that `B[`$-y$`]` can be used.

As we can see, for `LMT` = 256 multiplications occur very seldom in the version of algorithm computing u in the internal loop (recall that multiplications $x \cdot u_a$

n	EBGCD					EPMBGCD				
	$m \le 2$	$m \le 8$	$m > 8$	(C3+C4)/n	M	$m \le 2$	$m \le 8$	$m > 8$	(C8+C9)/n	M
160	24.20	58.10	2.84	0.3809	0	16.99	60.81	4.54	0.4084	0
180	27.25	65.46	3.23	0.3816	0	19.12	68.58	5.10	0.4094	0
210	31.65	76.35	3.80	0.3817	1	22.24	80.05	5.99	0.4097	1
230	34.65	83.69	4.15	0.3819	0	24.37	87.73	6.57	0.4100	0
256	38.47	93.00	4.68	0.3816	1	27.10	97.70	7.32	0.4102	1
310	46.63	112.82	5.68	0.3823	0	32.74	118.24	8.94	0.4103	0
330	49.67	120.14	6.05	0.3824	1	34.79	125.92	9.53	0.4104	0
410	61.55	149.27	7.56	0.3825	0	43.19	156.53	11.88	0.4107	0
430	64.62	156.59	7.94	0.3826	1	45.29	164.14	12.49	0.4107	4
510	76.57	185.80	9.44	0.3828	1	53.59	194.68	14.84	0.4108	6
512	76.80	186.40	9.49	0.3826	2	53.81	195.39	14.91	0.4107	2
530	79.51	193.04	9.84	0.3828	1	55.79	202.41	15.42	0.4110	3
610	91.51	222.23	11.32	0.3828	1	64.16	232.94	17.80	0.4110	1
630	94.49	229.58	11.70	0.3829	3	66.19	240.47	18.42	0.4109	4
710	106.54	258.80	13.19	0.3830	1	74.61	271.36	20.70	0.4113	1
730	109.33	265.99	13.56	0.3829	0	76.63	278.62	21.38	0.4109	2
768	115.08	279.77	14.31	0.3829	1	80.74	293.41	22.45	0.4112	2
810	121.38	295.14	15.08	0.3830	1	85.06	309.30	23.72	0.4111	2
830	124.27	302.43	15.46	0.3830	4	87.19	317.02	24.30	0.4112	2
910	136.35	331.64	16.96	0.3830	1	95.55	347.58	26.70	0.4113	3
930	139.35	338.89	17.36	0.3830	2	97.55	355.11	27.28	0.4111	4
1010	151.28	368.14	18.85	0.3831	2	105.99	385.74	29.65	0.4112	4
1024	153.15	373.08	19.09	0.3829	1	107.43	391.12	30.03	0.4112	6
1030	154.33	375.35	19.26	0.3831	1	108.10	393.52	30.20	0.4113	2

Table 2. The average number of m-shifts in a single call of a given algorithm

are not performed there). Results for 'internal loop' version, presented in Table 3, are extended for another `LMT`'s in appendix A.

6. Conclusions

Based on simple observation ($u + yB$, $v - yA$ are divisible by 2^k, for some y such that $0 \le y < 2^k$), we made two binary inverting algorithms noticeably faster. Furthermore, presented solution can be applied also on processors with relatively slow multiplication (slow in compare with addition)—precomputed array should be then used.

n	Pr(not(`B[ ]`)) in the EPMBGCD internal loop LMT =4	internal loop LMT =256	external loop LMT =4	external loop LMT =256	n	Pr(not(`B[ ]`)) in the EPMBGCD internal loop LMT =4	internal loop LMT =256	external loop LMT =4	external loop LMT =256
160	0.2074	0.00254	0.4695	0.0302	610	0.2079	0.00259	0.4737	0.0310
180	0.2070	0.00253	0.4695	0.0301	630	0.2082	0.00261	0.4741	0.0311
210	0.2076	0.00252	0.4703	0.0303	710	0.2078	0.00258	0.4739	0.0310
230	0.2074	0.00250	0.4709	0.0303	730	0.2083	0.00259	0.4744	0.0312
256	0.2076	0.00251	0.4715	0.0304	768	0.2079	0.00258	0.4738	0.0311
310	0.2077	0.00259	0.4725	0.0308	810	0.2080	0.00260	0.4742	0.0312
330	0.2075	0.00255	0.4722	0.0306	830	0.2081	0.00259	0.4744	0.0311
410	0.2079	0.00258	0.4729	0.0309	910	0.2082	0.00260	0.4745	0.0312
430	0.2080	0.00258	0.4731	0.0309	930	0.2082	0.00257	0.4745	0.0312
510	0.2080	0.00261	0.4736	0.0310	1010	0.2082	0.00262	0.4746	0.0312
512	0.2081	0.00257	0.4738	0.0310	1024	0.2081	0.00260	0.4745	0.0312
530	0.2080	0.00253	0.4734	0.0309	1030	0.2082	0.00258	0.4745	0.0312

Table 3. Probabilities of performing multiplication $y \cdot B$ in a single iteration

And last, but not least: our observation allows to extend the integer accelerated gcd algorithm—as a result we gain a very fast inverting method.

A. Probability of avoiding multiplication, in a single iteration of the EPMBGCD algorithm's internal loop

Below, under plausible assumptions, we estimate the probability of having $y = 0$ or using (for $y > 0$) elements from the array `B[ ]` in a single iteration of the EPM-BGCD algorithm computing u in the internal loop. Assume that elements `B[`$-y$`]` (see Subsection 3.1) can be used, and $2 \le k \le 31$.

For convenience we denote by Pr(`B[ ]`) the probability we are looking for.

First of all let us see, how often a fixed y_0 appears, where y is a result of the substitution

$$\begin{aligned} y &:= (|u_b| - |u_a|) \cdot (-B^{-1}) \pmod{2^k}, \text{ or} \\ y &:= (|u_b| + |u_a|) \cdot (-B^{-1}) \pmod{2^k}. \end{aligned} \tag{A.1}$$

Let $L = \max(2, \ell)$, where ℓ is such that $2^{\ell-1} \le y_0 < 2^\ell$ for $y_0 > 0$, and $\ell = 1$ for $y_0 = 0$. Thus ℓ equals the least positive exponent of 2-power greater than y_0. In a given iteration the substitution (A.1) is fixed. Obviously, y_0 appears only if $k \ge L$.

Assume that the differences $(|u_b| - |u_a|) \pmod{2^k}$ and sums $(|u_b| + |u_a|) \pmod{2^k}$ behave randomly, i.e., for a fixed k the probability of obtaining a fixed y_0 is equal to $\frac{1}{2^k}$. The assumption that $k \leq 31$ means that the 31-bit (the most significant bit in the least significant computer word) of the integer t_3 is set to one if only all younger bits are set to zero. Thus the probability of $k = 31$ equals $\frac{1}{2^{29}}$ instead of $\frac{1}{2^{k-1}}$ as we suppose for $2 \leq k < 31$. Therefore the probability that y obtained in a given iteration is equal to y_0, can be described as follows:

$$\Pr(y = y_0) = \sum_{k=L}^{30} \frac{1}{2^k} \cdot \frac{1}{2^{k-1}} + \frac{1}{2^{31}} \cdot \frac{1}{2^{29}}$$

$$= \frac{1}{2^{60}} + 2 \cdot \sum_{k=L}^{30} \frac{1}{4^k} = \frac{1}{3} \cdot \left(\frac{1}{2^{60}} + \frac{1}{2^{2L-3}} \right).$$

For $L = 2$ we get about $\frac{1}{6}$, and for $L = 3$ about $\frac{1}{24}$.

Since $\Pr(y = y_0)$ depends on L, the probabilities of obtaining y's look like stairs: the next, lower step $(2^{L-1} \leq y < 2^L)$ is so wide as the whole stairs before $(0 \leq y < 2^{L-1})$. Moreover, the next step is about four times lower than the current step. Therefore elements of the array `B[ ]`, belonging to the further steps of our stairs, are rarely obtained during the algorithm's processing time. Hence even small array `B[ ]` is intensively used in the algorithm EPMBGCD computing u in the inner loop.

Let us return to $\Pr(\texttt{B[ ]})$. It is clear that a multiplication can occur if $2^k \geq 2 \cdot \texttt{LMT}$, i.e., $k \geq 1 + \lceil \log_2 \texttt{LMT} \rceil$. Let $D = 1 + \lceil \log_2 \texttt{LMT} \rceil$. We remember that the array `B[ ]` contains $\texttt{LMT} - 1$ elements: $B, 2B, \ldots, (\texttt{LMT} - 1) \cdot B$. A multiplication is not performed when $\texttt{B}[y]$ or $\texttt{B}[-y]$ can be used, or if $y = 0$.

Thus

$$\Pr(\texttt{B[ ]}) = \Pr(k < D) + \sum_{k=D}^{30} \frac{2(\texttt{LMT} - 1) + 1}{2^k} \cdot \frac{1}{2^{k-1}} + \frac{2(\texttt{LMT} - 1) + 1}{2^{31}} \cdot \frac{1}{2^{29}}$$

$$= 1 - \frac{1}{2^{D-2}} + \frac{2\texttt{LMT} - 1}{3} \cdot \left(\frac{1}{2^{60}} + \frac{1}{2^{2D-3}} \right).$$

By subtracting the above probabilities for different `LMT`'s we can observe how often elements from the tailing part of the array `B[ ]` are used.

Assume that $\texttt{LMT} = 2^s$, hence $D = s + 1$. The reader can easily check that then

$$\Pr(\texttt{B[ ]}) \approx 1 - \frac{2^s + 1}{3 \cdot 2^{2s-1}}.$$

Thus for $s = 2$ $\quad \Pr(\texttt{B[ ]}) = 1 - \frac{5}{3 \cdot 8} = 1 - \frac{5}{24} \approx 1 - 0.2083,$

and for $s = 8$ $\quad \Pr(\texttt{B[ ]}) = 1 - \frac{257}{3 \cdot 2^{15}} = 1 - \frac{257}{98304} \approx 1 - 0.0026,$

which basically agrees with results presented in Table 3.

References

[1] Bojańczyk, A.W., Brent, R.P., A systolic algorithm for extended GCD computation. Comput. Math. Applic. 14 (1987), 233–238.

[2] Brent, R.P., Analysis of the binary Euclidean algorithm. In: Algorithms and Complexity. New Directions and Recent Results (ed. by J.F. Traub), 321–355. Academic Press, New York 1976.

[3] —, Twenty years' analysis of the Binary Euclidean Algorithm. In: Millennial Perspectives in Computer Science: Proceedings of the 1999 Oxford-Microsoft Symposium in honour of Professor Sir Antony Hoare, 41–53. Palgrave, New York 2000.

[4] Brent, R.P., Kung, H.T., Luk, F.T., Some Linear-Time Algorithms for Systolic Arrays. In: Information Processing 83, 865–876. North Holland, Amsterdam 1984.

[5] Cohen, H., A Course in Computational Algebraic Number Theory. Springer, Berlin 1993.

[6] Jebelean, T., Systolic Multiprecision Arithmetic. Ph.D. Thesis. RISC, Johannes Kepler University, Linz, Austria, March 1994.

[7] Knuth, D., The Art of Computer Programming, vol. 2: Seminumerical Algorithms, 2nd ed. Addison-Wesley, Reading 1981.

[8] A C++ Library For Computational Number Theory. `http://www.informatik.tu-darmstadt.de/TI/LiDIA/`

[9] Vallée, B., Dynamics of the binary Euclidean algorithm: functional analysis and operators. Algorithmica 22 (1998), 660–685.

[10] Weber, K., The accelerated integer GCD algorithm. ACM Trans. Math. Software 21 (1995), 111–122.

Stochastic kleptography detection*

Daniel Kucner and Mirosław Kutyłowski

Abstract. We consider kleptographic attack introduced by Young and Yung to steal secrets by a manufacturer of cryptographic black-box devices. This attack is particularly dangerous, since the attacker need only to eavesdrop the ciphertexts. Moreover, the attack possibility is confined to the manufacturer, even reverse engineering the black box device cannot enable an attack for the third party (which is not the case for the classical subliminal channel techniques). Young and Yung prove that their kleptographic protocols cannot be detected by the user provided that the analysis is based on the input-output data of the device.

In this paper, we present a method of testing black-box devices against kleptography. We show that in most cases a simple stochastic analysis of computation time of such devices might easily detect kleptographic code. This includes modifications of ElGamal, DSS, Diffie–Hellman protocol and RSA proposed by Young and Yung. The test can be used in practice.

1. Introduction

Using tamper-proof devices is very often recommended as a remedy for all threats against security of cryptographic computations. Such devices cannot be attacked from outside, since essentially there is no way to change the code executed. Typical dangers of an "open" system do not exist for black-box tamper-proof devices.

The problem is that we have to trust a manufacturer of the device. It is quite easy to put extra code into the device and since it is tamper-proof, the user cannot detect the extra code inside. Therefore, there is a growing concern about using such devices. One obvious scenario of this kind is using subliminal channels inside a device. However, this would enable anyone who opens the device to read subliminal channel messages. Far more dangerous technique is a so called kleptography discovered by Young and Yung [5, 9, 10]. This method enables the manufacturer of the device to steal secrets merely by eavesdropping ciphertexts generated by a contaminated device. Interesting (and dangerous) features of this technique are the following:

- no additional bit strings are sent by the device (such an extra communication would indicate illegal actions);

* Partially supported by Komitet Badań Naukowych, grant 8 T11C 032 15.

- only the manufacturer may perform the attack designed, even reverse engineering the device and reading its hidden secrets does not enable a kleptographic attack;
- the output of a contaminated device is polynomially indistinguishable from the output of an uncontaminated device.

At the first look, the conditions above seem to be impossible to fulfil. The attacker must get some information bits, but on the other hand all bits are used by the protocol executed. However, random numbers used by the protocols executed by the device need not to be really random, but obtained by public key cryptography. Then the owner of the public key may recover strings used for generating random keys, while with the knowledge of the public key only this is infeasible.

Kleptography might have a tremendous impact on applied cryptography. However, we show that despite the claims above, kleptography is easily detectable (at least in the present shape). The test we propose is based on "side-channel" information generated by a cryptographic device. The first test of this kind was discovered by P.C. Kocher [4] and called *timing attack.*

Our test examines stochastic characteristics of the response time of the device, namely the variation coefficient of the time (which is independent of the actual speed of the processing unit). As shown by our experiments, it works astonishingly well for the original protocols of Young and Yung.

Organization of the paper. In Section 2 we recall the kleptographic protocols. In Section 3 we describe the underlying idea of our tests. In Section 4 we present the experimental results.

2. Kleptography protocols

In order to explain the idea of the attack, we have to recall kleptography protocols developed by A. Young and M. Yung [9, 10, 11]. They called their protocols SETUP as an acronym for Secretly Embedded Trapdoor with Universal Protection.

SETUP for Diffie–Hellman protocol. Recall that if Alice and Bob are using insecure communication channel and wish to agree upon a secret key, they may use the Diffie–Hellman key exchange protocol [1]. First, they fix a large prime p, for which computing discrete logarithms modulo p is infeasible, and a generator g in $\mathbb{Z}_p^*$. Then they are executing the following protocol:

Alice	**Bob**
1. generate random $a < p-1$	1. generate random $b < p-1$
2. $m_A := g^a \bmod p$	2. $m_B := g^b \bmod p$
3. send m_A to Bob	3. send m_B to Alice
4. $k := m_B^a \bmod p$	4. $k := m_A^b \bmod p$

Now, let us recall the SETUP mechanism for this protocol [11]. Let X be the private attacker's key and let $Y = g^X \bmod p$ be his public key. Assume that $H : \mathbb{Z}_p^* \to \mathbb{Z}_{p-1}$ and let W, a, b be fixed odd integers. The contaminated Diffie–Hellman protocol executed by Alice (Bob) looks as follows [11]:

1. for the first usage, generate random $c_1 \in \mathbb{Z}_{p-1}$
2. return $m_1 = g^{c_1} \bmod p$
3. store c_1 in memory for the next usage;
4. for the second usage, generate random $t \in \{0, 1\}$
5. $z := g^{c_1 - Wt} \cdot Y^{-ac_1 - b} \bmod p$
6. $c_2 := H(z)$
7. return $m_2 = g^{c_2} \bmod p$

The code above corresponds to two rounds of Diffie–Hellman key generation protocol. (In fact, the first part must include some extra code to slow down the response time. Otherwise, the user may detect that the device is faster every second time.) To see how kleptography works, assume that the attacker has got m_1 and m_2. Now he can calculate c_2 as follows:

1. $r := m_1^a \cdot g^b \bmod p$
2. $z_1 := m_1 / r^X \bmod p$
3. if $m_2 := g^{H(z_1)} \bmod p$, then return $H(z_1)$
4. $z_2 := z_1 / g^W \bmod p$
5. if $m_2 := g^{H(z_2)} \bmod p$, then return $H(z_2)$

(Note that the private key of the attacker is necessary in this computation.) Having the value of c_2, the attacker can easily calculate the secret key generated in the second key exchange.

ElGamal SETUP. Let p be an appropriate large prime and let g be a generator in $\mathbb{Z}_p^*$. Let x be a private key and $y = g^x \bmod p$ the corresponding public key. ElGamal algorithm [3] encrypts a message M as follows:

1. generate a random number $k < p$
2. $m_1 := g^k \bmod p$
3. $m_2 := M \cdot y^k \bmod p$
4. return (m_1, m_2)

Having (m_1, m_2) and x, one may decipher M, since $m_2 (m_1^x)^{-1} = M \bmod p$. The contaminated ElGamal encryption scheme [10] looks as follows:

1. for the first usage, generate a random number $k_1 < p$
2. $a_1 := g^{k_1} \bmod p$
3. $b_1 := M y^{k_1} \bmod p$
4. return (a_1, b_1)
5. for the second usage, generate a random number $t \in \{0, 1\}$
6. $z := g^{k_1 - Wt} \cdot Y^{-ak_1 - b} \bmod p$
7. $k_2 := H(z)$
8. $a_2 := g^{k_2} \bmod p$

9. $b_2 := My^{k_2} \bmod p$
10. return (a_2, b_2)

(Note that the first part of the protocol exploits the same mechanism as in the case of Diffie–Hellman.) For the further details on the attack mechanism see [10].

DSA SETUP. First let us recall briefly the DSA algorithm [5]. Let p be a large prime such that $512 \leq \log_2 p \leq 1024$ and let q be a 160 bit prime factor of $p-1$. Let g be a generator in $\mathbb{Z}_p^*$ and let SHA be Secure Hash Function. A private key is a number $x < q$ and the corresponding public is $y = g^x \bmod p$. In order to compute the signature of a message M, the owner of x proceeds as follows:

1. generate a random number $k < q$
2. $r := (g^k \bmod p) \bmod q$
3. $s := k^{-1}(\mathrm{SHA}(M) + x \cdot r) \bmod q$
4. return (s, r)

The signature is verified by checking that $r = (g^{s^{-1}\mathrm{SHA}(M)} \cdot y^{s^{-1}r} \bmod p) \bmod q$. SETUP for DSA is based on the same idea as for ElGamal, the contaminated code looks as follows [11]:

1. for the first usage generate a random number $k_1 < q$
2. $r_1 := (g^{k_1} \bmod p) \bmod q$
3. $s_1 := k_1^{-1} \cdot (\mathrm{SHA}(M_1) + x \cdot r_1) \bmod p$
4. return (r_1, s_1)
5. for the second usage, generate a random number $t \in \{0, 1\}$
6. $z := g^{k_1 - Wt} \cdot Y^{-ak_1 - b} \bmod p$
7. $k_2 := H(z)$
8. $r_2 := (g^{k_2} \bmod p) \bmod q$
9. $s_2 := k_2^{-1}(\mathrm{SHA}(M_2) + x \cdot r_2) \bmod p$
10. return (r_2, s_2)

SETUP in RSA Key Generation. To use the RSA algorithm [6] one has to generate public modulus n, a public exponent e and a private exponent d. Now we describe a SETUP for RSA key generation [11]. Let X be the private key of an attacker and $Y = g^X \bmod P$ his public key where P is an appropriate large prime and g is a generator in $\mathbb{Z}_P^*$. Let G be a pseudorandom function having two arguments, data and key. Finally, let $M = \lfloor \log_2 P \rfloor$ and let K be a fixed symmetric key for G.

1. generate a random number $c_1 < P - 1$
2. solve for z in $Y^{ac_1+b} \cdot g^{Wt}z = g^{c_1} \bmod P$
3. remove the bias of z to get $z' := \mathrm{PBRM}(z)$ by a *probabilistic bias removal* algorithm PBRM (see below for details)
4. $z'' := H(z')$ and set the lowest bit of z'' to 1,
5. $p := \min\{r : r > z'' \text{ and } r \text{ is (pseudo)prime}\}$
6. for $i = 0$ to $B_1 - 1$ do
   (a) $U := G(p, K + i)$

   (b) generate a random number V with binary representation of length M
   (c) $X := U \cdot 2^M + V$
   (d) $q := X$ div p
   (e) if q is not a (pseudo)prime, then goto 1
7. take $n = pq$ and proceed to generate a pair of RSA keys e and d for RSA modulus n.

PBRM (Probabilistic Bias Removal Method) [11] looks as follows: Having value x uniformly distributed in set $\{1, \ldots, R\}$ we obtain uniformly distributed $x' \in \{1, \ldots, S\}$, where $\frac{S}{2} < R < S$:

1. generate random $z \in \{0, 1\}$
2. if $x < (S - R)$, then
   (a) if $z = 1$, then $x' = x$
   (b) else $x' = S - x$
3. else
   (a) if $z = 1$, then $x' = x$
   (b) else goto 1

To find out the private key, the attacker takes the public key generated by the protocol and reverses the whole computation (for details see [11]).

3. Test mechanism

Our test uses variation coefficient of the computation time for the protocol considered. All protocols that we consider involve exponentiations. Moreover, this is the most costly operation. The other operations are mainly modular addition, multiplication and, in some cases, modular inverse. Exponentiation is clearly much more difficult than addition. Exponentiation takes longer than multiplication, since, for the number of interest, exponentiation algorithms are implemented by performing hundreds of multiplications (see below). Computing modular inverse can be performed by binary GCD, which again is much faster than exponentiation. So, the probability distribution of computation time is determined mostly by the exponentiation time.

Below we sketch the methods of modular exponentiations to provide some intuition why the computation times have large variation (for a detailed survey on fast exponentiation algorithms see [2]).

Modular exponentiation. While implementing public key algorithms, the major concern is to execute exponentiation in a fast enough way. Exponentiation time depends on the size of numbers used and thereby states a trade-off between efficiency (smaller numbers preferred) and security (larger numbers preferred). Due to this problem, there are a lot of efforts to design efficient exponentiation algorithms and even accelerating a computation by a factor close to 1 might be interesting from the practical point of view.

The basic method is BSM (binary square and multiply): for computing $g^x \bmod n$ we take $x_l x_{l-1} \ldots x_0$, the binary representation of x and compute a sequence $v_{l+1} = 1, v_l, \ldots, v_0$ satisfying

$$v_i = \begin{cases} v_{i+1}^2 \bmod n & \text{if } x_i = 0, \\ v_{i+1}^2 \cdot g \bmod n & \text{if } x_i = 1. \end{cases}$$

Then obviously, $v_0 = g^x \bmod n$. The number of multiplications and squaring performed is l plus the number of ones in the binary representation of x. The main advantage is that very often only squaring suffices. However, it follows that the computation time depends on the number of ones in the binary representation of x.

The method BSM can be tuned in may ways. The first one is the m-ary method. We consider an m-ary representation of x and instead of multiplying v_{i+1} by 1 or g, we multiply it by one of the precomputed values $1, g, g^2, \ldots, g^{m-1}$, and instead of squaring we compute v_{i+1}^m. This method reduces the number of phases, but in each phase the chance of avoiding multiplication is smaller than in the case of BSM. Therefore the method shows to be most effective for small m. In this case, the computation time depends on the number of zeroes in the m-ary representation of x.

The window method, another variant of BSM, takes advantage of existence of blocks of zeroes in the binary representation of the exponent, the next window is not the block of the next m digits (as in the case of m-ary method), but the next nonzero-block of m digits.

Special strategies have been developed for the case when the exponent is fixed. For instance, if we use a pair of RSA keys for a long time, it might be an advantage to optimize the exponentiation for these particular keys. The addition chain method builds a chain $s_0, s_1, \ldots, s_l$ such that $x = s_l$ and for each $i \leq l$, the number s_i is a sum $s_j + s_k$ for some $j, k < i$. Then we may use equality $g^{s_i} = g^{s_j} \cdot g^{s_k}$ to compute all numbers g^{s_i}, and hence also g^x. The length of an addition chain used (and hence the number of multiplications) depends on the exponent and the search method—practically suboptimal solutions are used.

Another method of this kind is BWHT. It finds coefficients $a_i \leq h$ such that $x = \sum_{i=1}^{l} a_i x_i$, where each g^{x_i} is precomputed. Let $c_d = \prod_{i:a_i=d} g^{x_i}$. Then $g^x = \prod_{i=1}^{h} c_i^i$. The key property leading to a fast computation is the equality $g^x = c_h(c_h c_{h-1})(c_h c_{h-1} c_{h-2}) \ldots$. Once all c_i are known it requires $l + h - 2$ multiplications. However, computing numbers c_i involve the number of multiplications that depends on the numbers a_i, and hence on the exponent x.

We conclude that the number of multiplications required to perform an exponentiation depends on x. It is obvious from the context that deviations from the average number of multiplications are substantial.

Variation of computation times. The first idea to recognize a SETUP hidden in a cryptographic algorithm would be to inspect computation time. Since a SETUP requires extra operations, the computation time should be generally longer. However, we are never sure about the speed of the processor and such a detection method is questionable. Our proposal is to use characteristics of the computation time that are independent from the actual speed of the processor.

The parameter that we test is variation coefficient

$$R[T] = \frac{\sigma[T]}{E[T]}$$

of the computation time T of the protocol under consideration. In the above formula, $E[T]$ stands for the expected value of T, $\sigma[T]$ is the standard deviation of T defined as $\sqrt{\sigma^2[T]}$, and $\sigma^2[T]$ stands for variance of T defined as

$$E[(T - E[T])^2].$$

Obviously, $R[T]$ is independent of the unit we use for measuring T. Indeed, if we decrease the unit α times, then $E[T]$ and T increase α times. Hence $\sigma^2[T]$ increases α^2 times and σ increases α times. Taking the quotient to compute $R[T]$ we see that its value remains unchanged.

As discussed above, the computation time for modular exponentiation for a random exponent is a random variable with a quite large variation. Recall also the exponentiation is the most time-consuming operation within Diffie–Hellman, ElGamal, DSS and RSA protocols.

Now let us consider the Diffie–Hellman protocol and investigate what is the effect of contaminating it with the kleptographic code on the value of $R[T]$. When a single m_i is generated, the contaminated device is performing the following exponentiations:

$$g^{c_{i-1}-Wt}, \quad Y^{-ac_{i-1}-b}, g^{c_i}.$$

If the device is not contaminated, then only one exponentiation is required, namely, g^{c_i}.

Assume that computation time for exponentiation (for numbers of the size used by the protocol) has variation σ^2 and expected value T_0. If we assume that the three exponentiations executed during contaminated Diffie–Hellman key generation are stochastically independent, then $\sigma^2[T]$ equals the sum of variations for the three exponentiations, $\sigma^2[T] = 3 \cdot \sigma^2$. (Of course, the exponentiations are not independent, however their computation times seem to behave in this way.) Therefore, $\sigma[T] = \sqrt{3} \cdot \sigma$, and

$$R[T] \approx \tfrac{\sqrt{3} \cdot \sigma}{3T_0} = \tfrac{1}{\sqrt{3}} \cdot \tfrac{\sigma}{T_0} \approx 0.58 \cdot \tfrac{\sigma}{T_0}.$$

In the case, when Diffie–Hellman protocol is uncontaminated, we have $\sigma^2[T] = \sigma^2$ and $E[T] = T_0$, hence

$$R[T] = \tfrac{\sigma}{T_0}.$$

We see that the value of $R[T]$ should be bigger by a factor of $\sqrt{3} \approx 1.73$ in the case of the uncontaminated protocol. This difference should be easily detected provided that we make enough trials to estimate the value of $R[T]$.

The difference we need between contaminated and uncontaminated Diffie–Hellman protocols is that the number of exponentiations executed while generating a single key differs in both cases. A similar situation occurs for the other kleptographic protocols proposed by Young and Yung.

For the ElGamal uncontaminated encryption algorithm, we have to compute g^k and y^k for a randomly chosen k. The contaminated protocol executes three exponentiations at the same time: the extra one is generating k by formula $k = Y^{-k'} \bmod p$, where k' is the previous key. Hence in the contaminated case we have $R[T] \approx 0.58 \cdot \frac{\sigma}{T_0}$, while in the uncontaminated case we expect

$$R[T] \approx \frac{\sqrt{2} \cdot \sigma}{2T_0} = \frac{1}{\sqrt{2}} \cdot \frac{\sigma}{T_0} \approx 0.71 \cdot \frac{\sigma}{T_0}.$$

They differ by a factor of $\sqrt{3}/\sqrt{2} \approx 1.22$.

In the case of ElGamal signature algorithm, in addition to computing g^{k_i}, the contaminated device must compute $k_i = Y^{-k_{i-1}} \bmod p$. Again the values of $R[T]$ should differ by a factor of about 1.73.

The case of DSA is similar. In addition to computing $g^k \bmod p$, the kleptographic procedure computes k as in the case of Diffie–Hellman contaminated protocol, i.e. with two extra exponentiations. So $R[T] \approx \frac{\sigma}{T_0}$ in the case of uncontaminated code, and $R[T] \approx \frac{1}{\sqrt{3}} \cdot \frac{\sigma}{T_0}$ in the case of contaminated code.

The situation for the RSA SETUP is much more difficult. The reason is that during the protocol a lot of candidates for prime numbers are tested.

The SETUP procedure for RSA key generation includes two exponentiations

$$Y^{-ac_1-b}, \quad g^{c_1-Wt}$$

in the second line of the pseudocode. The other operations that need a lot of computing time are testing primality in lines 5 and 6.e. If they are implemented by the Fermat test, then each test requires k exponentiations, where k is a constant security parameter (the number of actually executed exponentiations may vary—since in each trial we may find a witness for non-primality of a candidate). Since the way in which the consecutive candidates for p and q are generated seem to have no substantial influence on the probability that they are prime, the number of trials should be similar to the case of regular RSA key generation procedure.

Let p_i denote the probability that a (pseudo)prime p of a given length is generated after i exponentiations. Let σ^2 be the variation of a single exponentiation time for the numbers of the length considered. Then, in the case of the regular RSA key generation, the variation of the computation time is approximately

$$2 \cdot \sum_i p_i \cdot (i \cdot \sigma^2) = 2\sigma^2 \cdot \sum_i (p_i \cdot i).$$

For the contaminated code, the variation is approximately equal to

$$2\sigma^2 + 2\sum_i p_i \cdot (i \cdot \sigma^2) = \sigma^2 \cdot (2 + \sum_i (p_i \cdot i)).$$

since there are two extra exponentiations in the second line of the pseudocode. Note that $\sum_i (p_i \cdot i)$ is the expected number of exponentiations for finding a pseudoprime in a sequence of Fermat tests over independent candidates. If u is a probability that a single candidate is a pseudoprime, then the expected number of candidates required to find a pseudoprime is $1/u$. Hence the expected number of exponentiations equals c/u for some constant c. We conclude that the variance is approximately $2\sigma^2 \cdot c/u$ for the uncontaminated protocol and $2\sigma^2 \cdot (c/u + 1)$ for the contaminated protocol. Therefore, $R[T]$ can be approximated by $\frac{\sigma}{\sqrt{2c/u \cdot T_0}}$ in the uncontaminated case, and by $\frac{\sigma}{\sqrt{2(c/u+1) \cdot T_0}}$ in the contaminated case. The difference is by a factor of $\sqrt{c/(c+u)}$. This is a constant close to 1, but nevertheless a difference might be detected by examination of a statistics for a large number of key generations. Indeed, we can easily estimate u since by Prime Number Theorem there are approximately $\frac{1}{\ln n}$ prime numbers not exceeding n. For the numbers of the size considered, this shows that u is not very small and the difference might be visible.

4. Experimental results

Our goal is to compare computation time of contaminated and uncontaminated devices. The heuristic argument in the previous section is a good motivation for such tests. Once the tests confirm this behavior, we have a detection method, even if there is no rigorous mathematical proof that our assumptions well approximate the real situation.

We have implemented contaminated and uncontaminated versions of ElGamal, DSS and Diffie–Hellman protocols in ANSI C and linked with RSAREF 2.0 library using GNU C v. 2.7 compiler. All tests were run on Linux system using a computer with a Pentium II processor (350 MHz) and 64 Mb memory. Computation time for a single protocol was measured in $10^{-2}s$.

ElGamal

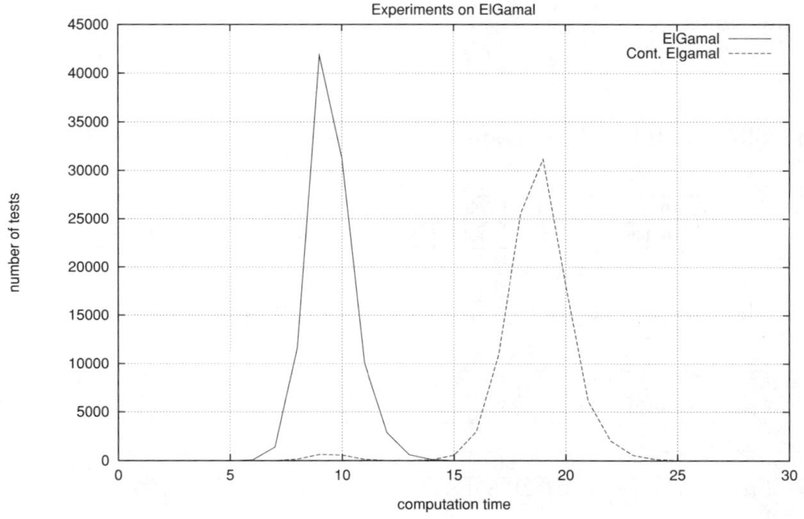

The following table shows computation times for the regular ElGamal:

time	5	6	7	8	9	10	11	12
# of tests	9	73	1399	11629	41874	31226	10095	2895

time	13	14	15	16	17	18	19
# of tests	608	113	22	1	2	1	1

The following table shows the results for the contaminated ElGamal:

time	6	7	8	9	10	11	12	13	14	15	16
# of tests	3	13	161	646	566	132	23	25	81	571	3000

time	17	18	19	20	21	22	23	24	25	26
# of tests	10979	25580	31175	18148	6098	2031	568	150	33	10

For the uncontaminated ElGamal algorithm we have obtained $R[T] = 0.1065$, while $R[T] = 0.0917$ for the contaminated version. The values obtained differ by a factor of 1.161, which is roughly as expected despite inaccuracy in measuring computation time.

It is interesting that the curve representing the contaminated implementation has a small peak at the same value of computation time where the correct implementation has its only peak. The same feature will be clearly visible at the next figures. To explain it let us look at SETUP for ElGamal. There are two different parts which occur every second call to device. The first one is identical to original ElGamal protocol and exactly this part is presented on the small peak. The disproportion between two peaks of curve representing contaminated implementation is clearly visible. The reason is that for practical usage after the first part of the protocol, (i.e. lines 1–4) device repeats the second part (i.e. lines 5–10) not once but many times.

DSS

The following table shows computation times for the regular DSS:

time	1	2	3	4	5	6	7
# of tests	84	720	79343	19796	46	2	1

The following table shows the results for the contaminated DSS:

time	2	3	4	5	11	13	14	15	16	17
# of tests	136	1140	249	10	5	14	34	262	1448	5924

time	18	19	20	21	22	23	24	25	26	27
# of tests	17528	38707	21085	8509	3431	1089	296	109	17	10

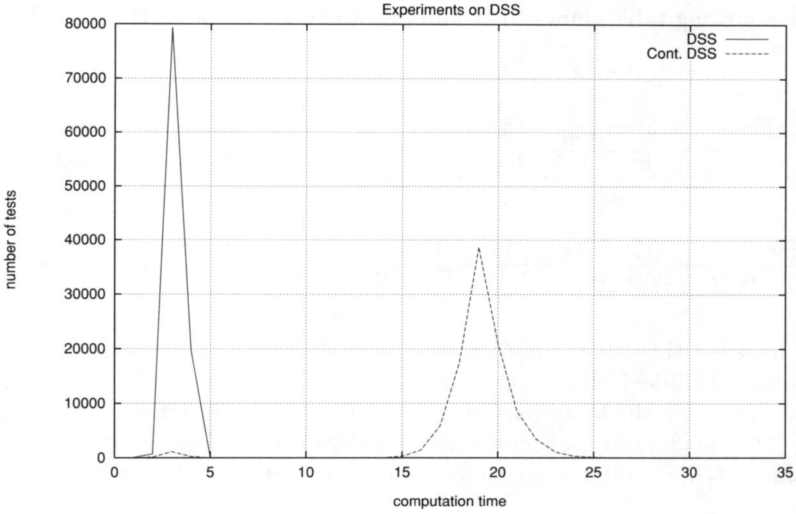

For standard DSS algorithm we gained $R[T] = 0.130042$ and for contaminated version $R[T] = 0.114013$ so they differ by the factor of 1.140, which as the previous result is close to our theoretical results.

Diffie–Hellman

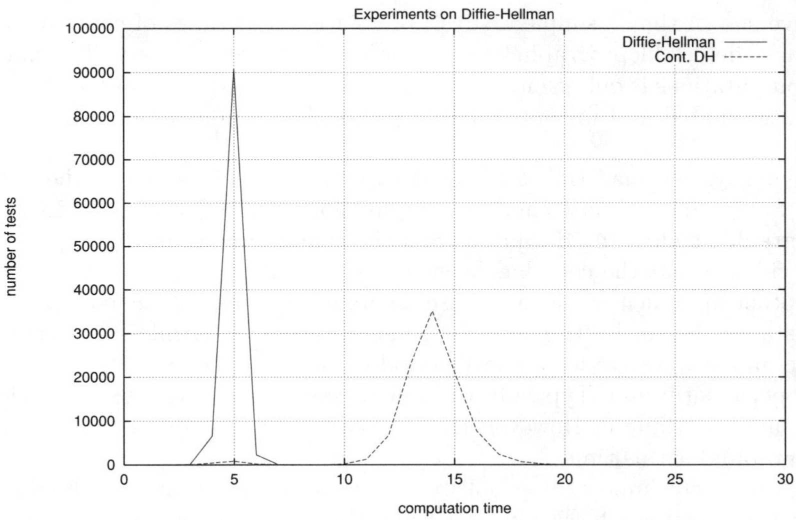

The following table shows computation times for the regular Diffie–Hellman protocol:

time	1	2	3	4	5	6	7
# of tests	28	45	106	6579	90873	2293	46

The following table shows the results for the contaminated Diffie–Hellman protocol:

time	3	4	5	6	7	8	9	10	11	12
# of tests	49	417	833	200	36	5	15	158	1283	6734

time	13	14	15	16	17	18	19	20	21	22
# of tests	22944	35145	21145	7740	2381	691	167	35	16	3

We have got the value of $R[T] = 0.1475$ for the Diffie–Hellman protocol. For the contaminated Diffie–Hellman protocol $R[T] = 0.102748$, so the factor they differ by is 1.435. If we skip the lower part of computations time of contaminated algorithm, i.e. the interval $[3; 7]$ (this effect should be hidden!) we will get $R[T] = 0.077$. Hence the values of $R[T]$ differ by a factor of 1.9 what is very close to our theoretical results.

5. Conclusions and final remarks

We have shown that a simple analysis of computation times of contaminated devices may detect kleptographic code. The test is extremely easy. The number of trial computations is quite small (say, for a cryptanalysis). All this happens despite the contaminated and uncontaminated protocols are polynomially indistinguishable.

Obviously, one may try to build black-box devices immune to the tests proposed. Such a device must emulate computation times of regular devices without kleptographic code—any deviation from the characteristics of the regular devices is an evidence that the code has been manipulated.

Another approach would be to use kleptographic part of the code very rarely. For instance, it can be triggered by some special but harmless ciphertext from outside and stay active for a short period of time. In this case, the user is again in a hopeless situation. Typically what he or she can do is to test the device after buying and installing it. However, the attacker could be wise enough to deactivate kleptography for this time.

An extremely important problem for the future of security technology is to answer the question whether kleptography that is not based on exponentiation is possible. Such protocols would be immune against detection tests such as proposed in this paper.

Acknowledgment. We thank Przemek Kubiak and members of Algorithms and Complexity Seminar at University of Wrocław for many helpful comments.

References

[1] Diffie, W., Hellman, M., New directions in cryptography. IEEE Trans. Information Theory 22 (1976), 644–654.

[2] Gordon, D., A survey of fast exponentiation methods. J. Algorithms 27 (1998), 129–146.

[3] ElGamal, T., A public-key cryptosystem and a signature scheme based on discrete logarithms. In: Advances in Cryptology (CRYPTO '84) (ed. by G.R. Blakley, D. Chaum; Lecture Notes in Comput. Sci. 196), 10–18. Springer, Berlin 1985.

[4] Kocher, P.C., Timing attacks on implementations of Diffie-Hellman, RSA, DSS, and other systems. In: Advances in Cryptology – CRYPTO '96 (ed. by N. Koblitz; Lecture Notes in Comput. Sci. 1109), 104–113. Springer, Berlin 1996.

[5] Proposed Federal Information Processing Standard for Digital Signature Standard (DSS). Federal Register 56, n. 169, 42980–42982, 1991.

[6] Rivest, R., Shamir, S., Adleman, L., A method for obtaining digital signatures and public-key cryptosystems. Comm. ACM 21 (1978), 120–126.

[7] de Rooij, P., Efficient exponentiation using precomputation and vector addition chains. In: Advances in Cryptology – EUROCRYPT '94 (ed. by A. De Santis; Lecture Notes in Comput. Sci. 950), 389–399. Springer, Berlin 1995.

[8] Schueth, J., Public communication in *sci.crypt*.

[9] Young, A., Yung, M., The dark side of black-box cryptography. In: Advances in Cryptology – CRYPTO '96 (ed. by N. Koblitz; Lecture Notes in Comput. Sci. 1109), 89–103. Springer, Berlin 1996.

[10] —, —, The prevalence of kleptographic attacks on discrete-log based cryptosystems. In: Advances in Cryptology – CRYPTO '97 (ed. by B.S. Kaliski; Lecture Notes in Comput. Sci. 1294), 264–275. Springer, Berlin 1997.

[11] —, —, Kleptography: using cryptography against cryptography. In: Advances in Cryptology – EUROCRYPT '97 (ed. by W. Fumy; Lecture Notes in Comput. Sci. 1233), 62–74. Springer, Berlin 1997.

References

[1] [illegible]

[2] Gordon, [illegible] A survey of [illegible]

[3] [illegible]

[4] Kohl, [illegible]

[5] [illegible] Standard for [illegible]

[6] [illegible] ACM [illegible]

[7] [illegible]

[8] [illegible]

[9] [illegible]

[10] [illegible]

[11] [illegible] Springer [illegible]

An overview of the XTR public key system

Arjen K. Lenstra and Eric R. Verheul

Abstract. XTR is a new method to represent elements of a subgroup of a multiplicative group of a finite field. Application of XTR in cryptographic protocols leads to substantial savings both in communication and computational overhead without compromising security. This paper describes and explains the techniques and properties that are relevant for the XTR cryptosystem and its implementation. It is based on the material from [10, 11, 12, 28].

1. Introduction

XTR stands for 'ECSTR', which is an abbreviation for Efficient and Compact Subgroup Trace Representation. It is a novel method that makes use of traces to represent and calculate powers of elements of a subgroup of a finite field. XTR is not the first method to do so. The LUC cryptosystem (cf. [25], and also [13, 17, 18, 22]) uses the trace over $\mathrm{GF}(p)$ to represent elements of the order $p+1$ subgroup of $\mathrm{GF}(p^2)^*$. Compared to the traditional representation this leads to a factor 2 reduction in the representation size. The variant described in [7] uses the subgroup of order p^2+p+1 of $\mathrm{GF}(p^3)^*$ instead, but as a result sizes are reduced by only a factor 1.5. XTR uses the trace over $\mathrm{GF}(p^2)$ to represent elements of the order p^2-p+1 subgroup of $\mathrm{GF}(p^6)^*$, thereby achieving a factor 3 size reduction. Also, the resulting calculations are appreciably faster than using the standard representation. The factor 3 size reduction was first achieved—at much lower speed than XTR, but using the same subgroup—in the predecessor paper [3].

From a security point of view, XTR is a traditional discrete logarithm system: for its security it relies on the difficulty of solving discrete logarithm related problems in the multiplicative group of a finite field. Thus, XTR is not based on any new primitive or new allegedly hard problem—on the contrary, it is based on the primitive underlying the very first public key cryptosystem, the Diffie–Hellman key agreement protocol. Other advantages of XTR are its very fast parameter and key selection (much faster than RSA, orders of magnitude faster than ECC), small key sizes (much smaller than RSA, comparable with ECC for current security settings), and speed (overall comparable with ECC for current security settings). Combined with its very easy programmability, this makes XTR an excellent public key system

Public-Key Cryptography and Computational Number Theory
K. Alster, J. Urbanowicz and H. C. Williams (Eds.)

for a very wide variety of environments, ranging from smart cards to web servers, without the need to share system parameters with other users (as is often the case in ECC systems). For actual implementation results and comparisons with other cryptosystems, see [10, Subsection 4.4].

The purpose of this paper is to present a unified description of the XTR results obtained so far (cf. [10, 11, 12, 28]). Proofs are included only inasmuch they are required to implement XTR; all other proofs can be found in the original papers on which this survey is based. An outline of the paper is as follows:

- Section 2 introduces the mathematics of XTR including the basic parameters and the fundamental algorithms to calculate traces of powers.
- Section 3 describes the selection of XTR parameters and keys.
- Section 4 contains XTR-based encryption schemes, providing confidentiality services.
- Section 5 focuses on XTR-based digital signature schemes, providing authentication and non-repudiation services, and includes methods to reduce public key sizes for these applications.
- Section 6 describes how to efficiently verify that data exchanged during XTR-based protocols are correctly formatted. Such checks can be used to prevent so-called subgroup based attacks.
- Section 7 summarizes the most relevant security properties of XTR. This includes a result providing evidence that the XTR group is more secure than an algebraically isomorphic supersingular elliptic curve over $\mathrm{GF}(p^2)$, thereby settling a problem posed by [16].

2. Fundamentals of XTR

2.1. XTR supergroup and XTR group

Many cryptographic protocols used to be based on a generator of the full multiplicative group of a finite field. Schnorr introduced the idea to replace this generator by the generator of a relatively small subgroup of sufficiently large prime order q (cf. [24]). This same idea is used in XTR in a specific setting, namely XTR uses a subgroup of prime order q of the order p^2-p+1 subgroup of $\mathrm{GF}(p^6)^*$. The latter group is referred to as the *XTR supergroup* and the order q subgroup $\langle g\rangle$ generated by g is referred to as the *XTR (sub)group*. The XTR supergroup is not contained in any proper subfield of $\mathrm{GF}(p^6)$ (cf. [9]). Combined with the choice of q it follows that computing discrete logarithms in $\langle g\rangle$ is as hard, in general, as it is in $\mathrm{GF}(p^6)^*$ (cf. [10, Section 5]).

The reason that XTR uses this specific subgroup $\langle g\rangle$ is not just that it provides the full $\mathrm{GF}(p^6)^*$ security, but also that the elements of the XTR supergroup, and thus of $\langle g\rangle$, allow a very efficient representation, at a small cost: if one is willing to give up the distinction between elements and their conjugates over $\mathrm{GF}(p^2)$, then not only elements of the XTR supergroup can be represented using an element of $\mathrm{GF}(p^2)$ as opposed to $\mathrm{GF}(p^6)$ (i.e., just a third of the usual number of bits). But also calculations take place in $\mathrm{GF}(p^2)$ instead of $\mathrm{GF}(p^6)$ and can thus be performed much faster than usual. As shown in Subsection 2.3 below, this is all a consequence of the particularly nice form of the minimal polynomial over $\mathrm{GF}(p^2)$ of the elements of the XTR supergroup. First, however, it is described how computations in $\mathrm{GF}(p^2)$ can be done efficiently.

2.2. Arithmetic operations in $\mathrm{GF}(p^2)$

As set forth in Subsection 2.1 above, a representation of $\mathrm{GF}(p^2)$ is needed that allows efficient arithmetic operations, where p is a prime such that $p^2 - p + 1$ has a sufficiently large prime factor q. Here it is indicated how this can be done, irrespective of the existence of q. Selection of p (and q) is described in Section 3.

Let p be a prime that is $2 \bmod 3$. It follows that $(X^3-1)/(X-1) = X^2+X+1$ is irreducible over $\mathrm{GF}(p)$ and that the roots α and α^p form an optimal normal basis for $\mathrm{GF}(p^2)$ over $\mathrm{GF}(p)$, i.e., $\mathrm{GF}(p^2) \cong \{x_1\alpha + x_2\alpha^p\colon x_1, x_2 \in \mathrm{GF}(p)\}$. With $\alpha^i = \alpha^{i \bmod 3}$ it follows that

$$\mathrm{GF}(p^2) \cong \{x_1\alpha + x_2\alpha^2\colon \alpha^2 + \alpha + 1 = 0 \text{ and } x_1, x_2 \in \mathrm{GF}(p)\}.$$

Note that in this representation of $\mathrm{GF}(p^2)$ an element t of $\mathrm{GF}(p)$ is represented as $-t\alpha - t\alpha^2$, e.g. 3 is represented as $-3\alpha - 3\alpha^2$. The cost of arithmetic operations in $\mathrm{GF}(p^2)$ follows from Lemma 2.2.1 below.

Lemma 2.2.1 ([10, Lemma 2.1.1]). *Let $x, y, z \in \mathrm{GF}(p^2)$ with $p \equiv 2 \bmod 3$. Not counting additions or subtractions in $\mathrm{GF}(p)$:*

(i) *computing x^p is for free;*

(ii) *computing x^2 takes two multiplications in $\mathrm{GF}(p)$;*

(iii) *computing $x \cdot y$ takes three multiplications in $\mathrm{GF}(p)$;*

(iv) *computing $x \cdot z - y \cdot z^p$ takes four multiplications in $\mathrm{GF}(p)$.*

Proof. Let $x = x_1\alpha + x_2\alpha^2$, $y = y_1\alpha + y_2\alpha^2$, $z = z_1\alpha + z_2\alpha^2 \in \mathrm{GF}(p^2)$. To prove (i) observe that

$$x^p = x_1^p\alpha^p + x_2^p\alpha^{2p} = x_2\alpha + x_1\alpha^2.$$

To prove (ii), write

$$(x_1\alpha + x_2\alpha^2)^2 = x_2(x_2 - 2x_1)\alpha + x_1(x_1 - 2x_2)\alpha^2.$$

Under the reasonable assumption that a squaring in $\mathrm{GF}(p)$ takes 80% of the time of a multiplication in $\mathrm{GF}(p)$ (cf. [5]), this is faster than the three squarings in $\mathrm{GF}(p)$ that would result if x^2 were computed using the Karatsuba-like approach that is used for (iii). To compute

$$x \cdot y = (x_2y_2 - x_1y_2 - x_2y_1)\alpha + (x_1y_1 - x_1y_2 - x_2y_1)\alpha^2,$$

compute $x_1 \cdot y_1$, $x_2 \cdot y_2$, and $(x_1 + x_2) \cdot (y_1 + y_2)$, so that $x_1 \cdot y_2 + x_2 \cdot y_1$ and thus $x \cdot y$ follow using four subtractions. This proves (iii). Finally, (iv) follows from

$$x \cdot z - y \cdot z^p = (z_1(y_1 - x_2 - y_2) + z_2(x_2 - x_1 + y_2))\alpha \\ + (z_1(x_1 - x_2 + y_1) + z_2(y_2 - x_1 - y_1))\alpha^2. \qquad \square$$

2.3. Traces

The *conjugates* over $\mathrm{GF}(p^2)$ of $h \in \mathrm{GF}(p^6)$ are h, h^{p^2}, and h^{p^4}. The sum of the conjugates over $\mathrm{GF}(p^2)$ of $h \in \mathrm{GF}(p^6)$ is known as the *trace* $\mathrm{Tr}(h) = h + h^{p^2} + h^{p^4}$ over $\mathrm{GF}(p^2)$ of h. From this definition and the fact that $h^{p^6} = 1$ it follows that $\mathrm{Tr}(h)^{p^2} = \mathrm{Tr}(h)$, so that $\mathrm{Tr}(h) \in \mathrm{GF}(p^2)$. Furthermore,

$$\mathrm{Tr}(h_1 + h_2) = \mathrm{Tr}(h_1) + \mathrm{Tr}(h_2) \text{ and } \mathrm{Tr}(c \cdot h_1) = c \cdot \mathrm{Tr}(h_1)$$

for $h_1, h_2 \in \mathrm{GF}(p^6)$ and $c \in \mathrm{GF}(p^2)$. That is, the trace over $\mathrm{GF}(p^2)$ is $\mathrm{GF}(p^2)$-linear. Unless specified otherwise, conjugates and traces in this paper are over $\mathrm{GF}(p^2)$.

Now let $g \in \mathrm{GF}(p^6)^*$ be of order > 3 and dividing $p^2 - p + 1$. As argued in Subsection 2.1 the subgroup $\langle g \rangle$ is as secure, with respect to the discrete logarithm related problems, as the full multiplicative group $\mathrm{GF}(p^6)^*$, assuming a sufficiently large prime divides the order of g. For that reason g is later chosen as an element of order q dividing $p^2 - p + 1$. For the present purposes, however, q is not needed and it suffices to take g of order > 3 and dividing $p^2 - p + 1$.

Because $p^2 \equiv p - 1 \bmod p^2 - p + 1$ and $p^4 \equiv -p \bmod p^2 - p + 1$, the conjugates of g are g, g^{p-1} and g^{-p}, so that $\mathrm{Tr}(g) = g + g^{p-1} + g^{-p}$. It follows that the product of the conjugates equals 1, so that the polynomial $(X - g)(X - g^{p-1})(X - g^{-p})$ has the form $X^3 - \mathrm{Tr}(g)X^2 + uX - 1$, where

$$u = g \cdot g^{p-1} + g \cdot g^{-p} + g^{p-1} \cdot g^{-p} = g^p + g^{1-p} + g^{-1} = \mathrm{Tr}(g)^p \in \mathrm{GF}(p^2)$$

(the last equality follows from $1 - p \equiv -p^2$ and $-1 \equiv p^2 - p$, both modulo $p^2 - p + 1$). Thus

$$(X - g)(X - g^{p-1})(X - g^{-p}) = X^3 - \mathrm{Tr}(g)X^2 + \mathrm{Tr}(g)^p X - 1 \in \mathrm{GF}(p^2)[X]$$

is actually the minimal polynomial of g over $\mathrm{GF}(p^2)$, and this polynomial—and thereby g's conjugates—is fully determined by $\mathrm{Tr}(g)$. This is the fundamental observation underlying XTR. The same holds for any power of g: for any integer n the conjugates of g^n are the roots of $X^3 - \mathrm{Tr}(g^n)X^2 + \mathrm{Tr}(g^n)^p X - 1 \in \mathrm{GF}(p^2)[X]$, and the latter polynomial is fully determined by $\mathrm{Tr}(g^n)$.

This observation is useful for cryptographic purposes if there is a way to efficiently compute $\mathrm{Tr}(g^n)$ given $\mathrm{Tr}(g)$: in cryptographic protocols $g^n \in \mathrm{GF}(p^6)$ can then be replaced by $\mathrm{Tr}(g^n) \in \mathrm{GF}(p^2)$, thereby obtaining a saving of a factor 3 in the representation size. It is shown in Algorithm 2.3.5 below that $\mathrm{Tr}(g^n)$ can indeed be computed quickly given $\mathrm{Tr}(g)$—it turns out that this computation can be done much faster than computing g^n given g, so that a considerable speed advantage is obtained as well.

Definition 2.3.1 ([10, Definition 2.3.1]). For $c \in \mathrm{GF}(p^2)$ define

$$F(c,X) = X^3 - cX^2 + c^pX - 1 \in \mathrm{GF}(p^2)[X],$$

and define $\tau(c,n) = h_0^n + h_1^n + h_2^n$ for $n \in \mathbf{Z}$, where h_0, h_1, h_2 are the (not necessarily distinct) roots of $F(c,X)$ in $\mathrm{GF}(p^6)$. The shorthand c_n is used for $\tau(c,n)$.

The definition of $F(c,X)$ for all c in $\mathrm{GF}(p^2)$ is more general than that implied by the argument before Definition 2.3.1. There only c are considered of the form $\mathrm{Tr}(g)$ for g of order > 3 and dividing $p^2 - p + 1$. The more general definition allows the application in Section 3. For the present purposes, if $c = \mathrm{Tr}(g)$ then $c_n = \mathrm{Tr}(g^n)$ (as argued above), which makes fast computation of c_n relevant. This can be done based on the following properties of c_n and $F(c,X)$.

Lemma 2.3.2 ([10, Lemmas 2.3.2 and 2.3.4]).

(i) $c = c_1$.

(ii) $c_{-n} = c_{np} = c_n^p$ *for* $n \in \mathbf{Z}$.

(iii) $c_n \in \mathrm{GF}(p^2)$ *for* $n \in \mathbf{Z}$.

(iv) $c_{u+v} = c_u \cdot c_v - c_v^p \cdot c_{u-v} + c_{u-2v}$ *for* $u, v \in \mathbf{Z}$.

(v) *Either all* h_j *have order dividing* $p^2 - p + 1$ *and* > 3 *or all* $h_j \in \mathrm{GF}(p^2)$. *In particular,* $F(c,X)$ *is irreducible if and only if its roots have order dividing* $p^2 - p + 1$ *and* > 3.

(vi) $F(c,X)$ *is reducible over* $\mathrm{GF}(p^2)$ *if and only if* $c_{p+1} \in \mathrm{GF}(p)$.

Corollary 2.3.3 ([10, Corollary 2.3.5]). *Let* c, c_{n-1}, c_n, *and* c_{n+1} *be given.*

(i) *Computing* $c_{2n} = c_n^2 - 2c_n^p$ *takes two multiplications in* $\mathrm{GF}(p)$.

(ii) *Computing* $c_{n+2} = c \cdot c_{n+1} - c^p \cdot c_n + c_{n-1}$ *takes four multiplications in* $\mathrm{GF}(p)$.

(iii) *Computing* $c_{2n-1} = c_{n-1} \cdot c_n - c^p \cdot c_n^p + c_{n+1}^p$ *takes four multiplications in* $\mathrm{GF}(p)$.

(iv) *Computing* $c_{2n+1} = c_{n+1} \cdot c_n - c \cdot c_n^p + c_{n-1}^p$ *takes four multiplications in* $\mathrm{GF}(p)$.

Proof. Use Lemmas 2.2.1 and 2.3.2. □

Definition 2.3.4 ([10, Definition 2.3.6]). Let $S_n(c) = (c_{n-1}, c_n, c_{n+1}) \in \mathrm{GF}(p^2)^3$.

Algorithm 2.3.5 (Computation of $S_n(c)$ given n and c [10, Algoritm 2.3.7]).

- If $n < 0$, apply this algorithm to $-n$ and c, and apply Lemma 2.3.2.(ii) to the resulting value.
- If $n = 0$, then $S_0(c) = (c^p, 3, c)$ (cf. Lemma 2.3.2.(ii)).
- If $n = 1$, then $S_1(c) = (3, c, c^2 - 2c^p)$ (cf. Corollary 2.3.3.(i)).
- If $n = 2$, use Corollary 2.3.3.(ii) and $S_1(c)$ to compute c_3 and thereby $S_2(n)$.
- Otherwise, to compute $S_n(c)$ for $n > 2$ define $\bar{S}_i(c) = S_{2i+1}(c)$ and let $\bar{m} = n$. If $\bar{m}$ is even, then replace $\bar{m}$ by $\bar{m} - 1$. Let $\bar{m} = 2m + 1$, $k = 1$, and compute $\bar{S}_k(c) = S_3(c)$ using Corollary 2.3.3.(ii) and $S_2(c)$.
  Let $m = \sum_{j=0}^{r} m_j 2^j$ with $m_j \in \{0,1\}$ and $m_r = 1$. For $j = r-1, r-2, \ldots, 0$ in succession do the following:

— If $m_j = 0$ then use

$$\bar{S}_k(c) = (c_{2k}, c_{2k+1}, c_{2k+2}) \text{ to compute } \bar{S}_{2k}(c) = (c_{4k}, c_{4k+1}, c_{4k+2})$$

(using Corollary 2.3.3.(i) for c_{4k} and c_{4k+2} and Corollary 2.3.3.(iii) for c_{4k+1}).

— If $m_j = 1$ then use

$$\bar{S}_k(c) = (c_{2k}, c_{2k+1}, c_{2k+2}) \text{ to compute } \bar{S}_{2k+1}(c) = (c_{4k+2}, c_{4k+3}, c_{4k+4})$$

(using Corollary 2.3.3.(i) for c_{4k+2} and c_{4k+4} and Corollary 2.3.3.(iv) for c_{4k+3}).

Comment. The great similarity between the computation for $m_j = 0$ and $m_j = 1$ makes this algorithm much less susceptible to environmental attacks than usual exponentiation routines.

— Replace k by $2k + m_j$.

After this iteration $k = m$ and $S_{\bar{m}}(c) = \bar{S}_m(c)$. If n is even use

$$S_{\bar{m}}(c) = (c_{\bar{m}-1}, c_{\bar{m}}, c_{\bar{m}+1}) \quad \text{to compute} \quad S_{\bar{m}+1}(c) = (c_{\bar{m}}, c_{\bar{m}+1}, c_{\bar{m}+2})$$

(using Corollary 2.3.3.(ii)) and replace $\bar{m}$ by $\bar{m}+1$. As a result $S_n(c) = S_{\bar{m}}(c)$.

Theorem 2.3.6 ([10, Theorem 2.3.8]). *Given the sum c of the roots of $F(c, X)$, the sum c_n of the nth powers of the roots of $F(c, X)$ can be computed in $8\log_2(n)$ multiplications in* GF(p).

Thus, given the representation $\text{Tr}(g) \in \text{GF}(p^2)$ of the conjugates of g, the representation $\text{Tr}(g^n) \in \text{GF}(p^2)$ of the conjugates of the nth power of g can be computed at the cost of $8\log_2(n)$ multiplications in GF(p), for any integer n. This compares quite favorably to the speed of the computation of $g^n \in \text{GF}(p^6)$ given $g \in \text{GF}(p^6)$ (cf. [10, Subsection 2.4]).

Before cryptographic applications of this alternative representation of the elements of $\langle g \rangle$ can be discussed, it remains to show how p, q, and $\text{Tr}(g)$ are chosen.

3. XTR parameter and key selection

3.1. Selection of p and q

As indicated in Subsection 2.1, primes p and q have to be selected in such a way that q divides $p^2 - p + 1$, and such that the resulting fields and subgroups are large enough to withstand known attacks. Furthermore, in order to be able to use the fast GF(p^2) arithmetic described in Subsection 2.2, the prime p should be 2 mod 3. Primes p that are 1 mod 3 can be used as well, but they may not always achieve the same speed.

Let P and Q denote the bit lengths of the primes p and q to be generated, respectively. The prime p should be such that the field GF(p^6) cannot be effectively attacked using the Discrete Logarithm variant of the Number Field Sieve, and the prime q should be such that an order q subgroup cannot be effectively attacked using Pollard's rho method. For current security levels, a choice where $6P$ is close to 1024 and Q is close to 160 is acceptable. Choosing P much smaller than Q cannot be recommended given current cryptanalytic methods.

In principle p may also be a non-trivial prime power (cf. [10, Section 6]). This is, however, incompatible with the first two methods presented in this subsection, and makes selection of a proper p and q in general much harder. It can be used in conjunction with Algorithm 3.1.3, but not efficiently if q must be much smaller than intended there.

A detailed analysis of the run times of the algorithms in this subsection is straightforward and left to the reader.

Algorithm 3.1.1 (Selection of q and 'nice' p [10, Algorithm 3.1.1]).

1. Find $r \in \mathbf{Z}$ such that $q = r^2 - r + 1$ is a Q-bit prime.
2. Find $k \in \mathbf{Z}$ such that $p = r + k \cdot q = kr^2 + (1-k)r + k$ is a P-bit prime that is $2 \bmod 3$.

Algorithm 3.1.1 is very fast and can be used to find primes p that satisfy a second degree polynomial with small coefficients. Such p lead to fast arithmetic operations in $\mathrm{GF}(p)$. In particular if the search for k is restricted to $k = 1$ (i.e., search for an r such that both $r^2 - r + 1$ and $r^2 + 1$ are prime and such that $r^2 + 1 \equiv 2 \bmod 3$, thereby slowing down Algorithm 3.1.1 considerably) the primes p have a very nice form; note that in this case r must be even and $p \equiv 1 \bmod 4$.

On the other hand, such 'nice' p may be undesirable from a security point of view because they may make application of the Discrete Logarithm variant of the Number Field Sieve easier. Another method to generate p and q that does not have this disadvantage (and thus neither the advantage of fast arithmetic modulo p) and that is about equally fast, is the following.

Algorithm 3.1.2 (Selection of q and p [10, Algorithm 3.1.2]).

1. First, select a Q-bit prime $q \equiv 7 \bmod 12$.
2. Find the roots r_1 and r_2 of $X^2 - X + 1 \bmod q$.
   Comment. It follows from $q \equiv 1 \bmod 3$ and quadratic reciprocity that r_1 and r_2 exist. Since $q \equiv 3 \bmod 4$ they can be found using a single $\frac{q+1}{4}^{\text{th}}$ powering modulo q.
3. Find $k \in \mathbf{Z}$ such that $p = r_i + k \cdot q$ for $i = 1$ or 2 is a P-bit prime that is $2 \bmod 3$.

In Section 3.5 a fast algorithm is given to verify that an element $c \in \mathrm{GF}(p^2)$ is the trace of an element of the XTR supergroup: according to Theorem 3.5.5 this can be done at the cost of about $1.8 \log_2(p)$ multiplications in $\mathrm{GF}(p)$. Verifying XTR subgroup membership, however, amounts to checking that $c_q = 3$. This costs $8 \log_2(q)$ multiplications in $\mathrm{GF}(p)$ (cf. Algorithm 2.3.5), and is thus substantially more expensive than XTR supergroup membership verification. In some applications (and in particular to avoid so-called subgroup attacks, cf. Subsection 6.1), XTR subgroup membership verification is required. In that case it turns out to be practical to choose the size Q of the XTR subgroup close to the size of the XTR supergroup: the amount of damage can in general be bounded by $2P - Q$, so that only limited damage can be done if $2P - Q$ is small and the 'cheap' XTR supergroup test is carried out instead of the expensive XTR subgroup test. Furthermore, by using 'short exponents' (e.g. of size 170 bits, cf. [23]) in this setting, the use of a large XTR subgroup does not have a negative impact on the speed of cryptographic operations. The best that can be achieved (given that $p = 2 \bmod 3$) is to choose p such that $(p^2 - p + 1)/3$ is prime (and equal to q). The following straightforward algorithm determines satisfactory primes p and q for such applications.

Algorithm 3.1.3 (Selection of large q and p). Select a P-bit prime p until $p^2 - p + 1$ is of the form $q \cdot s$ where q is a prime number and s is small.

Algorithm 3.1.3 can be improved by choosing p in such a way that some fixed number of small primes does not divide $(p^2 - p + 1)/3$.

An alternative way to render subgroup attacks mostly ineffective, is by choosing p and q such that $(p^2 - p + 1)/q$ is a small multiple of a prime of the same order of magnitude as q. Assuming that Q is only slightly smaller than P, finding such p and q can be achieved by the following simple (but on average considerably slower) adaptation of Algorithm 3.1.2.

Algorithm 3.1.4 (Selection of subgroup attack resistant q and p).

1. First, select a Q-bit prime $q \equiv 7 \bmod 12$.
2. Find the roots r_1 and r_2 of $X^2 - X + 1 \bmod q$.
3. Find $k \in \mathbf{Z}$ such that $p = r_i + k \cdot q$ for $i = 1$ or 2 is a P-bit prime that is $2 \bmod 3$ and $(p^2 - p + 1)/q$ is of the form $s \cdot q'$ for a small s and prime q' of at least Q bits.

3.2. Basic subgroup selection

Given p and $q > 3$, it remains to find an element $c \in \mathrm{GF}(p^2)$ such that $c = \mathrm{Tr}(g)$ for an element $g \in \mathrm{GF}(p^6)$ of order q dividing $p^2 - p + 1$. Note that finding $\mathrm{Tr}(g)$ suffices and that g itself is not needed. But, given $\mathrm{Tr}(g)$, a generator g of the XTR subgroup (cf. Subsection 2.1) can be found by determining any root of $F(\mathrm{Tr}(g), X)$ (cf. Definition 2.3.1).

According to Lemma 2.3.2.(v), if $c \in \mathrm{GF}(p^2)$ is such that $F(c, X)$ is irreducible, then c is the trace of an element $h \in \mathrm{GF}(p^6)^*$ of order > 3 dividing $p^2 - p + 1$. Thus, if furthermore $c_{(p^2-p+1)/q} \neq 3$ (cf. Definition 2.3.1), which can be verified using Algorithm 2.3.5 at the cost of $8 \log_2((p^2 - p + 1)/q)$ multiplications in $\mathrm{GF}(p)$, then $c_{(p^2-p+1)/q}$ is the trace of an element of order q. Consequently, $\mathrm{Tr}(g)$ can be defined as $c_{(p^2-p+1)/q}$. It remains to find $c \in \mathrm{GF}(p^2)$ such that $F(c, X)$ is irreducible. According to the following lemma, this can be done by randomly picking c's until $F(c, X)$ is irreducible.

Lemma 3.2.1 ([10, Lemma 3.2.1]). *For a randomly selected $c \in \mathrm{GF}(p^2)$ the probability that $F(c, X) \in \mathrm{GF}(p^2)[X]$ is irreducible is about one third.*

As shown in Subsections 3.3 and 3.5, testing $F(c, X)$ for irreducibility for a randomly selected $c \in \mathrm{GF}(p^2)$ can be done very fast, but those methods require additional code. The following method, based on Lemma 2.3.2.(vi), requires only Algorithm 2.3.5 and thus hardly any additional code.

Algorithm 3.2.2 (Computation of $\mathrm{Tr}(g)$ [10, Algorithm 3.2.2]).

1. Apply Algorithm 2.3.5 to $n = p + 1$ and a random $c \in \mathrm{GF}(p^2) \setminus \mathrm{GF}(p)$ to compute c_{p+1}.
2. If $c_{p+1} \in \mathrm{GF}(p)$ then return to Step 1.
3. Apply Algorithm 2.3.5 to $n = (p^2 - p + 1)/q$ and c to compute $d = c_{(p^2-p+1)/q}$.
4. If $d = 3$, then return to Step 1.
5. Let $\mathrm{Tr}(g) = d$.

Theorem 3.2.3 ([10, Theorem 3.2.3]). *Algorithm 3.2.2 computes an element of $\mathrm{GF}(p^2)$ that equals $\mathrm{Tr}(g)$ for some $g \in \mathrm{GF}(p^6)$ of order q. It can be expected to require $\frac{3q}{q-1}$ applications of Algorithm 2.3.5 with $n = p + 1$ and $\frac{q}{q-1}$ applications with $n = (p^2 - p + 1)/q$.*

The next subsection contains a faster method to find the trace of a generator of the XTR subgroup, based on a more direct method to test $F(c, X)$ for irreducibility.

3.3. Subgroup selection using an irreducibility test

The roots of a third-degree equation can be computed directly by means of one of Cardano's classical formulas, more precisely Scipione del Ferro's method (cf. [19, page 559]).

Algorithm 3.3.1 ((Scipione del Ferro, ~1465–1526)). To compute the roots of $f(X) = aX^3 + bX^2 + dX + e$ in a field of characteristic $p \neq 2, 3$, do the following.

1. Compute the polynomial

$$\frac{f(X - \frac{b}{3a})}{a} = X^3 + f_1 X + f_0$$

with

$$f_1 = \frac{3ad - b^2}{3a^2} \text{ and } f_0 = \frac{27a^2e - 9abd + 2b^3}{27a^3}.$$

2. Compute the discriminant $\Delta = f_0^2 + 4(\frac{f_1}{3})^3$ of the polynomial $X^2 + f_0 X - (\frac{f_1}{3})^3$, and compute its roots $r_{1,2} = \frac{-f_0 \pm \sqrt{\Delta}}{2}$ using the standard method to compute the roots of a second-degree equation.
3. If $r_1 = r_2 = 0$, then let $u = v = 0$. Otherwise, let $r_1 \neq 0$, compute a cube root u of r_1, and let $v = -\frac{f_1}{3u}$ be a cube root of r_2.

4. The roots of $f(X)$ are
$$u+v-\frac{b}{3a}, \quad uw+vw^2-\frac{b}{3a}, \quad uw^2+vw-\frac{b}{3a},$$
where $w \in \mathrm{GF}(p^2)$ is a non-trivial cube root of unity, i.e., $w^3=1$ and $w^2+w+1=0$.

Lemma 3.3.2. *With Algorithm* 3.3.1 *applied to* $f(X)=F(c,X) \in \mathrm{GF}(p^2)[X]$:

(i) $\Delta \in \mathrm{GF}(p)$ (*cf.* [11, *Lemma* 3.3]).

(ii) Δ *is a quadratic residue in* $\mathrm{GF}(p)$ *if and only if either* $F(c,X)$ *is irreducible in* $\mathrm{GF}(p^2)[X]$ *or all roots in* $\mathrm{GF}(p^2)$ *of* $F(c,X)$ *have order dividing* $p+1$ (*cf.* [11, *Lemma* 3.6]).

(iii) $F(c,X) \in \mathrm{GF}(p^2)[X]$ *is reducible over* $\mathrm{GF}(p^2)$ *if and only if* r_1 *satisfies* $r_1^{p(p+1)/3} = r_1^{(p+1)/3}$ (*i.e.,* r_1 *is a cube in* $\mathrm{GF}(p^2)$) (*cf.* [11, *Corollary* 3.4]).

Algorithm 3.3.1 and Lemma 3.3.2 lead to the following irreducibility test.

Algorithm 3.3.3 (Irreducibility test [11, Algorithm 3.5]). To test $F(c,X)$ in $\mathrm{GF}(p^2)[X]$ for irreducibility over $\mathrm{GF}(p^2)$ with $p \neq 2,3$, do the following.

1. Compute
$$f_0 = \frac{-27+9c^{p+1}-2c^3}{27} \text{ and } f_1 = c^p - \frac{c^2}{3} \in \mathrm{GF}(p^2).$$

2. If $\Delta = f_0^2 + 4(\frac{f_1}{3})^3 \in \mathrm{GF}(p)$ (cf. Lemma 3.3.2.(i)) is a quadratic non-residue in $\mathrm{GF}(p)$ then $F(c,X)$ is reducible (cf. Lemma 3.3.2.(ii)).
   Comment. This step requires the computation of a Jacobi symbol.
3. Otherwise, compute $r_1 = \frac{-f_0+\sqrt{\Delta}}{2} \in \mathrm{GF}(p^2)$.
4. Compute $y = r_1^{(p+1)/3} \in \mathrm{GF}(p^2)$, then $F(c,X)$ is irreducible if and only if $y \neq y^p$ (cf. Lemma 3.3.2.(iii)).

Algorithm 3.3.4 (Computation of $\mathrm{Tr}(g)$).

1. Pick a random $c \in \mathrm{GF}(p^2) \setminus \mathrm{GF}(p)$ and use Algorithm 3.3.3 to test $F(c,X)$ for irreducibility.
2. If $F(c,X)$ is reducible, then return to Step 1.
3. Apply Algorithm 2.3.5 to $n=(p^2-p+1)/q$ and c to compute $d = c_{(p^2-p+1)/q}$.
4. If $d=3$, then return to Step 1.
5. Let $\mathrm{Tr}(g)=d$.

Theorem 3.3.5 ([11, Theorem 3.7]). *Finding the trace of a generator of the XTR group, using Algorithm* 3.3.4 *takes an expected number*
$$\frac{q}{q-1}(7.2\log_2(p) + 8\log_2((p^2-p+1)/q))$$
plus a small constant number of multiplications in $\mathrm{GF}(p)$.

Proof. According to Lemma 3.2.1, Algorithm 3.3.3 is called, on average, $\frac{3q}{q-1}$ times. For half the calls, on average, Δ in Step 2 is a quadratic non-residue, and the cost is a small constant number of multiplications in $\mathrm{GF}(p)$. For the other calls, first $\sqrt{\Delta}$ is computed at an expected cost of $\log_2(p)$ squarings and $0.5\log_2(p)$ multiplications in $\mathrm{GF}(p)$ (for a total of $1.3\log_2(p)$ multiplications in $\mathrm{GF}(p)$, cf. assumption in the proof of Lemma 2.2.1). This is followed by the computation of y at an expected cost of $\log_2(p)$ squarings and $0.5\log_2(p)$ multiplications in $\mathrm{GF}(p^2)$ (for a total of $3.5\log_2(p)$ multiplications in $\mathrm{GF}(p)$, cf. Lemma 2.2.1). Thus, for random c application of Algorithm 3.3.3 costs

$$\frac{1.3+3.5}{2}\log_2(p) = 2.4\log_2(p)$$

plus a small constant number of multiplications in $\mathrm{GF}(p)$, on average. □

Thus, Algorithm 3.3.4, based on Scipione del Ferro's method in Algorithm 3.3.1, is more than 50% faster than Algorithm 3.2.2. Though useful, Algorithm 3.3.4 and Theorem 3.3.5 are just a side result of a more important consequence of Algorithm 3.3.1, namely the key recovery method from Subsection 5.5. The next two sections contain two even faster methods to find $\mathrm{Tr}(g)$. The first method poses the additional restriction that $p \equiv 2$ or $5 \bmod 9$ (i.e., $p \not\equiv 8 \bmod 9$).

3.4. Subgroup selection when $p \not\equiv 8 \bmod 9$

If $p \not\equiv 8 \bmod 9$, then $(Z^9-1)/(Z^3-1) = Z^6+Z^3+1 \in \mathrm{GF}(p)[Z]$ is irreducible over $\mathrm{GF}(p)$, so that $\mathrm{GF}(p^6)$ can be represented as $\mathrm{GF}(p)(\zeta)$ where $\zeta^6+\zeta^3+1=0$. This representation of $\mathrm{GF}(p^6)$ allows symbolic calculation, i.e., irrespective of the value of p, of the trace of the $(p^6-1)/(p^2-p+1)^{\text{th}}$ power of elements of the form $\zeta + a$, for random $a \in \mathrm{GF}(p)$. This follows from a more general argument due to H.W. Lenstra, Jr. In particular (cf. [11, Proposition 4.3]),

$$\mathrm{Tr}\big((\zeta+a)^{(p^6-1)/(p^2-p+1)}\big) = \frac{-3}{a^6-a^3+1}\left((a^2-1)^3\alpha + a^3(a^3-3a+1)\alpha^2\right),$$

where it is shown that $a^6-a^3+1 \neq 0$. It follows that, for any $a \in \mathrm{GF}(p)^*$, $a \neq \pm 1$, the right hand side expression is the trace of an element of order dividing p^2-p+1 (cf. [11, Corollary 4.4]). With $a=2$ and $a=1/2$, this leads to the following very fast method to initialize $\mathrm{Tr}(g)$—obviously any $a \in \mathrm{GF}(p)^*$, $a \neq \pm 1$, can be used instead.

Algorithm 3.4.1 ((Computation of $\mathrm{Tr}(g)$ [11, Algorithm 4.5])).

1. Let $c = \frac{27\alpha+3\alpha^2}{19} \in \mathrm{GF}(p^2)$.
2. Apply Algorithm 2.3.5 to $n=(p^2-p+1)/q$ and c to compute $d = c_{(p^2-p+1)/q}$.
3. If $d \neq 3$, then let $\mathrm{Tr}(g) = d$ and return success.
4. Otherwise, if $d=3$, then replace c by $\frac{-27\alpha-24\alpha^2}{19} \in \mathrm{GF}(p^2)$.
5. Apply Algorithm 2.3.5 to $n=(p^2-p+1)/q$ and c and recompute $d = c_{(p^2-p+1)/q}$.
6. If $d \neq 3$, then let $\mathrm{Tr}(g) = d$ and return success.
7. Otherwise, if $d=3$, then return failure.

The probability of failure of Algorithm 3.4.1 may be expected to be q^{-2}, i.e., negligibly small. Its expected cost is about $8\log_2((p^2-p+1)/q)$ multiplications in GF(p).

Although Algorithm 3.4.1 is a very fast method to find a proper Tr(g), it is less general than the method from Subsection 3.3, and in particular does not provide a faster $F(c,X)$ irreducibility test. Because fast irreducibility testing has other applications than just Tr(g)-initialization, improving the test from Subsection 3.3 is relevant. In Subsection 3.5 below this is done by reformulating the third-degree GF(p^2)-irreducibility test as a third-degree GF(p)-irreducibility test, and by carefully analyzing the cost of the latter.

3.5. Subgroup selection using a faster irreducibility test

Definition 3.5.1 ([12, Definition 2.1]). For $c \in \mathrm{GF}(p^2)$ let

$$P(c,X) = X^3 + (c^p + c)X^2 + (c^{p+1} + c^p + c - 3)X + c^{2p} + c^2 + 2 - 2c^p - 2c.$$

It easily follows that $P(c,X)$ is a polynomial in GF(p). The following result indicates why it is relevant to consider $P(c,X)$.

Corollary 3.5.2 ([12, Corollary 2.5]). *$F(c,X)$ is irreducible over* GF(p^2) *if and only if $P(c,X)$ is irreducible over* GF(p).

As shown in [12, Section 3], efficient application of Algorithm 3.3.1 to $P(c,X)$ requires an equivalent of Algorithm 2.3.5 for traces over GF(p). It is well known how that is done; the details are given in Algorithm 3.5.3 below.

Algorithm 3.5.3 ([12, Algorithm 3.4]). To compute the trace $\mathrm{Tr}(y^n) \in \mathrm{GF}(p)$ over GF(p) of $y^n \in \mathrm{GF}(p^2)$, given an integer $n > 0$ and the trace $\mathrm{Tr}(y) \in \mathrm{GF}(p)$ over GF(p) of $y \in \mathrm{GF}(p^2)$ of order dividing $p+1$. This algorithm takes $1.8\log_2(p)$ multiplications in GF(p) (cf. assumption in the proof of Lemma 2.2.1).

1. Let $n = \sum_{i=0}^{k} n_i 2^i$ with $n_i \in \{0,1\}$ and $n_k \neq 0$ and let $v = \mathrm{Tr}(y) \in \mathrm{GF}(p)$.
2. Compute $w = (v^2 - 2) \in \mathrm{GF}(p)$.
3. For $i = k-1, k-2, \ldots, 0$ in succession, do the following.
   - If $n_i = 1$, then first replace v by $vw - \mathrm{Tr}(y)$ and next replace w by $w^2 - 2$.
   - If $n_i = 0$, then first replace w by $vw - \mathrm{Tr}(y)$ and next replace v by $v^2 - 2$.
4. Return $\mathrm{Tr}(y^n) = v$.

Algorithm 3.5.4 ([12, Algorithm 3.5]). To test $P(c,X) = X^3 + p_2X^2 + p_1X + p_0 \in \mathrm{GF}(p)[X]$ for irreducibility over GF(p) with $p \neq 2,3$, do the following.

1. Compute

$$f_0 = \frac{27p_0 - 9p_2p_1 + 2p_2^3}{27}, \quad f_1 = p_1 - \frac{p_2^2}{3} \in \mathrm{GF}(p).$$

2. If $\Delta = f_0^2 + 4(\frac{f_1}{3})^3 \in \mathrm{GF}(p)$ is a quadratic residue in GF(p), then $P(c,X)$ is reducible (cf. [12, Lemma 3.2]).
   Comment. This step requires the computation of a Jacobi symbol.

3. Otherwise, compute $s = 2\frac{f_0^2+\Delta}{f_0^2-\Delta} \in \mathrm{GF}(p)$.
   Comment. According to [12, Lemma 3.3], s is the trace of r_1^{p-1} over $\mathrm{GF}(p)$, where $r_1 = \frac{-f_0+\sqrt{\Delta}}{2}$.
4. Apply Algorithm 3.5.3 to $n = \frac{p+1}{3}$ and $\mathrm{Tr}(y) = s$ to compute $\mathrm{Tr}(y^{(p+1)/3})$. If $\mathrm{Tr}(y^{(p+1)/3}) = 2$, then $P(c, X)$ is reducible.
   Comment. If the trace over $\mathrm{GF}(p)$ of $(r_1^{p-1})^{(p+1)/3}$ equals 2, then r_1 is a cube in $\mathrm{GF}(p^2)$ and thus, according to [12, Lemma 3.2], $P(c, X)$ is not irreducible.
5. Otherwise, Δ is a quadratic non-residue and r_1 is not a cube in $\mathrm{GF}(p^2)$ so that, according to [12, Lemma 3.2], $P(c, X)$ is irreducible over $\mathrm{GF}(p)$.

In the worst case Algorithm 3.5.4 costs $1.8\log_2(p)$ plus a small constant number of multiplications in $\mathrm{GF}(p)$. For half the random $c \in \mathrm{GF}(p^2)$, however, Δ is a quadratic residue, and the cost is just a small constant number of multiplications in $\mathrm{GF}(p)$. The proof of Theorem 3.5.5 below and its corollary follow.

Theorem 3.5.5 ([12, Theorem 3.6]). *For $c \in \mathrm{GF}(p^2)$ the irreducibility of the polynomial $F(c, X) = X^3 - cX^2 + c^pX - 1$ over $\mathrm{GF}(p^2)$ can be tested at the cost of at most $m + 1.8\log_2(p)$ multiplications in $\mathrm{GF}(p)$, for some small constant m.*

Corollary 3.5.6 ([12, Corollary 3.7]). *Finding the trace of a generator of the XTR group can be expected to take about*

$$\frac{q}{q-1}(2.7\log_2(p) + 8\log_2((p^2 - p + 1)/q))$$

multiplications in $\mathrm{GF}(p)$ (cf. Theorem 3.3.5).

The result of Corollary 3.5.6 is only about $2.7\log_2(p)$ multiplications in $\mathrm{GF}(p)$ slower than Algorithm 3.4.1, but is more general since it applies to all $p \equiv 2 \bmod 3$ and not only to $p \equiv 2, 5 \bmod 9$.

Remark 3.5.7. Algorithm 3.5.4 provides an efficient way to test if $c \in \mathrm{GF}(p^2)$ is the trace of an element of the XTR supergroup. That is, by choosing the size of the XTR group close to the XTR supergroup, one obtains an efficient way to determine XTR group membership modulo a small error. In Algorithm 3.1.3 it is explained how such XTR parameters p and q can be found. In Section 6 it is shown that this has an application in the prevention of 'subgroup based attacks'.

4. XTR cryptographic schemes for confidentiality services

In any cryptosystem that relies on the (subgroup) discrete logarithm problem the ordinary representation of subgroup elements can be replaced by the XTR representation of subgroup elements of a multiplicative group of equivalent security. This section contains a description of some applications of XTR that provide confidentiality services: Diffie–Hellman key agreement in Subsection 4.1 and ElGamal encryption in Subsection 4.2. In both schemes random exponents modulo q are used. If q is chosen to be close to $p^2 - p + 1$ (cf. Algorithm 3.1.3), then much shorter random exponents, say of 170 bits, can be used instead if that is desirable for computational efficiency (cf. [23]).

4.1. XTR-DH

4.1.1. XTR-DH key agreement. Let p, q, $\mathrm{Tr}(g)$ be shared XTR public key data. If Alice and Bob want to agree on a secret key K they do the following.

1. Alice selects a random integer $a \in [2, q-3]$, applies Algorithm 2.3.5 to $n = a$ and $c = \mathrm{Tr}(g)$ to compute
$$S_a(\mathrm{Tr}(g)) = \left(\mathrm{Tr}(g^{a-1}), \mathrm{Tr}(g^a), \mathrm{Tr}(g^{a+1})\right) \in \mathrm{GF}(p^2)^3,$$
and sends $\mathrm{Tr}(g^a) \in \mathrm{GF}(p^2)$ to Bob.
2. Bob receives $\mathrm{Tr}(g^a)$ from Alice, selects a random integer $b \in [2, q-3]$, applies Algorithm 2.3.5 to $n = b$ and $c = \mathrm{Tr}(g)$ to compute
$$S_b(\mathrm{Tr}(g)) = \left(\mathrm{Tr}(g^{b-1}), \mathrm{Tr}(g^b), \mathrm{Tr}(g^{b+1})\right) \in \mathrm{GF}(p^2)^3,$$
and sends $\mathrm{Tr}(g^b) \in \mathrm{GF}(p^2)$ to Alice.
3. Alice receives $\mathrm{Tr}(g^b)$ from Bob, applies Algorithm 2.3.5 to $n = a$ and $c = \mathrm{Tr}(g^b)$ to compute
$$S_a(\mathrm{Tr}(g)^b) = \left(\mathrm{Tr}(g^{(a-1)b}), \mathrm{Tr}(g^{ab}), \mathrm{Tr}(g^{(a+1)b})\right) \in \mathrm{GF}(p^2)^3,$$
and determines K based on $\mathrm{Tr}(g^{ab}) \in \mathrm{GF}(p^2)$ (but see Remark 7.1.3).
4. Bob applies Algorithm 2.3.5 to $n = b$ and $c = \mathrm{Tr}(g^a)$ to compute
$$S_b(\mathrm{Tr}(g)^a) = \left(\mathrm{Tr}(g^{a(b-1)}), \mathrm{Tr}(g^{ab}), \mathrm{Tr}(g^{a(b+1)})\right) \in \mathrm{GF}(p^2)^3,$$
and determines K based on $\mathrm{Tr}(g^{ab}) \in \mathrm{GF}(p^2)$.
Comment. The 'neighboring' elements $\mathrm{Tr}(g^{(a-1)b})$ and $\mathrm{Tr}(g^{(a+1)b})$ computed by Alice are in general different from $\mathrm{Tr}(g^{a(b-1)})$ and $\mathrm{Tr}(g^{a(b+1)})$, the neighboring elements computed by Bob.

The communication and computational overhead of XTR-DH key agreement 4.1.1 are both about one third of traditional implementations of the Diffie–Hellman protocol that are based on subgroups of multiplicative groups of finite fields, and that achieve the same level of security (cf. Section 7.1).

4.2. XTR-ElGamal encryption

4.2.1. XTR-ElGamal encryption (cf. [6]). Let p, q, $\mathrm{Tr}(g)$ be XTR public key data, either owned (and made public) by Alice or shared by all parties. Furthermore, let $\mathrm{Tr}(g^k)$ be a value computed and made public by Alice, for some integer k selected (and kept secret) by Alice. Given $(p, q, \mathrm{Tr}(g), \mathrm{Tr}(g^k))$, Bob can encrypt a message M intended for Alice as follows.

1. Bob selects at random $b \in [2, q-3]$ and applies Algorithm 2.3.5 to $n = b$ and $c = \mathrm{Tr}(g)$ to compute
$$S_b(\mathrm{Tr}(g)) = \left(\mathrm{Tr}(g^{b-1}), \mathrm{Tr}(g^b), \mathrm{Tr}(g^{b+1})\right) \in \mathrm{GF}(p^2)^3.$$
2. Bob applies Algorithm 2.3.5 to $n = b$ and $c = \mathrm{Tr}(g^k)$ to compute
$$S_b(\mathrm{Tr}(g^k)) = \left(\mathrm{Tr}(g^{(b-1)k}), \mathrm{Tr}(g^{bk}), \mathrm{Tr}(g^{(b+1)k})\right) \in \mathrm{GF}(p^2)^3.$$
3. Bob determines a symmetric encryption key K based on $\mathrm{Tr}(g^{bk}) \in \mathrm{GF}(p^2)$.
4. Bob uses an agreed upon symmetric encryption method with key K to encrypt M, resulting in the encryption E.
5. Bob sends $(\mathrm{Tr}(g^b), E)$ to Alice.
Comment. The message sent by Bob consists of an 'overhead part' $\mathrm{Tr}(g^b)$ and a message part E. The length of the former is independent of the length of M, but the length of the latter depends strongly on the length of M and the type of symmetric encryption used.

4.2.2. XTR-ElGamal decryption. Using her knowledge of k, Alice decrypts the message $(\mathrm{Tr}(g^b), E)$ encrypted using XTR-ElGamal encryption 4.2.1 as follows.

1. Alice applies Algorithm 2.3.5 to $n = k$ and $c = \mathrm{Tr}(g^b)$ to compute
$$S_k(\mathrm{Tr}(g^b)) = (\mathrm{Tr}(g^{b(k-1)}), \mathrm{Tr}(g^{bk}), \mathrm{Tr}(g^{b(k+1)})) \in \mathrm{GF}(p^2)^3.$$
2. Alice determines symmetric encryption key K based on $\mathrm{Tr}(g^{bk}) \in \mathrm{GF}(p^2)$.
3. Alice uses the agreed upon symmetric encryption method with key K to decrypt E, resulting in the encryption M.

The communication and computational overhead of XTR-based ElGamal encryption 4.2.1 and decryption 4.2.2 (with communication overhead as explained in Step 5 of 4.2.1) are both about one third of traditional implementations of the ElGamal encryption and decryption protocols that are based on subgroups of multiplicative groups of finite fields, and that achieve the same level of security (cf. Section 7.1).

Remark 4.2.3. The type of encryption described in 4.2.1 is commonly referred to as 'hybrid encryption', because the key K is used in conjunction with an agreed upon symmetric key encryption method. In the non-hybrid version the message is restricted to the key space and 'encrypted' using an invertible operation that takes place in the key space, such as multiplication by the key. In 4.2.1 and with $K = \mathrm{Tr}(g^{bk})$ this would amount to requiring that $M \in \mathrm{GF}(p^2)$ and computing E as $K \cdot M \in \mathrm{GF}(p^2)$. Compared to non-hybrid traditional ElGamal encryption, non-hybrid XTR-ElGamal encryption saves a factor three on the length of both parts of the encrypted message, for messages that fit in the key space (of one third of the 'traditional' size).

Remark 4.2.4. As is customary it is implicitly assumed in the decryption that the first component of an ElGamal encrypted message represents a conjugate of a power of g. In some situations this should be verified explicitly. A value, say c, can be checked by verifying that $c \in \mathrm{GF}(p^2) \setminus \mathrm{GF}(p)$ (implying, in particular, that $c \neq 3$), by applying Algorithm 2.3.5 to $n = q$ and c to compute $S_q(c) = (c_{q-1}, c_q, c_{q+1})$, and by verifying that $c_q = 3$. Other and more efficient techniques are discussed in Section 6.

5. XTR cryptographic schemes for non-repudiation services

5.1. Introduction

In this section two XTR applications are described that provide non-repudiation services: Nyberg–Rueppel message recovery digital signatures in Subsection 5.3 and XTR-DSA in Subsection 5.4. Both schemes require computation of the product of two powers of g. For the standard representation this can easily be done using well known multi-exponentiation techniques in substantially less time than required for two separate exponentiations. But if traces are used it is a relatively complicated operation. In Subsection 5.2 below it is described how this computation may be carried out in common cryptographic applications such as the ones in Subsections 5.3 and 5.4.

5.1.1. XTR public key data for signature verification. As in Subsection 4.2, Alice's XTR public key data for digital signatures consist of p, q, $\mathrm{Tr}(g)$, and $\mathrm{Tr}(g^k)$ for a secret integer k that is known only to Alice. However, in addition it is assumed that not only $\mathrm{Tr}(g^k)$ but also $\mathrm{Tr}(g^{k-1})$ and $\mathrm{Tr}(g^{k+1})$ (and thus $S_k(\mathrm{Tr}(g))$) are available to the verifier. These additional $\mathrm{GF}(p^2)$ elements are either part of the public key, or they are reconstructed by the verifier. As shown in Subsection 5.5, $\mathrm{Tr}(g^{k-1})$ (or $\mathrm{Tr}(g^{k+1})$) can be reconstructed from p, q, $\mathrm{Tr}(g)$, $\mathrm{Tr}(g^k)$, and $\mathrm{Tr}(g^{k+1})$ (or $\mathrm{Tr}(g^{k-1})$) using an explicit and easily computed formula. Reconstruction of $\mathrm{Tr}(g^{k+1})$ (or $\mathrm{Tr}(g^{k-1})$) given just $(p, q, \mathrm{Tr}(g), \mathrm{Tr}(g^k))$ requires additional assumptions and a slightly more involved computation (cf. Subsection 5.5).

5.2. Computing the trace of a product

Let $\mathrm{Tr}(g) \in \mathrm{GF}(p^2)$ and $S_k(\mathrm{Tr}(g)) \in \mathrm{GF}(p^2)^3$ be given for some secret integer k with $0 < k < q$. In Algorithm 5.2.7 below it is shown that $\mathrm{Tr}(g^a \cdot g^{bk})$ can be computed efficiently for any $a, b \in \mathbf{Z}$ given $\mathrm{Tr}(g)$ and $S_k(\mathrm{Tr}(g)$, i.e., without knowing k.

Definition 5.2.1 ([10, Definition 2.4.1]). Let $C(V)$ denote the center column of a 3×3 matrix V and let

$$A(c) = \begin{pmatrix} 0 & 0 & 1 \\ 1 & 0 & -c^p \\ 0 & 1 & c \end{pmatrix} \text{ and } M_n(c) = \begin{pmatrix} c_{n-2} & c_{n-1} & c_n \\ c_{n-1} & c_n & c_{n+1} \\ c_n & c_{n+1} & c_{n+2} \end{pmatrix}$$

be 3×3-matrices over $\mathrm{GF}(p^2)$ with c and c_n as in Definition 2.3.1.

Lemma 5.2.2 ([10, Lemma 2.4.2]). *For $n, m \in \mathbf{Z}$*

$$S_n(c) = S_m(c) \cdot A(c)^{n-m} \text{ and } M_n(c) = M_m(c) \cdot A(c)^{n-m}.$$

Corollary 5.2.3 ([10, Corollary 2.4.3]).

$$c_n = S_m(c) \cdot C(A(c)^{n-m}).$$

Lemma 5.2.4 ([10, Lemma 2.4.4]). *The determinant of $M_0(c)$ equals*

$$D = c^{2p+2} + 18c^{p+1} - 4(c^{3p} + c^3) - 27 \in \mathrm{GF}(p).$$

If $D \neq 0$ then

$$M_0(c)^{-1} = \frac{1}{D} \cdot \begin{pmatrix} 2c^2 - 6c^p & 2c^{2p} + 3c - c^{p+2} & c^{p+1} - 9 \\ 2c^{2p} + 3c - c^{p+2} & (c^2 - 2c^p)^{p+1} - 9 & (2c^{2p} + 3c - c^{p+2})^p \\ c^{p+1} - 9 & (2c^{2p} + 3c - c^{p+2})^p & (2c^2 - 6c^p)^p \end{pmatrix}.$$

Lemma 5.2.5 ([10, Lemma 2.4.5]).

$$\det(M_0(\mathrm{Tr}(g))) = (\mathrm{Tr}(g^{p+1})^p - \mathrm{Tr}(g^{p+1}))^2 \neq 0.$$

Combination of these results leads to the following corollary.

Corollary 5.2.6 ([10, Lemma 2.4.6, Corollary 2.4.7]). *Given $\mathrm{Tr}(g)$ and $S_n(\mathrm{Tr}(g))$,*

$$C(A(\mathrm{Tr}(g))^n) = M_0(\mathrm{Tr}(g))^{-1} \cdot (S_n(\mathrm{Tr}(g)))^T$$

can be computed at the cost of a small constant number of operations in $\mathrm{GF}(p^2)$.

Algorithm 5.2.7 (Computation of $\mathrm{Tr}(g^a \cdot g^{bk})$ [10, Algorithm 2.4.8]).
Let $\mathrm{Tr}(g)$, $S_k(\mathrm{Tr}(g))$ (for unknown k), and $a, b \in \mathbf{Z}$ with $0 < a, b < q$ be given.

1. Compute $e = a/b \bmod q$.
2. Apply Algorithm 2.3.5 to $n = e$ and $c = \mathrm{Tr}(g)$ to compute $S_e(\mathrm{Tr}(g))$.
3. Use Corollary 5.2.6 to compute $C(A(\mathrm{Tr}(g))^e)$ based on $\mathrm{Tr}(g)$ and $S_e(\mathrm{Tr}(g))$.
4. Use Corollary 5.2.3 to compute $\mathrm{Tr}(g^{e+k}) = S_k(\mathrm{Tr}(g)) \cdot C(A(\mathrm{Tr}(g))^e)$.
5. Apply Algorithm 2.3.5 to $n = b$ and $c = \mathrm{Tr}(g^{e+k})$ to compute $S_b(\mathrm{Tr}(g^{e+k}))$.
6. Return $\mathrm{Tr}(g^{(e+k)b}) = \mathrm{Tr}(g^a \cdot g^{bk})$.

Theorem 5.2.8 ([10, Theorem 2.4.9]). *Given $M_0(\mathrm{Tr}(g))^{-1}$, $\mathrm{Tr}(g)$, and*

$$S_k(\mathrm{Tr}(g)) = (\mathrm{Tr}(g^{k-1}), \mathrm{Tr}(g^k), \mathrm{Tr}(g^{k+1}))$$

the trace $\mathrm{Tr}(g^a \cdot g^{bk})$ of $g^a \cdot g^{bk}$ can be computed at a cost of

$$8 \log_2(a/b \bmod q) + 8 \log_2(b) + 34$$

multiplications in $\mathrm{GF}(p)$.

Remark 5.2.9. Assuming that $M_0(\mathrm{Tr}(g))^{-1}$ is computed once and for all (at the cost of a small constant number of operations in $\mathrm{GF}(p^2)$), it follows that $\mathrm{Tr}(g^a \cdot g^{bk})$

can be computed at a cost of $16\log_2(q)+34$ multiplications in $\mathrm{GF}(p)$. This is still substantially faster than traditional computation of $g^a \cdot g^{bk}$ in $\mathrm{GF}(p^6)$ using multi-exponentiation, but the speed-up factor drops from 3 (for a single exponentiation) to about 1.75.

5.3. XTR-Nyberg–Rueppel signatures

This subsection contains a description of the XTR version of the Nyberg–Rueppel (NR) message recovery signature scheme. XTR can in a similar way be used in other 'ElGamal-like' signature schemes as illustrated in Subsection 5.4 below.

5.3.1. XTR-NR signature generation. To sign a message M containing an agreed upon type of redundancy using the XTR version of the NR protocol, Alice does the following:

1. Alice selects a random integer $u \in [2, q-3]$, and applies Algorithm 2.3.5 to $n = u$ and $c = \mathrm{Tr}(g)$ to compute
$$S_u(\mathrm{Tr}(g)) = (\mathrm{Tr}(g^{u-1}), \mathrm{Tr}(g^u), \mathrm{Tr}(g^{u+1})) \in \mathrm{GF}(p^2)^3.$$
2. Alice determines a symmetric encryption key K based on $\mathrm{Tr}(g^u) \in \mathrm{GF}(p^2)$.
3. Alice uses an agreed upon symmetric encryption method with key K to encrypt M, resulting in the encryption E.
4. Alice computes the (integer valued) hash h of E.
5. Alice computes $s = (k \cdot h + u) \bmod q \in \{0, 1, \ldots, q-1\}$.
6. Alice's resulting signature on M is (E, s).

5.3.2. XTR-NR signature verification. It is assumed that Alice's XTR public key is as described in 5.1.1 and thus contains $S_k(\mathrm{Tr}(g))$. To verify Alice's signature (E, s) and to recover the signed message M, verifier Bob does the following.

1. Bob checks that $0 \le s < q$; if not failure.
2. Bob computes the hash h of E.
3. Bob replaces h by $-h \bmod q \in \{0, 1, \ldots, q-1\}$.
4. Bob applies Algorithm 5.2.7 to $\mathrm{Tr}(g)$, $S_k(\mathrm{Tr}(g))$ (with k unknown to Bob), $a = s$, and $b = h$ to compute $\mathrm{Tr}(g^s \cdot g^{hk})$ (which equals $\mathrm{Tr}(g^u)$).
5. Bob determines a symmetric encryption key K based on $\mathrm{Tr}(g^s \cdot g^{hk}) \in \mathrm{GF}(p^2)$.
6. Bob uses the agreed upon symmetric encryption method with key K to decrypt E resulting in M.
7. The signature is accepted if and only if M contains the agreed upon redundancy.

XTR-NR signature generation 5.3.1 and verification 5.3.2 are both considerably faster than traditional implementations of the NR scheme that are based on subgroups of multiplicative groups of finite fields of the same security level: XTR-NR signature generation 5.3.1 is about three times faster than traditional NR signature generation, and XTR-NR signature verification 5.3.2 is about 1.75 faster than the traditional method (cf. Remark 5.2.9). The length of the signature is identical to other variants of the hybrid version of the NR scheme (cf. Remark 4.2.3): an overhead part of length depending on the desired security (i.e., the subgroup size) and a message part of length depending on the message itself and the agreed upon redundancy and symmetric encryption.

5.4. XTR-DSA signatures

This subsection contains a description of the XTR version of the DSA signature scheme. As in the original standard, [21], we assume that the size of q is 160 bits, the same size as the SHA-1 secure hash [20].

5.4.1. XTR-DSA signature generation. To sign a message M using the XTR version of DSA, Alice does the following:

1. Alice selects a random integer $u \in [2, q-3]$.
2. Alice applies Algorithm 2.3.5 to $n = u$ and $c = \mathrm{Tr}(g)$ to compute
$$S_u(\mathrm{Tr}(g)) = \big(\mathrm{Tr}(g^{u-1}), \mathrm{Tr}(g^u), \mathrm{Tr}(g^{u+1})\big) \in \mathrm{GF}(p^2)^3.$$
3. Alice writes $\mathrm{Tr}(g^u) = x_1\alpha + x_2\alpha^2$ and computes $r = (x_1 + p \cdot x_2) \bmod q$. If $r = 0$, then Alice goes back to Step 1.
4. Alice computes $u^{-1} \bmod q$.
5. Alice computes the hash h of M.
Comment. The secure hash function SHA-1 (cf. [20]) is used in the Digital Signature Standard (cf. [21]).
6. Alice computes $s = u^{-1}(h + k \cdot r) \bmod q$. If $s = 0$, then Alice goes back to Step 1.
7. Alice's resulting signature on M is (r, s).

5.4.2. XTR-DSA signature verification. It is assumed that Alice's XTR public key is as described in 5.1.1 and thus contains $S_k(\mathrm{Tr}(g))$. To verify Alice's signature (r, s) on message M, verifier Bob does the following.

1. Bob checks that $0 < r, s < q$; if not failure.
2. Bob computes $w = s^{-1} \bmod q$.
3. Bob computes the hash h of M.
4. Bob computes $u_1 = w \cdot h \bmod q$ and $u_2 = r \cdot w \bmod q$.
5. Bob applies Algorithm 5.2.7 to $\mathrm{Tr}(g)$, $S_k(\mathrm{Tr}(g))$ (with k unknown to Bob), $a = u_1$, and $b = u_2$ to compute $v_0 = \mathrm{Tr}(g^{u_1} \cdot g^{k \cdot u_2})$ (which equals $\mathrm{Tr}(g^u)$).
6. Bob writes $v_0 = z_1\alpha + z_2\alpha^2$ and computes $v = (z_1 + p \cdot z_2) \bmod q$.
7. Bob accepts the signature if and only if $v = r$.

Remark 5.4.3. Note that if (r,s) is a valid signature on message M, then so are $(r, s\cdot p^2 \bmod q)$ and $(r, s\cdot p^4 \bmod q)$. This is similar to the property of the ECC-DSA signature scheme (cf. [1]) that if (r,s) is a valid signature, then so is $(r, -s \bmod q)$. In applications where this may cause problems, uniqueness can be achieved by selecting the signature for which $s\cdot p^{2i} \bmod q$, with $i = 0,1,2$, is minimal. Of course, the verification should then check this as well.

5.5. Key size reduction

In this subsection it is shown that $\mathrm{Tr}(g^{k+1})$ and $\mathrm{Tr}(g^{k-1})$ can be derived from $\mathrm{Tr}(g)$ and $\mathrm{Tr}(g^k)$, if the private key k is properly chosen. Throughout this section let $c = \mathrm{Tr}(g)$ and $c_n = \mathrm{Tr}(g^n)$ for $n \in \mathbf{Z}$. First of all, c_{k-1} (or c_{k+1}) can be computed from c, c_k, and c_{k+1} (or c_{k-1}) in a small constant number of operations in $\mathrm{GF}(p)$. This follows from Theorem 5.5.1 and Algorithm 5.5.2 below.

Theorem 5.5.1 ([11, Theorem 5.1]).

1. *If* $k \neq p, 1-p \bmod p^2-p+1$ *then* $c^p c_{k-1} - c c_k \neq 0$ *and*

$$c_{k+1} = \frac{c_k^p(c^2-3c^p) - c_{k-1}^p(c^{2p}-3c) - c_{k-1}^2 c + c_k^2(c^p - c^2) + c_k c_{k-1} c^{p+1}}{c^p c_{k-1} - c c_k}.$$

2. *If* $k \neq -p, p-1 \bmod p^2-p+1$ *then* $c c_{k+1} - c^p c_k \neq 0$ *and*

$$c_{k-1} = \frac{c_k^p(c^{2p}-3c) - c_{k+1}^p(c^2-3c^p) - c_{k+1}^2 c^p + c_k^2(c - c^{2p}) + c_k c_{k+1} c^{p+1}}{c c_{k+1} - c^p c_k}.$$

Algorithm 5.5.2 (Inversion in $\mathrm{GF}(p^2)$ [11, Algorithm 5.2]).
Let $x = x_1\alpha + x_2\alpha^2 \in \mathrm{GF}(p^2)$. Compute $t = (x_1x_2 + (x_1-x_2)^2)^{-1} \in \mathrm{GF}(p)$, then $1/x = t(x_2\alpha + x_1\alpha^2) \in \mathrm{GF}(p^2)$.

It follows that under a mild restriction on the private key k, if c and c_k are part of the public key, then either c_{k+1} or c_{k-1} suffices to compute c_{k-1} or c_{k+1}, in order to reconstruct the value $S_k(\mathrm{Tr}(g))$ required for signature verification. It is not hard to see that neither c_{k+1} nor c_{k-1} must in principle be part of the public key: they can both be computed by multiplying or dividing the roots of $F(c,X)$ and $F(c_k,X)$, leading to 3 possible representations c_{k+1} (and c_{k-1}). Two bits in the public key suffice to indicate which of the representations is the correct one. In general, finding the roots of $F(c,X)$ and $F(c_k,X)$ requires a more substantial computation than is acceptable for the recipient of the public key. If $p \not\equiv 8 \bmod 9$, then this idea can be made to work much more efficiently, however, and this can be done in such a way that the two additional bits are not even required. This is shown in the remainder of this subsection. The description focuses on reconstructing c_{k+1} from c and c_k, but works in a very similar way for c_{k-1}.

Roughly speaking, Algorithm 3.3.1 is used to compute explicit representations of g and g^k in $\mathrm{GF}(p^6) = \mathrm{GF}(p)[X]/(X^6+X^3+1)$ (cf. Section 3.3) based on their representations c and c_k, respectively. The value of c_{k+1} then follows as the trace over $\mathrm{GF}(p^2)$ of $g \cdot g^k \in \mathrm{GF}(p^6)$, where k is chosen in such a way that c_{k+1} is 'minimal' in some sense. How this is done is shown first, in 5.5.3 below, after which the reconstruction of c_{k+1} is described in 5.5.4.

5.5.3. Selecting *k*. After selecting the private key k, its owner applies Algorithm 2.3.5 to $n = k$ and $c = \mathrm{Tr}(g)$ to compute $S_k(\mathrm{Tr}(g)) = (c_{k-1}, c_k, c_{k+1})$. Because g^k, g^{kp^2}, and g^{kp^4} are conjugates, the same c_k is obtained for $n = k$, $n = kp^2 \bmod q$, and $n = kp^4 \bmod q$. The side result c_{k+1} obtained from the computation of c_k, however, is in general not the same for $n = k$, $n = kp^2 \bmod q$, and $n = kp^4 \bmod q$, simply because $\mathrm{Tr}(g^{k+1})$, $\mathrm{Tr}(g^{kp^2+1})$, and $\mathrm{Tr}(g^{kp^4+1})$ are, in general, not the same. Namely, unless $k = 0 \bmod q$, the elements g^{k+1}, g^{kp^2+1}, and g^{kp^4+1} are not conjugates over $\mathrm{GF}(p^2)$, despite the fact that g^k, g^{kp^2}, and g^{kp^4} are conjugates over $\mathrm{GF}(p^2)$.

It follows that for any pair (c, c_k) there are in principle three different possible values for c_{k+1}: one that corresponds to the selected private key k, and two that correspond to the related but 'wrong' values $kp^2 \bmod q$ and $kp^4 \bmod q$. This ambiguity, which will have to be resolved by any method to recover c_{k+1} from (c, c_k), is avoided in the following simple manner: the owner of k computes all three values $\mathrm{Tr}(g^{k+1})$, $\mathrm{Tr}(g^{kp^2+1})$, and $\mathrm{Tr}(g^{kp^4+1})$ for the k of his choice, and next replaces k by k, $kp^2 \bmod q$, or $kp^4 \bmod q$ depending on which of $\mathrm{Tr}(g^{k+1})$, $\mathrm{Tr}(g^{kp^2+1})$, and $\mathrm{Tr}(g^{kp^4+1})$ is the 'smallest'. As a consequence, c_{k+1} is the 'smallest' possibility given the pair (c, c_k). The security is not affected by changing the initially selected k in this way. It remains to define what is meant by 'smallest' and how $\mathrm{Tr}(g^{k+1})$, $\mathrm{Tr}(g^{kp^2+1})$, and $\mathrm{Tr}(g^{kp^4+1})$ are computed.

For $x \in \mathrm{GF}(p)$ let $\pi_0(x) \in \{0, 1, \ldots, p-1\}$ be the image of x under the 'natural' bijection between $\mathrm{GF}(p)$ and $\{0, 1, \ldots, p-1\}$. For $x = x_1\alpha + x_2\alpha^2 \in \mathrm{GF}(p^2)$ let $\pi(x) = \pi_0(x_1) + p \cdot \pi_0(x_2)$ (cf. Subsection 2.2) be a bijection from $\mathrm{GF}(p^2)$ to $\{0, 1, \ldots, p^2 - 1\}$. The mapping π induces an ordering on $\mathrm{GF}(p^2)$ and 'smallest' is defined as smallest with respect to this ordering.

To compute $\mathrm{Tr}(g^{k+1})$, $\mathrm{Tr}(g^{kp^2+1})$, and $\mathrm{Tr}(g^{kp^4+1})$ the owner of k can simply apply Algorithm 2.3.5 three times, namely with $n = k$, $n = kp^2 \bmod q$, and $n = kp^4 \bmod q$, at a total cost of about $24 \log_2(q)$ multiplications in $\mathrm{GF}(p)$ (cf. Theorem 2.3.6). A conceptually more complicated method that saves about $8 \log_2(q)$ multiplications in $\mathrm{GF}(p)$ is as follows. Compute (c_{k-1}, c_k, c_{k+1}) and $(c_{-p-1}, c_{-p}, c_{-p+1})$, at the cost of $16 \log_2(q)$ multiplications in $\mathrm{GF}(p)$, followed by $c_{k\pm 2}$ and $c_2 = c^2 - 2c^p$ (cf. Corollary 2.3.3.(ii) and (i)). Use these values to compute $\mathrm{Tr}(g^{kp^2+1})$ by observing that

$$\mathrm{Tr}(g^{kp^2+1}) = \mathrm{Tr}(g^{kp^2-p^3}) = \mathrm{Tr}(g^{(k-p)p^2}) = \mathrm{Tr}(g^{k-p})$$

and

$$\begin{pmatrix} \mathrm{Tr}(g^{k-p-1}) \\ \mathrm{Tr}(g^{k-p}) \\ \mathrm{Tr}(g^{k-p+1}) \end{pmatrix}^T = \begin{pmatrix} c_{-p-1} \\ c_{-p} \\ c_{-p+1} \end{pmatrix}^T \begin{pmatrix} c_2^p & c^p & 3 \\ c^p & 3 & c \\ 3 & c & c_2 \end{pmatrix}^{-1} \begin{pmatrix} c_{k-2} & c_{k-1} & c_k \\ c_{k-1} & c_k & c_{k+1} \\ c_k & c_{k+1} & c_{k+2} \end{pmatrix}$$

(cf. Lemmas 5.2.2 and 5.2.5). This takes a small constant number of multiplications in GF(p). With

$$\mathrm{Tr}(g^{kp^2-1}) = \mathrm{Tr}(g^{k+p})$$

a similar matrix identity involving (c_{p-1}, c_p, c_{p+1}) (obtained using $c_{-n} = c_n^p$, cf. Lemma 2.3.2(ii)) is used to compute $\mathrm{Tr}(g^{kp^2-1})$. The same method is then used to compute $\mathrm{Tr}(g^{kp^4+1})$ based on $(\mathrm{Tr}(g^{kp^2-1}), \mathrm{Tr}(g^{kp^2}), \mathrm{Tr}(g^{kp^2+1}))$ (where $\mathrm{Tr}(g^{kp^2}) = c_k$) and $(c_{-p-1}, c_{-p}, c_{-p+1})$.

5.5.4. Reconstructing c_{k+1}. Given (c, c_k) with k chosen as explained in 5.5.3, the correct (i.e., the 'smallest') c_{k+1} is reconstructed by means of repeated application of Algorithm 3.3.1. To get Algorithm 3.3.1 to work for this specific application, two auxiliary algorithms are needed.

Algorithm 5.5.5 (Exponentiation in GF(p^2) [11, Algorithm 5.3]). To compute $x^e \in \mathrm{GF}(p^2)$ given $x \in \mathrm{GF}(p^2)$ and positive integer e, do the following.

1. Let $e_0, e_1 \in \{0, 1, \ldots, p-1\}$ be such that $e_0 + e_1 p = e \bmod (p^2 - 1)$ and let $e_i = \sum_j e_{ij} 2^j$, with $e_{ij} \in \{0, 1\}$ for $i = 0, 1$ and $j \geq 0$, be the binary representations of e_0 and e_1.
2. Let n be the largest index such that $e_{in} \neq 0$ for $i = 0$ or 1.
3. Compute $x^{p+1} = x \cdot x^p \in \mathrm{GF}(p)$.
4. Let $y = 1$ in GF(p^2). For $j = n, n-1, \ldots, 0$ in succession do the following:
   - if $e_{0j} = 1$ and $e_{ij} = 1$, then replace y by $y \cdot x^{p+1}$;
   - if $e_{0j} = 1$ and $e_{1j} = 0$, then replace y by $y \cdot x$;
   - if $e_{0j} = 0$ and $e_{1j} = 1$, then replace y by $y \cdot x^p$;
   - if $j > 0$, then replace y by y^2.

   Comment. Note that this is similar to multi-exponentiation.
5. Return $y = x^e \in \mathrm{GF}(p^2)$.

Lemma 5.5.6 ([11, Lemma 5.4]). *The expected cost of Algorithm* 5.5.5 *is* $4\log_2(p)$ *multiplications in* GF(p).

Algorithm 5.5.7 ((Cube root in GF(p^2) if $p \not\equiv 8 \bmod 9$ [11, Algorithm 5.5])). To compute a cube root in GF(p^6) of $r \in \mathrm{GF}(p^2)$ perform the following steps.

1. If $p \equiv 2 \bmod 9$, then let $e = \frac{8p^2-5}{9}$, otherwise, if $p \equiv 2 \bmod 9$ then let $e = \frac{p^2+2}{9}$.
2. Apply Algorithm 5.5.5 to $x = r$ and e to compute $t = r^e \in \mathrm{GF}(p^2)$.
3. Compute $s = t^3 \in \mathrm{GF}(p^2)$ and determine $j = 0, 1$ or 2 such that $\alpha^j s = r$.
4. Return a cube root $\zeta^j t \in \mathrm{GF}(p^6)$ of r, where ζ is as in Subsection 3.3.
   Comment. The result is in GF(p^2) if $j = 0$.

Algorithm 5.5.8 ((Key recovery [11, Algorithm 5.6])). To compute the 'smallest' c_{k+1} corresponding to (c, c_k), do the following.

1. Apply Algorithm 3.3.1 to $f(X) = F(c, X)$ to compute a single root $g \in \mathrm{GF}(p^6) = \mathrm{GF}(p)(\zeta)$ of $F(c, X)$.

Comment. The representation of elements of $\mathrm{GF}(p^6)$ is explained in Subsection 3.3. In Step 3 of Algorithm 3.3.1 a cube root is computed using Algorithm 5.5.7. As a result, u is a $\mathrm{GF}(p^2)$-multiple of a power of ζ, so that Algorithm 5.5.2 can be used for the division by u in the same step.

2. Apply Algorithm 3.3.1 to $f(X) = F(c_k, X)$ to compute all three roots $y_1, y_2, y_3 \in \mathrm{GF}(p^6) = \mathrm{GF}(p)(\zeta)$ of $F(c_k, X)$
   Comment. Same comments as above. Furthermore, $w = \alpha$ in Step 4 of Algorithm 3.3.1.
3. For $i = 1, 2, 3$ compute the trace t_i over $\mathrm{GF}(p^2)$ of $gy_i \in \mathrm{GF}(p^6)$.
   Comment. The trace over $\mathrm{GF}(p^2)$ of $\sum_{j=0}^{5} a_j\zeta^j \in \mathrm{GF}(p^6)$ equals

$$3(a_3 - a_0)\alpha - 3a_0\alpha^2 \in \mathrm{GF}(p^2)$$

   (cf. [11, Lemma 4.1]).
4. Let c_{k+1} be the 'smallest' of t_1, t_2, and t_3, under the ordering induced by π as in 5.5.3.

Theorem 5.5.9 ([11, Theorem 5.7]). *Algorithm* 5.5.8 *can be expected to require* $10.6\log_2(p)$ *multiplications in* $\mathrm{GF}(p)$.

5.5.10. Key recovery summary. It follows from 5.5.3 and 5.5.4 that $\mathrm{Tr}(g^{k-1})$ and $\mathrm{Tr}(g^{k+1})$ do not have to be included in the XTR public key data for digital signature or authentication applications, as long as

1. the private key k is selected as explained in 5.5.3 above,
2. p, q, $\mathrm{Tr}(g)$, and $\mathrm{Tr}(g^k)$ are included in the public key,
3. the recipient of the public key is willing and able to perform Algorithm 5.5.8 to compute $\mathrm{Tr}(g^{k+1})$ followed by an application of Theorem 5.5.1 to compute $\mathrm{Tr}(g^{k-1})$.

Actually, there are three options for XTR public keys used for digital signatures or authentication: include one, two, or all three of the values $\mathrm{Tr}(g^{k-1})$, $\mathrm{Tr}(g^k)$, $\mathrm{Tr}(g^{k+1})$. In some applications, e.g. issuance of a certificate by a Certificate Authority, it may be required that the relative correctness of these components can be verified by a third party. A method to do this is described in Remark 6.2.10 below.

6. Correctness verification of XTR data formats

6.1. Subgroup based attacks against XTR

Security of cryptographic protocols may be endangered if elements of a certain group are exchanged but group membership is not properly verified by the recipient. Examples of such attacks can be found in [2, 4, 12, 14]. In XTR, subgroup attacks refer to attacks that take advantage of the omission to verify membership of the XTR (sub)group (cf. Subsection 2.1). As argued in [12], subgroup attacks can be rendered mostly ineffective if either

1. p and q are such that $(p^2 - p + 1)/q$ is small (cf. Algorithm 3.1.3), or
2. p and q are such that $(p^2 - p + 1)/q$ is a small multiple of a prime of the same order of magnitude as q (cf. Algorithm 3.1.4).

In either case, however, membership of the XTR supergroup, i.e., the order p^2-p+1 group containing the XTR group (cf. Subsection 2.1), still must be verified. In Subsection 6.2 below it is shown how that can be done. Note that if $(p^2-p+1)/q$ is small, short exponents may be used in the XTR versions of cryptographic protocols to maintain efficiency, as mentioned in the introduction to Section 4 (cf. [23]).

6.2. Prevention of subgroup attacks against XTR

Let G denote the order q XTR group and H the order p^2-p+1 XTR supergroup, as defined in Subsection 2.1. This section contains efficient methods to determine if an element $d \in \mathrm{GF}(p^2) \setminus \mathrm{GF}(p)$ is the trace of an element of H. As explained in Subsection 6.1 such methods are useful to prevent XTR subgroup attacks.

It follows from Lemma 2.3.2.(v) that it suffices to check that $F(d, X)$ is irreducible. According to Theorem 3.5.5 this can be done at the cost of $1.8 \log_2(p)$ plus a small constant number of multiplications in $\mathrm{GF}(p)$. Thus, checking membership of H can be done at the cost of a small overhead compared to the cost of the regular XTR cryptographic operations. As shown in the remainder of this section, the overhead can be reduced to just a small constant number of operations in $\mathrm{GF}(p)$, at the cost however of a small amount of additional communication: if d, the element to be checked, equals $\mathrm{Tr}(h)$ for $h \in H$, and $\mathrm{Tr}(h \cdot g)$ is sent along with d, then the fact that d is indeed the trace of an element of H can be verified in a small constant number of operations in $\mathrm{GF}(p)$ (cf. Corollary 6.2.8). Here it is assumed that the trace $\mathrm{Tr}(g)$ of an element $g \in H$ is known.

Definition 6.2.1 ([12, Definition 5.1 and Lemma 5.2]). For third-degree monic polynomials

$$R(X) = \prod_{i=0}^{2}(X - \alpha_i) \in \mathrm{GF}(p^2)[X] \text{ and } S(X) = \prod_{j=0}^{2}(X - \beta_j) \in \mathrm{GF}(p^2)[X]$$

with $\alpha_i, \beta_j \in \mathrm{GF}(p^6)^*$ for $0 \le i, j < 3$, the *root-product* $\Re(R, S)$ is the ninth-degree polynomial

$$\Re(R, S) = \prod_{i,j=0}^{2} (X - \alpha_i\beta_j) \in \mathrm{GF}(p^2)[X]$$

with non-zero constant term.

Lemma 6.2.2 ([12, Lemma 5.3]). *With $R(X)$ and $S(X)$ as in Definition* 6.2.1,

$$\Re(R, S) = (\beta_0 \cdot \beta_1 \cdot \beta_2)^3 R(X \cdot \beta_0^{-1}) \cdot R(X \cdot \beta_1^{-1}) \cdot R(X \cdot \beta_2^{-1}),$$

and if $S(X)$ is irreducible over $\mathrm{GF}(p^2)$ *then*

$$\Re(R,S) = \beta_0^{3(p^4+p^2+1)} R(X\cdot\beta_0^{-1})\cdot R(X\cdot\beta_0^{-p^2})\cdot R(X\cdot\beta_0^{-p^4}).$$

In the application of Lemma 6.2.2 the polynomial $F(c,X)$ plays the role of $S(X)$, where $c = \mathrm{Tr}(g)$ for some $g \in H$, so that $F(c,X) = S(X)$ is irreducible (cf. Lemma 2.3.2.(v)). This implies that $\mathrm{GF}(p^6)$ can be represented as $\mathrm{GF}(p^2)(g)$, and

$$\Re(R,F(c,X)) = R(X/g)\cdot R(X/g^{p-1})\cdot R(X/g^{-p})$$

(since $g^{3(p^4+p^2+1)} = 1$, $g^{p^2} = g^{p-1}$, and $g^{p^4} = g^{-p}$) can be computed using a constant number of operations in $\mathrm{GF}(p^6) = \mathrm{GF}(p^2)(g)$, if a representation of $g^p \in \mathrm{GF}(p^2)(g)$ is known. The following result shows how such a representation can be obtained. As before, $\mathrm{Tr}(g^i)$ is abbreviated to c_i.

Proposition 6.2.3 ([12, Proposition 5.4 and Corollary 5.5]). *Let $c = \mathrm{Tr}(g)$ for some $g \in H$ and let $c_{p-2} = \mathrm{Tr}(g^{p-2}) \in \mathrm{GF}(p^2)$ be given. Then $g^p = Kg^2 + Lg + M \in \mathrm{GF}(p^2)(g)$ can be computed in a small constant number of multiplications in* $\mathrm{GF}(p)$ *using*

$$\begin{pmatrix} M \\ L \\ K \end{pmatrix} = \begin{pmatrix} c_{-2} & c_{-1} & c_0 \\ c_{-1} & c_0 & c_1 \\ c_0 & c_1 & c_2 \end{pmatrix}^{-1} \cdot \begin{pmatrix} c_{p-2} \\ c \\ c^p \end{pmatrix},$$

where the inverse of the matrix on the right hand side exists (cf. Lemma 5.2.5) *and is given in Lemma* 5.2.4.

Theorem 6.2.4 ([12, Theorem 5.6]). *Let $R(X) \in \mathrm{GF}(p^2)[X]$ be a monic third-degree polynomial with non-zero constant term and let $c = \mathrm{Tr}(g)$ for some element $g \in H$. Given $\mathrm{Tr}(g^{p-2}) \in \mathrm{GF}(p^2)$, the root-product $\Re(R(X), F(c,X))$ can be computed at the cost of a small constant number of operations in* $\mathrm{GF}(p)$.

Proof. As argued above, this follows from Lemma 6.2.2 and Proposition 6.2.3. □

The value c_{p-2} plays an important role, so it could be precomputed and stored, independent of the value d to be checked. Note that $c_{p+1} = c\cdot c_p - c^p\cdot c_{p-1} + c_{p-2}$, $c_p = c^p$, and $c_{p-1} = c$, so that $c_{p-2} = c_{p+1}$. The following results shows that c_{p-2} can quickly be recovered from a single bit.

Proposition 6.2.5 ([12, Proposition 5.7]). *Let $c = \mathrm{Tr}(g)$ for some element $g \in H$. Then $\mathrm{Tr}(g^{p-2}) = c_{p-2}$ can be computed at the cost of a square-root computation in $\mathrm{GF}(p^2)$, assuming one bit of information to resolve the square-root ambiguity.*

Proof. With $c_{p-2} = x_1\alpha + x_2\alpha^2$ it simply follows that

$$(c_{p-2} - c_{p-2}^p)^2 = -3(x_1 - x_2)^2.$$

Combination with $c_{p-2} = c_{p+1}$, the identity for $(c_{p+1} - c_{p+1}^p)^2$ given in Lemmas 5.2.4 and 5.2.5, leads to

$$-3(x_1 - x_2)^2 = c^{2p+2} + 18c^{p+1} - 4(c^{3p} + c^3) - 27 \in \mathrm{GF}(p).$$

On the other hand $c_{p-2} + c_{p-2}^p = -(x_1 + x_2)$. With $c_{p-2} = g^{p-2} + g^{(p-2)p^2} + g^{(p-2)p^4} = g^{p-2} + g^{-2p+1} + g^{p+1}$, it follows that $c_{p-2}^p = g^{-p-1} + g^{-p+2} + g^{2p-1}$. Now,

$$\begin{aligned} c^{p+1} &= c \cdot c^p = (g + g^{p-1} + g^{-p})(g^p + g^{-1} + g^{-p+1}) \\ &= g^{p+1} + g^{p-2} + g^{-2p+1} + g^{-p-1} + g^{-p+2} + g^{2p-1} + 3 \\ &= c_{p-2} + c_{p-2}^p + 3. \end{aligned}$$

Thus, $x_1 + x_2 = 3 - c^{p+1} \in \mathrm{GF}(p)$. Combining the identities involving $x_1 - x_2$ and $x_1 + x_2$ it follows that c_{p-2} and its conjugate over $\mathrm{GF}(p)$ can be computed at the cost of a square-root calculation in $\mathrm{GF}(p^2)$. To distinguish $c_{p-2} = x_1\alpha + x_2\alpha^2$ from its conjugate $x_2\alpha + x_1\alpha^2$ over $\mathrm{GF}(p)$ a single bit that is on if and only if $x_1 > x_2$ suffices. □

Algorithm 6.2.6 ([12, Proof of Lemma 5.8]). Let $c = \mathrm{Tr}(g)$ for some element $g \in H$ and let $\mathrm{Tr}(g^{p-2}) = c_{p-2}$ be given. Given $d, d' \in \mathrm{GF}(p^2)$, to check if there exists an element $h \in H$ such that $d = \mathrm{Tr}(h)$ and $d' = \mathrm{Tr}(h \cdot g)$, do the following.

1. If $F(d, \alpha^i) = 0$ for either $i = 0$, 1, or 2, then the statement is not true.
2. Otherwise, if $F(d, \alpha^i) \neq 0$ for $i = 0, 1, 2$, then compute the root-product $\Re(F(d, X), F(c, X))$.
3. If $\Re(F(d, X), F(c, X))$ is divisible by $F(d', X)$, then the statement is true, otherwise it is false.

Lemma 6.2.7 ([12, Lemma 5.8]). *Algorithm* 6.2.6 *takes a small constant number of operations in* $\mathrm{GF}(p)$.

Corollary 6.2.8 ([12, Corollary 5.9]). *Let* $c = \mathrm{Tr}(g)$ *for some element* $g \in H$ *and let* $\mathrm{Tr}(g^{p-2})$ *be given. Given the trace values of an alleged element* $h \in H$ *and its 'successor'* $g \cdot h$, *it takes a small constant number of operations in* $\mathrm{GF}(p)$ *to verify that indeed* h *in* H.

Corollary 6.2.9 ([12, Corollary 5.10]). *Let* $c = \mathrm{Tr}(g)$ *where* g *is known to be a generator of the XTR group, let* d *be the trace of an element that is known to be in the XTR group* $\langle g \rangle$, *and let* d' *be some element of* $\mathrm{GF}(p^2)$. *Then it can efficiently be verified if* d *and* d' *are of the form* $\mathrm{Tr}(g^x)$ *and* $\mathrm{Tr}(g^{x+1})$, *respectively, for some integer* x, $0 < x < q$.

Remark 6.2.10. An XTR public key meant for digital signatures takes the form (p, q, c, d, d'), where p and q are primes satisfying the usual XTR conditions, $c = \mathrm{Tr}(g)$ for a generator g of the XTR group, $d = \mathrm{Tr}(g^k)$ for a secret key k, and $d' = \mathrm{Tr}(g^{k+1})$ (cf. Section 5). Corollary 6.2.9 implies that a Certificate Authority can efficiently verify the consistency of an XTR signature public key presented by a client, before issuing a certificate on it. More specifically, if a client provides a Certificate Authority with XTR public key data (p, q, c, d, d'), then the Certificate Authority checks that these data satisfy the conditions given above, using the following two step approach. First the Certificate Authority checks that p and q are well-formed and that $c, d \in \mathrm{GF}(p^2) \setminus \mathrm{GF}(p)$ are indeed traces of elements of the XTR group by verifying that $c_q = d_q = 3$ using standard XTR arithmetic (cf. Algorithm 2.3.5). Secondly, the Certificate Authority uses Corollary 6.2.9 to verify that d and d' are traces of consecutive powers of the generator corresponding to c. Note that the Certificate Authority does not obtain information about the secret key k.

7. Security of XTR

7.1. Security of the trace representation

In the XTR versions of 'subgroup discrete logarithm' based cryptographic protocols the subgroup elements are replaced by their traces. This implies that the security is no longer based on the regular and well known subgroup Discrete Logarithm (DL), Diffie–Hellman (DH), or Decision Diffie–Hellman (DDH) problems, but on their XTR counterparts. The *XTR-DH* problem is the problem of computing $\mathrm{Tr}(g^{xy})$ given $\mathrm{Tr}(g^x)$ and $\mathrm{Tr}(g^y)$. Given $\mathrm{Tr}(g^x)$ and $\mathrm{Tr}(g^y)$, the XTR-Diffie–Hellman value $\mathrm{Tr}(g^{xy})$ is denoted by $XDH(\mathrm{Tr}(g^x), \mathrm{Tr}(g^y))$. The *XTR-DDH* problem is the problem of determining whether $XDH(a, b) = c$ given $a, b, c \in \mathrm{Tr}(\langle g \rangle)$. The *XTR-DL* problem is to find $0 \leq x < q$ such that $a = \mathrm{Tr}(g^x)$ given $a \in \mathrm{Tr}(\langle g \rangle)$. Note that if x satisfies $a = \mathrm{Tr}(g^x)$, then so do $x \cdot p^2 \bmod q$ and $x \cdot p^4 \bmod q$.

Theorem 7.1.1 ([10, Theorem 5.2.1]). *The following equivalences hold:*

1. *The XTR-DL problem is $(1,1)$-equivalent to the DL problem in $\langle g \rangle$,*
2. *The XTR-DH problem is $(1,2)$-equivalent to the DH problem in $\langle g \rangle$,*
3. *The XTR-DDH problem is $(3,2)$-equivalent to the DDH problem in $\langle g \rangle$,*

*where $\mathcal{A}$ is (a,b)-*equivalent *to $\mathcal{B}$, if any instance of $\mathcal{A}$ (or $\mathcal{B}$) can be solved by at most a (or b) calls to an algorithm solving $\mathcal{B}$ (or $\mathcal{A}$).*

Remark 7.1.2. An algorithm solving DL, DH, or DDH with non-negligible probability can be transformed in an algorithm solving the corresponding XTR problem with non-negligible probability, and vice versa (cf. [10, Proof of Theorem 5.2.1]).

Despite the fact that, according to Theorem 7.1.1, XTR-DH is $(1,2)$-equivalent to ordinary DH, in many practical situations a single call to an XTR-DH solving algorithm would suffice to solve a DH problem. An example is DH key agreement where the resulting key is actually used after it has been established.

Remark 7.1.3. Theorem 7.1.1.2 states that determining the (small) XTR-DH key is as hard as determining the whole DH key in the representation group $\langle g \rangle$. From the results in [27] it actually follows that determining the image of the XTR-DH key under any non-trivial $\mathrm{GF}(p)$-linear function is also as hard as the whole DH key. This means that, for example, finding the coefficient of α or α^2 of the XTR-DH key is as hard as finding the whole DH key, implying that cryptographic applications may be based on just one of the coefficients. Note that in 4.1.1 both coefficients are used.

7.2. Relation between the XTR group and supersingular elliptic curves

The number of points over $\mathrm{GF}(p^2)$ (including the point at infinity) on an elliptic curve over $\mathrm{GF}(p^2)$ takes the form p^2-t+1 for some integer $-2p \leq t \leq 2p$. It is well known that there exist so-called supersingular elliptic curves over $\mathrm{GF}(p^2)$ where this order is equal to p^2-p+1 and that there exist efficiently computable (i.e., in polynomial time and space in the input length), injective homomorphisms based on the Weil pairing, of such curves onto the XTR supergroup. Such homomorphisms are known as MOV embeddings. See for instance [15] for further reference.

At the Crypto 2000 rump session (cf. [16]) it was suggested that the inverses of these homomorphisms might be efficiently computable too, and it was mentioned that this would imply that the XTR (sub)group is just an instance of the (sub)group of a supersingular elliptic curve. Thus, an attack affecting elliptic curve cryptosystems would affect XTR-based cryptosystems as well, implying that the security of XTR cryptosystems would not be not better than that of elliptic curve cryptosystems.

More precisely, the suggestion made at the Crypto 2000 rump session can be formulated as the following assumption:

X2C One can efficiently find a supersingular elliptic curve over $\mathrm{GF}(p^2)$, such that the group of points C over $\mathrm{GF}(p^2)$ (including the point at infinity) is of order p^2-p+1 and an efficiently computable, injective homomorphism from the XTR subgroup into C.

A similar problem is posed by N. Koblitz in [8, Remark on page 328]. It was shown in [28], however, that the suggested assumption is most likely false, because it would contradict several generally accepted hardness assumptions. The following is one of the results shown in [28].

Theorem 7.2.1. *Under the* **X2C** *assumption, the following problems are efficiently computable*:

1. *The Diffie–Hellman problem in the XTR subgroup.*
2. *The Diffie–Hellman problem in the group of points of order q on a supersingular elliptic curve over $\mathrm{GF}(p^2)$ of order p^2-p+1.*

This result gives evidence that the security provided by the XTR subgroup is better than that provided by the isomorphic group on supersingular elliptic curves. Additional evidence is provided by the fact that the Decision Diffie–Hellman problem is efficiently computable in the latter group, while this problem is believed to be hard in the XTR subgroup. See [28].

Note added in proof. The XTR exponentiation methods as presented in this paper can be improved considerably. The basis for the improvement is an alternative version of the XTR double exponentiation method from Algorithm 5.2.7 that is not only twice as fast on average, but that also does not require the matrix methods as introduced in Subsection 5.2. This improved version of Algorithm 5.2.7

can be used to formulate a new XTR single exponentiation method, improving Algorithm 2.3.5 of the present paper by more than 16%, or by more than 50% if a small one time precomputation is performed. As a result XTR signature applications become 50% faster and all other applications become significantly faster as well. For the details refer to [26].

References

[1] American National Standards Institute (ANSI) X9.62-1998, Public Key Cryptography for the Financial Services Industry: The Elliptic Curve Digital Signature Algorithm (ECDSA). 1998.

[2] Biehl, I., Meyer, B., Müller, V., Differential fault attacks on elliptic curve cryptosystems. In: Advances in Cryptology – CRYPTO 2000 (ed. by M. Bellare; Lecture Notes in Comput. Sci. 1880), 131–146. Springer, Berlin 2000.

[3] Brouwer, A.E., Pellikaan, R., Verheul, E.R., Doing more with fewer bits. In: Advances in Cryptology – ASIACRYPT '99 (ed. by K.Y. Lam, E. Okamoto, C. Xing; Lecture Notes in Comput. Sci. 1716), 321–332. Springer, Berlin 1999.

[4] Burmester, M.V.D., A remark on the efficiency of identification schemes. In: Advances in Cryptology – EUROCRYPT '90 (ed. by I.B. Damgård; Lecture Notes in Comput. Sci. 473), 493–495. Springer, Berlin 1991.

[5] Cohen, H., Miyaji, A., Ono, T., Efficient elliptic curve exponentiation using mixed coordinates. In: Advances in Cryptology – ASIACRYPT '98 (ed. by K. Ohta and D.Y. Pei; Lecture Notes in Comput. Sci. 1514), 51–65. Springer, Berlin 1998.

[6] ElGamal, T., A public key cryptosystem and a signature scheme based on discrete logarithms. IEEE Trans. Inform. Theory 31 (1985), 469–472.

[7] Gong, G., Harn, L., Public key cryptosystems based on cubic finite field extensions. IEEE Trans. Inform. Theory 45 (1999), 2601–2605.

[8] Koblitz, N., An elliptic curve implementation of the finite field digital signature algorithm. In: Advances in Cryptology – CRYPTO '98 (ed. by H. Krawczyk; Lecture Notes in Comput. Sci. 1462), 327–337. Springer, Berlin 1998.

[9] Lenstra, A.K., Using cyclotomic polynomials to construct efficient discrete logarithm cryptosystems over finite fields. In: Information Security and Privacy (ACISP '97) (ed. by V. Varadharajan, J. Pieprzyk, Y. Mu; Lecture Notes in Comput. Sci. 1270), 127–138. Springer, Berlin 1997.

[10] Lenstra, A.K., Verheul, E.R., The XTR public key system. In: Advances in Cryptology – CRYPTO 2000 (ed. by M. Bellare; Lecture Notes in Comput. Sci. 1880), 1–19. Springer, Berlin 2000; available from `www.ecstr.com`.

[11] —, —, Key improvements to XTR. In: Advances in Cryptology – ASIACRYPT 2000 (ed. by T. Okamoto; Lecture Notes in Comput. Sci. 1976), 220–233. Springer, Berlin 2000; available from `www.ecstr.com`.

[12] —, —, Fast irreducibility and subgroup membership testing in XTR. In: Public Key Cryptography (PKC 2001, Cheju Island, Korea) (ed. by K. Kim; Lecture Notes in Comput. Sci. 1992). Springer, Berlin 2001; available from `www.ecstr.com`.

[13] Lidl, R., Müller, W.B., Permutation polynomials in RSA-cryptosystems. In: Advances in Cryptology (CRYPTO '83) (ed. by D. Chaum), 293–301. Plenum Press, New York 1984.

[14] Lim, C.H., Lee, P.J., A key recovery attack on discrete log-based schemes using a prime order subgroup. In: Advances in Cryptology – CRYPTO '97 (ed. by B.S. Kaliski; Lecture Notes in Comput. Sci. 1294), 249–263. Springer, Berlin 1997.

[15] Menezes, A., Elliptic Curve Public Key Cryptosystems. Kluwer Academic Publishers, Boston 1993.

[16] Menezes, A., Vanstone, S., ECSTR (XTR): Elliptic Curve Singular Trace Representation. Rump session of CRYPTO 2000.

[17] Müller, W.B., Polynomial functions in modern cryptology. In: Contributions to General Algebra 3 (Vienna, 1984) (ed. by G. Eigenthaler et al.), 7–32. Hölder-Pichler-Tempsky, Vienna 1985.

[18] Müller, W.B., Nöbauer, R., Cryptanalysis of the Dickson-scheme. In: Advances in Cryptology – EUROCRYPT '85 (ed. by F. Pichler; Lecture Notes in Comput. Sci. 219), 50–61. Springer, Berlin 1986.

[19] Nicholson, W.K., Introduction to Abstract Algebra. PWS–Kent Publishing Company, Boston 1993.

[20] National Inst. of Standards and Technology, Secure Hash Standard (SHA-1). Federal Information Processing Standards Publication 180–1, 1995; available from `www.itl.nist.gov/fipspubs/fib180-1.htm`.

[21] —, Digital Signature Standard. Federal Information Processing Standards Publication 186, 1994.

[22] Nöbauer, R., Cryptanalysis of the Rédei scheme. In: Contributions to General Algebra 3 (Vienna, 1984) (ed. by G. Eigenthaler et al.), 255–264. Hölder-Pichler-Tempsky, Vienna 1985.

[23] van Oorschot, P.C.., Wiener, M.J., On Diffie-Hellman key agreement with short exponents. In: Advances in Cryptology – EUROCRYPT '96 (ed. by U. Maurer; Lecture Notes in Comput. Sci. 1070), 332–343. Springer, Berlin 1996.

[24] Schnorr, C.P., Efficient signature generation by smart cards. J. Cryptology 4 (1991), 161–174.

[25] Smith, P., Skinner, C., A public-key cryptosystem and a digital signature system based on the Lucas function analogue to discrete logarithms. In: Advances in Cryptology – ASIACRYPT '94 (ed. by J. Pieprzyk, R. Safavi-Naini; Lecture Notes in Comput. Sci. 917), 357–364. Springer, Berlin 1995.

[26] Stam, M., Lenstra, A.K., Speeding up XTR, submitted for publication; available from `www.ecstr.com`.

[27] Verheul, E., Certificates of recoverability with scalable recovery agent security. In: Public Key Cryptography (PKC 2000, Melbourne) (ed. by H. Imai, Y. Zhang; Lecture Notes in Comput. Sci. 1751), 258–275. Springer, Berlin 2000.

[28] —, Evidence that XTR is more secure than supersingular elliptic curve cryptosystems. In: Advances in Cryptology – EUROCRYPT 2001 (ed. by B. Pfitzmann; Lecture Notes in Comput. Sci. 2045), 195–210. Springer, Berlin 2001.

A survey of IND-CCA secure public-key encryption schemes relative to factoring

*Siguna Müller**

1. Introduction and summary

The most basic goal of encryption is to ensure that an adversary does not learn any useful information from given ciphertexts. In terms of a formalization this is usually captured by the notion of *indistinguishability* or *semantic security* [12]. On the other hand one considers any adversary under specific attack models, the strongest being the (*adaptive*) *chosen ciphertext attack* (CCA), in which it is given the ability to obtain plaintexts of ciphertexts of its choice. The combination of these security requirements leads to one of the strongest notions of security possible, known as IND-CCA (resp. IND-CCA2, when the 'adaptive' nature of the attack is stressed, [28]). For simplicity we will write IND-CCA to denote security against adaptive CCA. The strongest form of this notion informally captures the idea that the adversary does not learn anything about the challenge ciphertext even if it is allowed to query the decryption oracle with any ciphertext different from the challenge itself.

A commonly used security proving mechanism typically is based on specific hard problems (such as the Diffie–Hellman (DH) assumption or the factorization problem), which are widely believed to be difficult. If any attack on the system were possible then this should provably lead to the breaking of the underlying primitive.

Unfortunately for most of the public key encryption schemes it is not known whether breaking the scheme is indeed as difficult as breaking the problem it is based on. For instance, although RSA is based on the factorization intractability of large integers it is not known whether inverting RSA is computationally as difficult as factoring the modulus. Special cases have been found [11] for which inverting RSA is easier than factoring. On the contrary, it is known that obtaining the full plaintext of an ElGamal encrypted ciphertext is equivalent to solving the DH problem. Surprisingly, even the semantic security of ElGamal encryption (under a chosen plaintext attack, CPA [2]) is provably equivalent to the decisional DH problem.

In establishing security in the sense of IND-CCA, it seems to be easier to start off from systems based on ElGamal. In particular, when based on the decisional

* Supported by the Austrian Science Fund (FWF), P 14472-MAT.

Public-Key Cryptography and Computational Number Theory
K. Alster, J. Urbanowicz and H. C. Williams (Eds.)

DH problem, such proposals intrinsically already provide the exact same tools as required in the typical game play for CCA security. The idea is that when an adversary in the sense of CCA manages to determine the correct bit $b = 0$ or 1, when given the encryption of m_b, then this attacker can be converted to a statistical test for solving the underlying decisional DH problem (cf. e.g. [8]).

As ElGamal is by itself a probabilistic scheme, it seems difficult to generalize these methods to other primitives. Several suggestions have been made (cf. [23, 24]) based on other intractability assumptions, part of which are very specific and quite new (e.g., the higher residuosity problem), and rather strong (e.g., the decisional dependent RSA assumption, [25]).

However, not too many realizations based on the factorization primitive are known. In spite of efficient factorization techniques [31, 26], this problem still is considered to be difficult and yields a useful basis for various applications.

1.1. Previous schemes IND-CCA secure relative to factoring

The first suggestion based on the factorization problem $n = pq$ was made in [16], but the results can only be shown in the random oracle model (ROM) (this is due to some misunderstanding of an earlier version of [13]). Additionally, it requires that $p \equiv q \equiv 3 \pmod 4$. Being the first proposal of this kind, the method was merely of theoretical interest and is not very efficient.

A system based on the general factorization problem for arbitrary p and q is presented in [17] in terms of the Williams scheme. It is the only suggestion known without the ROM. This strong requirement necessitates rather involved design of the scheme with less practicality than the counterparts in the ROM.

Recently, by utilizing random oracles, a suggestion has been made in [15]. The underlying Kurosawa et al. scheme [14] is a special case of the Williams system ($e = 1$ of the Williams system), and certainly for this specific setting allows some simplification, as opposed to the general case for arbitrary exponents e. In [15] two strengthening methods for obtaining security against IND-CCA are proposed. The suggestion for short messages x essentially is the general OAEP scheme [4] introduced by Bellare–Rogaway for random oracles G, H,

$$f(x0^{k_1} \oplus G(r) \parallel r \oplus H(x0^{k_1} \oplus G(r))), \tag{1}$$

which they showed can be applied to any one-way permutation f. Kurosawa et al. proved that this requirement is not necessary and the analogous results can also be obtained for Rabin-type schemes (in particular, their scheme), which are not permutations but yield a $1 : k$ ambiguity in decryption, where in their case $k = 4$. As usual, unambiguity in decryption is achieved by two auxiliary bits in the ciphertext.

Unfortunately, it was very recently shown [30] that there are some serious flaws in the underlying proof of OAEP [4]. (Shoup also presents some correction, OAEP+.) In detail, the original proof of security in the sense of IND-CCA2 given for OAEP in [4] is not sufficient for obtaining the claimed result. Consequently, as any general scheme based on original OAEP, the Kurosawa et al. proposal is not provably secure against CCA. The gap in the proof stems from the fact that it is not sufficient to consider only oracle queries to both the random oracles G and H. Indeed, Shoup [30] described some classes of one-way trapdoor permutations for which an attack without any G-queries at all is shown to be successful. In some cases, the claimed security of OAEP can be obtained, when relying on stronger primitives such as partially trapdoor functions. Specifically, then the RSA based OAEP instantiation is shown to be secure [11]. However, it is stressed in [11] that the new security proof heavily relies on the random self-reducibility of RSA and cannot automatically be applied to other underlying functions (such as the ones in [15]). It is not known whether the proof of the Kurosawa et al. scheme for short messages [15] can be fixed by the methods in [11]. Certainly, any correction can be based on OAEP+ in an obvious way.

In [15], for $Y(r)$ the Kurosawa et al. encryption f, the suggestion due to [3],

$$x \oplus G(r) \,\|\, Y(r) \,\|\, H(x \,\|\, r) \tag{2}$$

is made for long messages x, but no proof of security is given in [15].

Pointcheval [24] showed that

$$g^r h^{H(m \,\|\, s)} \,\|\, G(r) \oplus (m \,\|\, s) \tag{3}$$

is secure in the sense of IND-CCA for random oracles G, H. The main disadvantage with this proposal is that it is based on the OU-encryption method [21] and as such requires the factorization primitive of the special numbers $n = p^2 q$.

1.2. Motivation for this paper

Originally, the motivation for this paper was to construct an efficient method which is IND-CCA secure relative to the general factorization primitive. We developed a scheme based on a proposal of Zheng–Seberry [36] and Baek et al. [1],

$$g^{H(m \,\|\, s)} \,\|\, G(g^{xH(m \,\|\, s)}) \oplus (m \,\|\, s) \tag{4}$$

which is known as the most efficient length-saving method [1] for security against CCA on the computational DH assumption.

The present research, independently driven from the Fujisaki and Okamoto [10] work, reaches a similar result as in [10], but for the factorization problem. Generally speaking, our system

$$E(s) \,\|\, G(s) \oplus (m \,\|\, r), \tag{5}$$

for $s = r \,\|\, H(m \,\|\, r)$ can be seen as a special instantiation of the hybrid scheme

$$\mathcal{E}^{\text{asym}}_{pk}(\sigma; H(\sigma, m)) \,\|\, \mathcal{E}^{\text{sym}}_{G(\sigma)}(m) \tag{6}$$

of [10] [1)]. Indeed, it seems as if a number of systems in this context have a structure like this, with the decrypted part of the first block yielding the secret quantity (resp., key), s, for the one-time padding via the random oracle function G in the

1) The author is grateful to the anonymous referee for this observation.

right block (compare (3), (4)). Thus, our proposed system is not entirely new. However, thanks to the specificities of our concrete system on special and suitable underlying functions, this has several advantages.

1.3. Specific advantages of the proposed system

- Instead of applying the general conversion method of [10], we establish the proof of security on the particular case of our underlying scheme, and thus, do not require notions like γ-uniformity, or security notions relative to symmetric encryption, but only the typical primitives in the public-key scenario. The proof of security was independently obtained from the one in [10] and for our case, is much shorter and yields more specific results than the ones for the general method in [10].

- The proposed system is not geared at a general method, but only at the factorization primitive. Our proof of security generalizes the one of [24] but allows a major relaxation of the underlying requirements.

  Pointcheval [24] needs as primitives partially trapdoor one-way functions f. Informally, this means that, when given $z = f(x, y)$, it is computationally infeasible to find a partial preimage x of z. We require only the *uninvertibility relative to factoring with respect to the entire preimage*. If s is a bit string depending on x and y, then, given $z = f(s) = f(x, y)$ it should be impossible to obtain the entire string s. This requirement is weaker, since, for example, if $s = x \| y$, then for some function f it might be possible that x leaks from $f(x \| y)$, whereas $x \| y$ does not leak in full from $f(x \| y)$. In that case such an f might be applicable to our setting, but not to Pointcheval's. As another example, RSA-based OAEP can be proven secure based on the stronger requirement of partial trapdoor one-wayness, but OAEP based on any trapdoor function is not secure [11, 30].

- Although Pointcheval's general suggestion [24] is low-cost for encryption, it appears to be heavy for decryption (the decryption process includes a complete re-encryption with the underlying encryption routine). This additional re-encryption will not be necessary in the proposed scheme.

- We achieve low message expansion, also for long messages. Our scheme yields an even more length-saving alternative than the (unproven) suggestion in [15] for long messages.

1.4. Comparison with other systems and extensions

The following table gives a short overview over all the existing proposals based on the factorization problem for which security in the sense of IND-CCA in the ROM is claimed. The first column contains the Kurosawa et al. system [15] based on OAEP, which was recently shown to be insecure [30]. However, the proposal of [15] can be validated secure on the basis of OAEP+. This describes the second column.

The third one considers Pointcheval's instantiation based on the OU-scheme [24], and the fourth one the proposed system for long messages in [15] which is based on Bellare and Rogaway's suggestion [4]. The symbol † indicates that the proof of security for this proposal is not given yet.

Here, $k \geq k_0 + k_1$ is the security parameter and k_0, k_1 are quantities with 2^{-k_0}, 2^{-k_1} negligibly small. Length of pt., resp. length of ct. denotes the length of the plaintext m, resp. the ciphertext, relative to k. The last two rows give an estimate on the number of computations which are additionally required in the security enhanced, as opposed to the underlying system (i.e., any system based on trapdoor functions, the OU-system, the Williams system).

Comparison with other factorization based variants					
	OAEP & KOMM	OAEP+	OU & Pointch.	BR & KOMM	Proposed system
Assumption	$n = pq$	$n = pq$	$n = p^2q$	$n = pq$	$n = pq$
Number of ROs	2	3	2	2	2
Length of pt.	$k - k_0 - k_1$	$k - k_0 - k_1$	$k - k_1$	$\|m\|$	$k - k_1$
Length of ct.	k	k	$k + \|n\|$	$\|m\| + k + k_0$	$k + \|n\|$
Comp. for encr.	$1G, 1H$	$1G, 1H, 1H'$	$1G, 1H$	$1G, 1H$	$1G, 1H$
Comp. for decr.	$1G, 1H$	$1G, 1H, 1H'$	$1G, 1H,\ 1f$	$1G, 1H$	$1G, 1H$
Security	IND-CPA	IND-CCA 2	IND-CCA 2	IND-CCA 2†	IND-CCA 2

As pointed out in [24], one-way trapdoor functions are very rare. We base our strengthening method on the general Williams scheme, which is one example of such a function. Contrary to the other schemes provably equivalent to factoring, this is the only one which imposes no restrictions on the primes or the structure of the modulus. Unfortunately it is not known whether factorizing the special moduli of the other factorization equivalent schemes, $n = p^2q$, or $n = pq$ with $p \equiv 3 \pmod 8$, $q \equiv 7 \pmod 8$, etc. is as difficult as factoring $n = pq$ with p, q arbitrarily. Thus, to be on the safe side, our basis will be the Williams system.

Evaluation of the Williams system is only slightly less efficient than evaluating power polynomials modulo n. If the degree of the underlying Lucas function is k, then its evaluation can be achieved with roughly 1.33 as many multiplications as evaluating $x^k \pmod n$. In the worst case this takes no longer than $2\log_2 k$ multiplications [34, 35]. Any further progress for its efficient evaluation would make the Williams scheme more applicable still.

Additionally, our enhanced scheme can be applied to any one-way trapdoor function and hence allows a number of possible extensions. Since practically the Williams system is less efficient than RSA, we can analogously obtain an IND-CCA secure Rabin-type RSA modification. As underlying basis other systems provably equivalent to factoring can be chosen. The suggestions in [32, 14] are more efficient than the general Williams scheme. Combined with the proposed method they establish another alternative for OAEP, even for long messages.

2. The proposed scheme

2.1. The underlying Williams system

Key generation. Let $n = pq$, where p and q are two large primes. Further, let $t, c \in \mathbb{Z}_n^*$ be chosen such that $p \equiv -\left(\frac{c}{p}\right) \pmod 4$ and $q \equiv -\left(\frac{c}{q}\right) \pmod 4$, $\gcd(t^2 - c, n) = 1$ and $\left(\frac{t^2-c}{n}\right) = -1$.

Let the encryption key e and the decryption key d with $\gcd(e, (p+1)(q+1)) = 1$ be determined according to $ed \equiv \frac{M+1}{2} \pmod m$, where $M = \frac{\left(p-\left(\frac{c}{p}\right)\right)\left(q-\left(\frac{c}{q}\right)\right)}{4}$.

Throughout, let m be the message and $b_1 = 1$, if $\left(\frac{m^2-c}{n}\right) = 1$, and $b_1 = -1$, if $\left(\frac{m^2-c}{n}\right) = -1$. Suppose $\gcd(m^2 - c, n) = 1$ and define

$$a = a(m) \equiv \frac{\widehat{m}^2 + c}{\widehat{m}^2 - c} \pmod n, \qquad b = b(m) \equiv \frac{2\widehat{m}}{\widehat{m}^2 - c} \pmod n, \tag{7}$$

where $\widehat{m} \equiv m \pmod n$, if $b_1 = 1$, and $\widehat{m} \equiv \frac{mt+c}{m+t} \pmod n$, if $b_1 = -1$.

The encryption and decryption functions are described by means of $X_i(a) = \frac{\alpha^i + \overline{\alpha}^i}{2} = V_i(2a, 1)/2$ and $Y_i(a, b) = b\frac{\alpha^i - \overline{\alpha}^i}{\alpha - \overline{\alpha}} = bU_i(2a, 1)$, where $\alpha, \overline{\alpha} = a \pm b\sqrt{c}$ and U_i, V_i denote the Lucas sequences of first and second kind, respectively, [35].

In order to minimize any problems concerning the existence of the above multiplicative inverses modulo n (cf. [18]) it turned out to be preferable to work with a slightly modified version of the original scheme. In the following we will exclusively be applying this modification which essentially consists of reversing numerator and denominator of the original encryption function $X_e(a)/Y_e(a, b) \pmod n$ of [33] and adapting the decryption scheme.

Williams encryption $E(m)$. The first step of the encryption process consists of calculating $a(m), b(m)$ from m by means of (7). Then m is encoded as

$$E(m) \equiv \frac{Y_e(a(m), b(m))}{X_e(a(m))} \pmod n.$$

The cryptogram C is the triple $[E(m), b_1, b_2]$, where b_1 is defined above and b_2 equals 0 or 1, depending on whether $a(m) \pmod n$ is even or odd.

Williams decryption $D(C)$. Upon receiving C the receiver firstly calculates

$$a_0 \equiv \frac{1 + E(m)^2 c}{1 - E(m)^2 c} \pmod n, \qquad b_0 \equiv \frac{2E(m)}{1 - E(m)^2 c} \pmod n. \tag{8}$$

The second step consists of determining $\sigma = (-1)^{b_2 - X_d(a_0)}$ and $a(m), b(m)$ by means of

$$a(m) \equiv \sigma X_d(a_0) \pmod n \quad \text{and} \quad b(m) \equiv \sigma Y_d(a_0, b_0) \pmod n.$$

Finally, the message m can be retrieved from $a(m)$ and $b(m)$ via

$$\widetilde{m} \equiv \frac{a(m) + 1}{b(m)} \pmod n, \qquad \text{and} \quad \begin{cases} m \equiv \widetilde{m} \pmod n, & \text{if } b_1 = 1, \\ m \equiv \frac{\widetilde{m}t - c}{t - \widetilde{m}} \pmod n, & \text{if } b_1 = -1, \end{cases} \tag{9}$$

provided $\gcd(b(m), n) = 1$ for $b_1 = 1$, $\gcd(a(m) + 1 - tb(m), n) = 1$ for $b_1 = -1$. (It was shown in [18] that the number of messages not fulfilling these gcd-conditions is negligible.)

By extending Rabin's result [27], Williams establishes the equivalence to factoring for the messages with $\left(\frac{m^2-c}{n}\right) = -1$.

Definition 1. Let $a'(w)$ and $b'(w)$ be chosen such that $a'(m), b'(m)$ correspond to $a(m), b(m)$ for $-b_1$ (instead of b_1). Further denote the *false encryption of m* by $E'(m) \equiv \frac{Y_e(a'(m),b'(m))}{X_e(a'(m))} \pmod{n}$, that is defined by following the formulas with respect to $b'_1 = -b_1$ (rather than b_1).

Lemma 1. *If* $\left(\frac{m^2-c}{n}\right) = -1$ *then* $E'(m) \equiv E(z) \pmod{n}$ *for some* $z \in \mathbb{Z}_n^*$ *with* $b'_1(w) = b_1(z)$. *Let* $u \equiv D(E(z)) \pmod{n}$ *where the parameters for decryption are* b'_1 *and* b_2. *Then* $\gcd(m - u, n)$ *gives the factorization of* n.

Crucial for this property is the value b_1, whereas b_2 plays no role in the equivalence to factoring. Indeed, suppose that $E(w)$ is being decrypted with respect to the wrong σ. Then the decryption obtained equals $w(-\sigma) = c\frac{1}{w} \pmod{n}$.

2.2. The enhanced system

Let $E : \{0,1\}^k \mapsto \mathbb{Z}_n^*$ be the Williams encryption function, where $n = pq$ is the product of two large primes with $|n| = k$ and k is the security parameter. Messages m are of length k_0, where $k_0 + k_1 \leq k$ and it should be the case that 2^{-k_0} and 2^{-k_1} are negligible quantities.

We need two functions, H and G, both assumed to be ideal random functions (cf. [3]),

$$H : \{0,1\}^k \mapsto \{0,1\}^{k_0}, \; G : \{0,1\}^{k_0+k_1} \mapsto \{0,1\}^k.$$

Key generation $\mathcal{K}$ is as in the Williams system.

Encryption $\mathcal{E}(m)$. Given a plaintext m, the encryption algorithm randomly chooses $r \in \{0,1\}^{k_1}$ and then computes $s \in \{0,1\}^{k_0+k_1}$, $\alpha \in \mathbb{Z}_n^*$, $\beta \in \{0,1\}^k$ as follows:

$$s = r \,\|\, H(m \,\|\, r), \tag{10}$$

$$\alpha = E(s) \pmod{n}, \tag{11}$$

$$\beta = G(s) \oplus (m \,\|\, r). \tag{12}$$

The ciphertext is $C = [\alpha, b_1, b_2, \beta]$, where b_1 and b_2 are as above.

Decryption $\mathcal{D}(C)$. Given a ciphertext $C = [\alpha, b_1, b_2, \beta]$ the decryption oracle computes $S \in \{0,1\}^{k_0+k_1}$, $R_1 \in \{0,1\}^{k_1}$, $H \in \{0,1\}^{k_0}$, $R_2 \in \{0,1\}^{k_1}$, $\mathcal{M} \in \{0,1\}^k$,

$M \in \{0,1\}^{k_0}$ as follows:

$$\begin{aligned} S &= D([\alpha, b_1, b_2]) \text{ by following the formulas of Section 2.1,} &(13)\\ R_1 &= S[0...k_1], &(14)\\ H &= S[k_1...k_0 + k_1], &(15)\\ \mathcal{M} &= G(S) \oplus \beta, &(16)\\ M &= \mathcal{M}[0...k_0], &(17)\\ R_2 &= \mathcal{M}[k_0...k_0 + k_1]. &(18) \end{aligned}$$

If $H(M) = H$ and $R_1 = R_2$ then the algorithm outputs the plaintext M; otherwise the algorithm *rejects* the ciphertext and outputs NULL.

3. Security analysis

3.1. Notions of security

For brevity we only informally recall the basic notions required below. The attack models considered are well-known. For the exact definitions we refer to [2].

Chosen plaintext attacks. An adversary $A = (A_1, A_2)$ defining security against CPA is usually described via the well-known game play [2]. During the 'find'-stage, A_1 is run on input the public key, pk. At the end of $A_1's$ execution it outputs a triple (m_0, m_1, st), where m_0, m_1 are messages of the same length and st is some state information. A random one of m_0 and m_1 is selected, say m_b, and a 'challenge' y is determined by encrypting m_b under pk. For the 'guess'-stage A_2 is given y but not m_b. It is now $A_2's$ job to decide, if y is the encryption of m_0 or of m_1. At the end of the game it outputs $b' \in \{0,1\}$, representing its "guess" of b. The adversary's advantage in this scenario is defined to be $2\Pr[b' = b] - 1$.

In public key cryptography such an attack is always possible, since any adversary has access to the encryption oracle, as pk is always publicly known.

Plaintext awareness (PA). The basic property of PA is that any party which generates a valid ciphertext also must "know" (be able to calculate) the underlying plaintext. The underlying idea is obvious. If any attacker is allowed to query the deciphering oracle then it will only obtain an output different from NULL when the input was a valid ciphertext. However, in that case the adversary must have known the plaintext to begin with. Hence, it does not gain advantage by any decryption queries and its success rate is the same as without access to the deciphering oracle.

Formalizing this requirement turned out to involve some subtleties. Indeed, the first definition of PA given in [4] required some modification, which finally results in the present definition of PA in [2] in the ROM. An adversary $\mathcal{A}$ in the sense of PA has access to the enciphering algorithm and to any random oracles present. It is allowed to create a list of valid ciphertexts via encryption or by any other means. Suppose at the end of its execution $\mathcal{B}$ outputs some ciphertext y which is not the result of some $\mathcal{E}$-query. $\mathcal{B}$ is plaintext aware, if for any valid ciphertext which it produces, it necessarily knows the corresponding plaintext. More precisely, to be PA, there must exist some algorithm K, the plaintext extractor, which can find the plaintext to y by just looking at $\mathcal{B}$'s interaction with its oracles. K can be seen as a simulator of the deciphering algorithm $\mathcal{D}$, and itself does not have access to $\mathcal{D}$. In particular, the extractor can see all of $\mathcal{B}$'s direct oracle queries and their answers. The system is PA, if the information that K sees can be shown to suffice to simulate $\mathcal{D}$ correctly in almost all cases.

Plaintext awareness seems to be the strongest notion of security which is known. It captures security in the sense of IND-CCA2 [2], non-malleability [9], and security against the one-more CCA [29].

Several systems have been published (for DH based, [29, 1], and others [24, 15]) which provably achieve some form of PA under the specific settings of these schemes. Below, we will show that also our proposed system also fulfils this strong attribute.

3.2. Proof of security

Pointcheval's partially trapdoor one-way functions are best suitable for systems based on DH (compare Section 3.2 in [24] which shows the intrinsic relationship between the DH problem and this particular class of functions).

In the proof for the semantic security he shows that if a CPA adversary is successful in finding the underlying message m_b, then with overwhelming probability this involves queries to the random oracle G. Thus, such a CPA adversary could be used to build another adversary that obtains a partial preimage r of $f(r,(m\,\|\,s))$ from the input to a suitable G-query. By his hypothesis on the partial trapdoorness of f, this contradiction implies that a CPA adversary cannot be successful.

The queries to G alone are not sufficient in retrieving the complete preimage of f. Nonetheless, we show that Pointcheval's ideas can be extended that way.

Theorem 1. *If there exists an adversary $\mathcal{A}$ attacking the proposed encryption scheme in the chosen-plaintext scenario, then we can construct an efficient adversary $\mathcal{B}$ that factorizes $n = pq$ with non-negligible probability.*

In particular, if $\mathcal{A}$ needs t_A steps for breaking the scheme under a CPA, then $\mathcal{B}$ needs $t_A + q_G q_H(T_E(k) + \lambda k) + T_{\text{gcd}}(k)$ steps for factorizing n, where λ is some constant, $T_E(k)$ denotes the time complexity of the Williams encryption $E_{pk}(x)$ and $T_{\text{gcd}}(k)$ denotes that of $\gcd(x, y)$. Additionally, if ϵ denotes the advantage of the CPA attacker, then $\mathcal{B}$ factorizes n with probability greater than $\epsilon - q_H/2^{k_1}$.

Proof. Let $\mathcal{A} = (A_1, A_2)$ be an adversary attacking the proposed system $\Pi = (\mathcal{K}, \mathcal{E}, \mathcal{D})$ under a CPA where A_1 denotes the "find"-stage and A_2 the "guess"-stage. Further, let ϵ be the advantage of $\mathcal{A}$ and assume that both G and H are random oracles with q_G and q_H the number of queries to G and H respectively.

We use adversary $\mathcal{A}$ to construct an adversary $\mathcal{B}$ that factorizes n. Firstly, $\mathcal{B}$ randomly chooses e, c, t according to the requirements of the key generation algorithm and sets $pk = (n, e, c, t)$. Consequently, $\mathcal{B}$ chooses randomly $\widehat{s}$ from $\mathbb{Z}_n^*$ with $\left(\frac{\widehat{s}^2 - c}{n}\right) = -1$ and evaluates the corresponding $b_1 = b_1(\widehat{s})$, $b_2 = b_2(\widehat{s})$. Then $\mathcal{B}$ calculates $z = E'(\widehat{s})$ by following the formulas for b_1' rather than b_1. It then follows from Lemma 1 (cf. [33]) that there exists some $s \in \mathbb{Z}_n^*$ such that $z = E(s)$, $C = [z, b_1, b_2]$ is a valid Williams-ciphertext, and $\gcd(\widehat{s} - s, n) = p$ or q. Our strategy will be to show that $\mathcal{B}$ can use $\mathcal{A}$ to find this value s. The machine $\mathcal{B}$ works as follows.

- It runs $\mathcal{A}$ on the public key pk where, as in Pointcheval's case, all queries to G and H are intercepted. $\mathcal{B}$ makes two lists, G-list and H-list, in the following way. When A_1 makes any new oracle query j to G, $\mathcal{B}$ chooses a random bit string G_j in $\{0,1\}^k$ as $G(j)$ and adds (j, G_j) to the G-list. The same way, for any new query l asked to H, $\mathcal{B}$ chooses a random H_l in $\{0,1\}^{k_0}$ as $H(l)$ and adds (l, H_l) to the H-list.

- A_1 finally outputs two messages m_0 and m_1. The adversary $\mathcal{B}$ chooses a random bit b, a random string $r \in_R \{0,1\}^{k_1}$, and a random string $T \in_R \{0,1\}^k$ for $G(s) \oplus (m_b \| r)$. (The adversary cannot really compute T because it does not know s.) It outputs $[\alpha, b_1, b_2, \beta] = [z, b_1, b_2, T]$ as a ciphertext.

- The attacker A_2 is fed with the ciphertext $[\alpha, b_1, b_2, \beta]$ and queries to the oracles are again intercepted and answered. Additionally to above, $\mathcal{B}$ performs the following.

(a) When A_2's query g to G is found in the G-list, return the value G_g for g in the G-list. Otherwise, for each entry (h, H_h) in the H-list, if

$$g[k_1 ... k_0 + k_1] = H_h, \tag{19}$$

then put $s = g$. If there exists such a h for which additionally

$$z = E(s), \;\; b_1(s) = b_1, \;\; b_2(s) = b_2, \tag{20}$$

then obtain p, q from $\gcd(\widehat{s} - s, n)$ and return $G_g = \beta \oplus h$. Otherwise, choose a random bit string G_g of length k, add (g, G_g) to the G-list and return G_g.

(b) If A_2 queries H with h and there is an entry (h, H_h) in the H-list, return H_h. Otherwise choose H_h, a random string of length k_0 and add the entry (h, H_h) to the H-list. For each entry (g, G_g) in G-list compute (19). If there exists g which also satisfies (20), then obtain p, q from $\gcd(\widehat{s} - s, n)$ and return H_h.

- When the attacker A_2 returns its answer b', the machine $\mathcal{B}$ returns both the G-list and the H-list. If $\mathcal{B}$ finds p and q it outputs them too.

Recall that $z = E(s)$. We write s as $s = r \| y$ where r and y are of length k_1 and k_0, respectively. By relating the corresponding b_1 and b_2, if such a pair (r, y) exists, it is unique.

On the other hand, if pk is a legal public key there always exists exactly one pair (g, H_h) which satisfies (19) and (20) and then $\gcd(\widehat{s} - s, n)$ always is a proper factor of n. Then, in particular, the above unique r and s fulfil $G(s) = \beta \oplus (m_b \| r)$.

To determine the success rate of the adversary $\mathcal{B}$ relative to that of $\mathcal{A}$, we consider a new game. Define the following events.

- Ask G, the query s was made to G,
- Ask H, the query $(\widetilde{m} \,\|\, r)$ was made to H, for the specific r chosen in the beginning, and for some $\widetilde{m}$.

In the first case, $\mathcal{A}$ knows s and thus can find the desired m_b from $G(s) \oplus \beta$. In the latter case $\mathcal{A}$ knows the r relative to the challenge and since s is one of $r \,\|\, H(m_0 \,\|\, r)$ or $r \,\|\, H(m_1 \,\|\, r)$, the correct one follows by comparison with $z = E(s)$. Thus, $\mathcal{A}$ also correctly distinguishes between m_0 and m_1. We say that $\mathcal{A}$ wins, if one of these events occur, or if at the end, the value b' returned by $\mathcal{A}$ as the guess of b, is equal to b. Then the advantage of the adversary in the new game is defined by

$$\mathrm{Adv} = 2\Pr[\mathrm{wins}] - 1. \tag{21}$$

With a random simulation of G and H, as described above, we have a perfect simulation of the real life, except for the unlikely case where s or $m_b \,\|\, r$ have already been asked to G or H during the find stage, i.e., before their assignment. But then the adversary wins, so that $\mathrm{Adv} \geq \epsilon$, where ϵ denotes $\mathcal{A}$'s advantage in the CPA game.

On the other hand, it is obvious that no advantage can be gained by the adversary without Ask G or Ask H, so that

$$2\Pr[\mathrm{wins}] \leq (1 + \Pr[\mathrm{Ask}\, G \vee \mathrm{Ask}\, H]). \tag{22}$$

By combining the above inequalities, and since $\mathrm{Adv} \geq \epsilon$, this leads to $\epsilon \leq \Pr[\mathrm{Ask}\, G \vee \mathrm{Ask}\, H]$. Our task is to lower bound $\Pr[\mathrm{Ask}\, G]$. We have

$$\begin{aligned}
\Pr[\mathrm{Ask}\, G \vee \mathrm{Ask}\, H] &= \Pr[\mathrm{Ask}\, G] + \Pr[\mathrm{Ask}\, H \wedge \neg\, \mathrm{Ask}\, G] \\
&= \Pr[\mathrm{Ask}\, G] + \Pr[\mathrm{Ask}\, H | \neg\, \mathrm{Ask}\, G] \Pr[\neg\, \mathrm{Ask}\, G] \\
&\leq \Pr[\mathrm{Ask}\, G] + \Pr[\mathrm{Ask}\, H | \neg\, \mathrm{Ask}\, G].
\end{aligned}$$

But without having asked s to G, the probability that the input to H has the last bits equal to r is unlikely. More precisely, $\Pr[\mathrm{Ask}\, H | \neg\, \mathrm{Ask}\, G] \leq q_H/2^{k_1}$, and therefore,

$$\Pr[\mathrm{Ask}\, G] \geq \epsilon - \frac{q_H}{2^{k_1}},$$

which means that with probability greater than $\epsilon - q_H/2^{k_1}$ the desired value s lies in the set of the queries asked to the oracle G. Thus, if the advantage of $\mathcal{A}$ is non-negligible, the probability that the query s is made to G is non-negligible, in which case $\mathcal{B}$ obtains the factorization of n as above. □

A widely applied method for establishing security in the sense of IND-CCA2 is to firstly establish semantic security against CPA, and successively showing that the scheme allows a plaintext extractor [1, 10, 15, 24].

Theorem 2. *Let $\Pi = (\mathcal{K}, \mathcal{E}, \mathcal{D})$ be the proposed system. Then there exists a plaintext extractor K and hence Π is secure in the sense of PA.*

Proof.

Game 1. Suppose that $\mathcal{A}$ is an adversary in the sense of PA. Let S_G and S_H denote the set of s, resp. (m, r) at which G, resp. H was evaluated by the adversary. Additionally, let C be a list of $\mathcal{A}$'s answers (ciphertexts) obtained as the result of some $\mathcal{E}$-queries and let $y = [\alpha, b_1, b_2, \beta] \notin C$ be another ciphertext produced by $\mathcal{B}$ which is not the result of some encryption. The knowledge extractor K works as follows:

- K considers all query-answer pairs (g, G_g) and (h, H_h), respectively and checks if there exist pairs such that $H_h = g[k_1...k_0 + k_1]$, $\alpha = E(g)$, and $\beta = G_g \oplus h$.
- At most one $\{(g, G_g), (h, H_h)\}$ may satisfy these conditions, and in that case, K outputs $m = h[0...k_0]$. Otherwise, it rejects (the ciphertext is considered an invalid one).

With this simulation, only valid ciphertexts will be decrypted. However, there is a chance that a valid ciphertext can be produced without asking queries to G or H. We now show that the probability for this is negligible.

Game 2. We modify the decryption oracle as follows (cf. [30]). Given a ciphertext, the new decryption oracle runs as the original one. If the old oracle rejects, so does the new one. But the new decryption oracle also rejects if $\mathcal{M} \notin S_H$. Additionally, we require that in such a case the new oracle does not evaluate $H(\mathcal{M})$. Then the decryption oracle never evaluates H at points other than those at which the adversary did.

Let F_1 be the event that a ciphertext is rejected by the new decryption oracle, but not the original one. Consider a ciphertext $y \notin C$ submitted to the decryption oracle and suppose that F_1 has not occurred up until this point. This y defines values r, s, m as above. If $r = r'$ and $m = m'$ for some r', m' relating to some ciphertext $y' \in C$ then $y = y'$. So we can assume that $r \neq r'$ or $m \neq m'$ for all $y' \in C$. Unless either the adversary or the encryption oracle made the query $H(m \| r)$, then this value is independent of all the other values available (directly or indirectly) to the adversary, and hence F_1 occurs with probability $1/2^{k_0}$. When obtaining C, the encryption oracle has made queries $H(m' \| r')$, but this will not "help" the adversary since $(m, r) \neq (m', r')$. It follows that in the course of the attack, $\Pr[F_1] \leq q_D/2^{k_0}$, which means that only with negligible probability the original decryption oracle does not reject a ciphertext whose creation did not involve the adversary's direct H-query with $(m \| r)$.

Game 3. Now we modify the decryption oracle just considered again. This new decryption oracle runs as the former one. If any of the previous decryption oracles rejects, so does the new one. But the new oracle also rejects if $s \notin S_G$.

Consider firstly the case that $s \notin S_G$ and $s = s'$ where s' relates to some $y' \in C$. Now, $s = s'$ and $y \neq y'$ implies that $(m, r) \neq (m', r')$ where the prime values correspond to the quantities relative to y'. On the other hand, if y is not rejected by the previous decryption oracle, then the condition $s = s'$ implies $H(m \| r) = H(m' \| r')$. The probability that such a collision can be found over the course of the attack is $q_H/2^{k_0}$.

Also, if $s \notin S_G$ and $s \neq s'$ for all $y' \in C$, then the oracle G was never evaluated at s by either the adversary, the encryption oracle, or the decryption oracle. Since $\mathcal{M} = \beta \oplus G(s)$, the fact that G is a random function implies that the value $\mathcal{M}$ is independent from all other values accessible by the adversary, directly or indirectly. It follows that the probability that $\mathcal{M} \in S_H$ is at most $q_H/2^k$. Over the course of the attack, the probabilities in this case sum to $q_D q_H/2^k$.

This shows that in each case the adversary's success probability without queries of H or G is negligible. With overwhelming probability the decryption oracle rejects all the adversary's queries which did not involve direct oracle queries asked by the adversary. In particular, these queries need to be contained in the oracle lists, and therefore they need to be known to the adversary. In other words, even if the adversary is given a list of valid ciphertexts and may access $\mathcal{D}$, it cannot produce a new valid ciphertext without knowing the secret underlying parameters. This constitutes the core property of plaintext awareness [2]. Hence, with overwhelming probability, any queries to $\mathcal{D}$ relative to unknown underlying parameters (e.g., such that are related to the challenge) will be rejected by $\mathcal{D}$. Consequently, the adversary does not gain advantage by having access to $\mathcal{D}$ which is the general argument used for establishing security against CCA. □

4. Extensions

The above method may be generalized to other basis schemes.

4.1. RSA based

We can apply the following Rabin-type alternative with unique decryption [19].

Key generation. Let p and q be primes with $p \equiv q \equiv 3 \pmod 4$, put $k = \frac{(p-1)(q-1)}{4}$ and let e and d be chosen according to $ed \equiv \frac{k+1}{2} \pmod k$. Set $Q \equiv qq^* - pp^* \pmod n$, where $n = pq$ and $q^* \equiv q^{-1} \pmod p$ and $p^* \equiv p^{-1} \pmod q$.

The public key, is (n, e) and the private key is (p, q, d, Q).

Fact-RSA encryption. Given a message $a \in \mathbb{Z}_n^*$, the encryption is obtained by $c \equiv a^{2e} \pmod n$. Additionally, let $B_1 = \left(\frac{a}{n}\right)$ and $B_2 \equiv a \pmod 2$ with $B_2 \in \{0,1\}$. The ciphertext is $[c, B_1, B_2]$.

Fact-RSA decryption. On receiving $[c, B_1, B_2]$ the designer firstly calculates

$$\begin{cases} K \equiv c^d \pmod n & \text{if } B_1 = 1, \\ K \equiv \frac{c^d}{Q} \pmod n & \text{if } B_1 = -1, \end{cases}$$

where $0 < K < n$. Finally, the correct message is $a \equiv K$ or $n - K \pmod n$ whichever satisfies $a \equiv B_2 \pmod 2$.

Obviously, [19], decrypting the above RSA modification and factorizing the modulus are computationally equivalent. As for all the proposals in the Rabin family, the situations is the same as for the Williams scheme above.

Proposition 1. *If E and D denote the encryption and decryption algorithms of this section then for any $a \in \mathbb{Z}_n^*$ with $\left(\frac{a}{n}\right) = -1$ one obtains a non-trivial factor of n by the determination of* $\gcd(D(E(a))-a, n)$, *where in the decryption procedure B_1 is assigned the value* 1 *instead of* -1.

Hence, by our IND-CCA strengthening method of Section 2.2, we obtain a fast method securing this scheme against CCA. The Williams encryption and decryption only need to be replaced by the Fact-RSA encryption and decryption, respectively.

4.2. One-way functions

The proposed strengthening method is applicable for any one-way trapdoor function f. That is, inverting f without the trapdoor information needs to be provably as difficult as the underlying functions. Additionally to the above Williams system and the Fac-RSA system, other such candidates are given in terms of Williams' RSA modification based on Blum integers [32], the Dickson based RSA modification [19], or any of the other factorization equivalent proposals.

Our method is not limited to the factorization problem but (as the schemes in [10]) only requires the provable one-wayness of the encryption function.

Acknowledgement. The author is grateful to the anonymous referees for their many insightful comments and for pointing out [10].

References

[1] Baek, J., Lee, B., Kim, K., Provably secure length-saving public-key encryption scheme under the computational Diffie-Hellman assumption. ETRI J. 22 (2000), 25–31.

[2] Bellare, M., Desai, A., Pointcheval, D., Rogaway, P., Relations among notions of security for public-key encryption schemes. Extended abstract in: Advances in Cryptology – CRYPTO '98 (ed. by H. Krawczyk; Lecture Notes in Comput. Sci. 1462), 26–45. Springer, Berlin 1998. Full version available at `www-cse.ucsd.edu/users/mihir/crypto-papers.html`.

[3] Bellare, M., Rogaway, P., Random oracles are practical: a paradigm for designing efficient protocols. Extended abstract in: Proceedings of the 1st ACM Conference on Computer and Communication Security (CCS '93), 62–73. ACM Press, New York 1993. Full paper available at `http://www.cs.ucdavis.edu/~rogaway/papers/`.

[4] Bellare, M., Rogaway, P., Optimal asymmetric encryption – how to encrypt with RSA. In: Advances in Cryptology – EUROCRYPT '94 (ed. by A. De Santis; Lecture Notes in Comput. Sci. 950), 92–111. Springer, Berlin 1995.

[5] Boneh, D., Venkatesan, R., Breaking RSA may not be equivalent to factoring. In: Advances in Cryptology – EUROCRYPT '98 (ed. by K. Nyberg; Lecture Notes in Comput. Sci. 1403), 59–71. Springer, Berlin 1998.

[6] Canetti, R., Goldreich, O., Halevi, S., The random oracle methodology, revisited. In: STOC '98, 209–218. ACM Press, New York 1999. `http://www.acm.org/pubs/contents/proceedings/stoc/276698/`

[7] Coppersmith, D., Franklin, M., Patarin, J., Reiter, M., Low-Exponent RSA with Related Messages, In: Advances in Cryptology – EUROCRYPT '96 (ed. by U. Maurer; Lecture Notes in Comput. Sci. 1070), 1–9. Springer, Berlin 1996.

[8] Cramer, R., Shoup, V., A practical public key cryptosystem provable secure against adaptive chosen ciphertext attack. In: Advances in Cryptology – CRYPTO '98 (ed. by H. Krawczyk; Lecture Notes in Comput. Sci. 1462), 13–25. Springer, Berlin 1998.

[9] Dolev, D., Dwork, C., Naor, M., Non-malleable cryptography. In: STOC '91, 542–552. ACM Press, New York 1991.

[10] Fujisaki, E., Okamoto, T., Secure integration of asymmetric and symmetric encryption schemes. In: Advances in Cryptology – CRYPTO '99 (ed. by M. Wiener; Lecture Notes in Comput. Sci. 1666), 537–554. Springer, Berlin 1999.

[11] Fujisaki, E., Okamoto, T., Pointcheval, D., Stern, J., RSA-OAEP is still alive!. Cryptology ePrint Archive 2000/061.

[12] Goldwasser, S., Micali, S., Probabilistic encryption. J. Comput. System Sci. 28 (1984), 270–299.

[13] Håstad, J., Näslund, M., The security of individual RSA bits. In: 39th Symposium on Foundations of Computer Science, 510–519. IEEE, 1998.

[14] Kurosawa, K., Ito, T., Takeuchi, M., Public key cryptosystem using a reciprocal number with the same intractability as factoring a large number. Cryptologia 12 (1988), 225–233.

[15] Kurosawa, K., Ogata, W., Matsuo, T., Makishima, S., IND-CCA public key schemes equivalent to factoring $n = pq$. In: Public Key Cryptography (PKC 2001, Cheju Island, Korea) (ed. by K. Kim; Lecture Notes in Comput. Sci. 1992). Springer, Berlin 2001.

[16] Müller, S., On the security of an RSA based encryption scheme. In: Information Security and Privacy – ACISP'99 (ed. by J. Pieprzyk, R. Safavi-Naini, J. Seberry; Lecture Notes in Comput. Sci. 1587), 135–148. Springer, Berlin 1999.

[17] —, On the security of a Williams based public key encryption scheme. In: Public Key Cryptography (PKC 2001, Cheju Island, Korea) (ed. by K. Kim; Lecture Notes in Comput. Sci. 1992), 1–18. Springer, Berlin 2001.

[18] —, Some remarks on Williams' public key crypto functions. Manuscript, University of Klagenfurt, 2001.

[19] Müller, S., Müller, W.B., The security of public key cryptosystems based on integer factorization. In: Information Security and Privacy – ACISP'98, (ed. by C. Boyd, E. Dawson; Lecture Notes in Comput. Sci. 1438), 7–23. Springer, Berlin 1998.

[20] Naor, M., Yung, M., Public-key cryptosystems provably secure against chosen ciphertext attacks. In: 22nd Annual ACM Symposium on Theory of Computing (1990), 427–437.

[21] Okamoto, T., Uchiyama, S., A new public-key cryptosystem as secure as factoring. In: Advances in Cryptology – EUROCRYPT '98 (ed. by K. Nyberg; Lecture Notes in Comput. Sci. 1403), 308–318. Springer, Berlin 1998.

[22] Okamoto, T., Uchiyama, S., Fujisaki, E., EPOC: Efficient Probabilistic Public-Key Encryption. Submission to P1363a (1998).

[23] Paillier, P., Pointcheval, D., Efficient public-key cryptosystems provably secure against active adversaries. In: Advances in Cryptology – ASIACRYPT '99 (ed. by K.Y. Lam, E. Okamoto, C. Xing; Lecture Notes in Comput. Sci. 1716), 165–179. Springer, Berlin 1999.

[24] Pointcheval, D., Chosen-ciphertext security for any one-way cryptosystem. In: Public Key Cryptography (PKC 2000, Melbourne) (ed. by H. Imai, Y. Zhang; Lecture Notes in Comput. Sci. 1751), 129–146. Springer, Berlin 2000.

[25] —, New public key cryptosystems based on the dependent-RSA problems. In: Advances in Cryptology – EUROCRYPT '99 (ed. by J. Stern; Lecture Notes in Comput. Sci. 1592), 239–254. Springer, Berlin 1999.

[26] Pomerance, C., The quadratic sieve factoring algorithm. In: Advances in Cryptology – EUROCRYPT '84 (ed. by T. Beth, N. Cot, I. Ingemarsson; Lecture Notes in Comput. Sci. 209), 169–182. Springer, Berlin 1985.

[27] Rabin, M.O., Digitalized signatures and public-key functions as intractable as factorization. MIT/LCS/TR-212, MIT Laboratory for Computer Science, 1979.

[28] Rackoff, C., Simon, D.R., Non-interactive zero-knowledge proof of knowledge and chosen ciphertext attack. In: Advances in Cryptology – CRYPTO '91 (ed. by J. Feigenbaum; Lecture Notes in Comput. Sci. 576), 433–444. Springer, Berlin 1992.

[29] Schnorr, C.P., Jakobsson, M., Security of signed ElGamal encryption. In: Advances in Cryptology – ASIACRYPT 2000 (ed. by T. Okamoto; Lecture Notes in Comput. Sci. 1976), 73–89. Springer, Berlin 2000.

[30] Shoup, V., OAEP reconsidered. Cryptology ePrint Archive 2000/060.

[31] The Development of the Number Field Sieve (ed. by A.K. Lenstra, H.W. Lenstra, Jr.; Lecture Notes in Math. 1554). Springer, Berlin 1993.

[32] Williams, H.C., A modification of the RSA public-key encryption procedure. IEEE Trans. Inform. Theory 26 (1980), 726–729.

[33] —, Some public-key crypto-functions as intractable as factorization. Cryptologia 9 (1985), 223–237.

[34] —, A $p+1$ method of factoring. Math. Comp. 39 (1982), 225–234.

[35] —, Édouard Lucas and Primality Testing (Canad. Math. Soc. Ser. Monogr. Adv. Texts 22). John Wiley & Sons, New York 1998.

[36] Zheng, Y., Seberry, J., Immunizing public key cryptosystems against chosen ciphertext attacks. IEEE J. on Selected Areas in Communications 11 (1993), 715–724.

Efficient point multiplication for elliptic curves over special optimal extension fields

Volker Müller

Abstract. We generalize an idea of Gallant, Lambert, Vanstone for fast multiplication of points on elliptic curves with efficient endomorphisms. We describe how this generalization improves point multiplication for elliptic curves defined over optimal extension fields. Finally we present practical results for the new algorithm compared to Gallant's algorithm.

Key words and phrases: Elliptic curve, point multiplication, endomorphisms, optimal extension fields.

1. Introduction

The dominant cost operation in elliptic curve cryptosystems (ECC) is *point multiplication*. Therefore a large number of different algorithms exist for this problem [3], often designed for special situations. For a particular elliptic curve, Koblitz showed how to use endomorphisms for point multiplication ([5], [7]), gaining a significant speed increase for this operation. Later his work was extended to a greater number of elliptic curves [10]. Both authors used the fact that the so called *Frobenius endomorphism* can be quickly evaluated (faster than a point doubling) if the elliptic curve is defined over a small field of characteristic two. This idea is also applicable in fields of small odd characteristic [12]. Then Gallant et al. presented a different algorithm for improved point multiplication with "efficient endomorphisms" [2]. Efficient endomorphisms are endomorphisms whose computation is more expensive than a point doubling, but still "reasonable fast" (say, faster than up to 5 point additions).

In this paper we extend the ideas of Gallant et al. We show that the new extension is especially useful for elliptic curves defined over optimal extension fields as defined in [1]. These finite fields allow very fast software implementation of field arithmetic and are therefore a good choice as basis for ECC. We will present a list of cryptographic secure elliptic curves defined over such fields, where the new idea can be used easily. In addition, we list practical data for the average number of point operations for random scalars for the new algorithm compared to [2].

Public-Key Cryptography and Computational Number Theory
K. Alster, J. Urbanowicz and H. C. Williams (Eds.)

Let E be an elliptic curve defined over a finite field $GF(q)$ with $q = p^r$ elements. In the following sections, we always want to determine $k \cdot P$ for some point $P \in E(GF(q))$ of order n and some integer $0 < k < n$. Let ψ be an endomorphism of E. If this endomorphism is of small norm and can be evaluated quickly for any point P (i.e. faster than doubling a point), then expansions for ψ can be used to speed up the computation of $k \cdot P$ (see [10], [12]). Therefore we assume that either the norm of ψ is "reasonably large" (say, bigger than 256), or ψ cannot be evaluated faster than a point doubling.

The remainder of this paper is organized as follows. In Section 2, we shortly review and generalize the idea of [2]. Section 3 explains the usage of this generalization for point multiplication. In Section 4, we present examples for cryptographic secure elliptic curves and endomorphisms that can be used with the new multiplication algorithm. Finally, Section 5 examines the problem of simultaneous multiple point additions, one basic part of the new algorithm. We describe how general parallel exponentiation algorithms can be modified for elliptic curves, and we report on practical running time results for the curves given before.

2. Parallel point multiplication with efficient endomorphisms

Let the action of the efficient endomorphism ψ on the point P be given as $\psi(P) = \lambda \cdot P$; the integer λ can be easily computed as a root of the characteristic polynomial of ψ modulo n. In [2], Gallant et al. describe an algorithm that for any integer $1 \leq k < n$ computes integers k_0, k_1 of size approximately $\sqrt{n}$ with $k \equiv k_0 + k_1 \cdot \lambda \bmod n$ (under a reasonable assumption on λ and n, which can be tested before). We denote such integers k_0, k_1 *a balanced length-two representation for* k. Then $k \cdot P$ can be computed as

$$k \cdot P = k_0 \cdot P + k_1 \cdot (\lambda \cdot P) = k_0 \cdot P + k_1 \cdot \psi(P).$$

The latter equation can be evaluated with parallel point multiplication and one evaluation of ψ. Since the bit length of k_0, k_1 is in general roughly half the bit length of k and the evaluation of ψ is "sufficiently fast", this algorithm should be faster than the standard multiplication algorithm (for statistical data on the number of point operations, see Table 3).

We shortly review their idea for finding k_0 and k_1. We define the (obviously linear) map

$$\kappa : \mathbb{Z}^2 \longrightarrow \mathbb{Z}/n\mathbb{Z},$$
$$(x_1, x_2) \longmapsto x_1 + x_2 \cdot \lambda \bmod n.$$

Suppose $\underline{b}_0, \underline{b}_1$ are linearly independent vectors in $\mathbb{Z}^2$ with $\kappa(\underline{b}_0) = \kappa(\underline{b}_1) = 0$. Any vector $\underline{v} \in \mathbb{Z}^2$ is "close" to some vector in the lattice $\mathbb{Z}[\underline{b}_0, \underline{b}_1]$. More precisely, for any $\underline{v}$ we can find a vector $\underline{w} \in \mathbb{Z}[\underline{b}_0, \underline{b}_1]$ such that

$$||\underline{v} - \underline{w}|| \leq \frac{1}{2} \max\big(||\underline{b}_0 - \underline{b}_1||, ||\underline{b}_0 + \underline{b}_1||\big)$$

and $\kappa(\underline{v}-\underline{w}) = \kappa(\underline{v})-\kappa(\underline{w}) = \kappa(\underline{v})$. Thus, if we choose $\underline{v} = (k, 0)$ and obtain $\underline{v}-\underline{w} = (k_0, k_1)$, then $k \equiv k_0 + k_1 \cdot \lambda \bmod n$. If the vectors $\underline{b}_0, \underline{b}_1$ were short, i.e. of length $O(\sqrt{n})$, then obviously k_0, k_1 are also of size $O(\sqrt{n})$. Note therefore that small coefficients k_0, k_1 exist for every scalar $0 \leq k < n$ if the lattice $\{\underline{x} \in \mathbb{Z}^2; \kappa(\underline{x}) = 0\}$ has a short basis. This condition can be checked beforehand.

As can be easily guessed, $\underline{w}$ can be found using "change of basis" and "rounding". Specifically, write $\underline{v} = \eta_0 \cdot \underline{b}_0 + \eta_1 \cdot \underline{b}_1$ with $\eta_0, \eta_1 \in \mathbb{Q}$, using standard change of basis techniques in the vector space $\mathbb{Q}^2$. Round η_0, η_1 to nearest integers $\lfloor \eta_0 \rceil, \lfloor \eta_1 \rceil$, and set $\underline{w} = \lfloor \eta_0 \rceil \cdot \underline{b}_0 + \lfloor \eta_1 \rceil \cdot \underline{b}_1$. See [2] for a detailed description why this choice of $\underline{w}$ is sufficient.

It remains to clarify how to find short basis vectors $\underline{b}_0, \underline{b}_1$. We can use the LLL lattice basis reduction algorithm, which in lattices of dimension two reduces to an extended gcd computation. We compute $\gcd(n, -\lambda)$ and its representation, i.e. we successively determine integers r_i, s_i, t_i such that $r_i \cdot n - t_i \cdot \lambda = s_i$ for $i \geq 0$. Obviously $\kappa(s_i, t_i) = 0$ for all $i \geq 0$. The basis determination algorithm stops for the first index i such that $|s_i| < \sqrt{n}$. An analysis of Euclid's extended gcd algorithm shows that then $|t_i| \leq \frac{|n|}{|s_i|}$. Thus, if s_i is not too far away from $\sqrt{n}$, the first short basis vector can be chosen as (s_i, t_i). Gallant et al. propose to choose the second basis vector as (s_{i-1}, t_{i-1}) since very often s_{i-1} is also close to $\sqrt{n}$. This suggestion is supported by practical computations, we almost never experienced a problem with this strategy.

The name "balanced length-2 representation" comes from the fact that both coefficients k_1, k_2 are of approximately the same size. Exactly the same ideas can also be used to compute *non-balanced length-2 representations*, i.e. representations k_1, k_2 where $k_1 = O(n^x)$ and $k_2 = O(n^{1-x})$ for some $0 < x < 1$.

Assume again that $\underline{b}_1 = (b_{11}, b_{12})$, $\underline{b}_2 = (b_{21}, b_{22})$ is a basis for the lattice $L = \{\underline{x} \in \mathbb{Z}^2, \kappa(\underline{x}) = 0\}$. Now we do not assume that these vectors form a short basis, but a basis in a special form. The first component of both basis vectors is expected to be of size $\approx n^x$, the second one of size $\approx n^{1-x}$. For any input vector $\underline{v} = (u_1, u_2) \in \mathbb{Z}^2$, we compute a translation vector $\underline{w} = (w_1, w_2) \in L$ such that $\underline{v} - \underline{w}$ lies in one of the fundamental parallelograms spanned by $\pm\underline{b}_1$ and $\pm\underline{b}_2$. A simple geometric argument shows that

$$|v_i - w_i| \leq \max\big(|b_{1i}|, |b_{2i}|, |b_{1i} + b_{2i}|\big), \; i = 1, 2.$$

Exactly as in balanced length-2 representations, we choose $\underline{v} = (k, 0)$ for the given scalar k, and determine $\underline{w}$ as above. Then the two components of $\underline{v} - \underline{w} = (k_1, k_2)$ satisfy the length conditions, i.e. $|k_1| = O(n^x)$ and $|k_2| = O(n^{1-x})$.

It remains to describe the computation of the given lattice basis. The idea is again as in the original algorithm. We compute the representation of $\gcd(n, -\lambda)$, i.e. integers s_i, r_i, t_i with $r_i \cdot n + t_i \cdot (-\lambda) = s_i$ for $i \geq 0$. If we want to find a non-balanced length-two representation, where the first component of the lattice basis is approximately n^x, then we stop the iteration of the extended gcd-algorithm if $|s_i| < n^x$. Then we get $|t_i| \leq n/|s_i|$, and if $|s_i|$ is not too far away from n^x, the second component will be of size $O(n^{1-x})$. For finding the second basis vector, we proceed as in the original algorithm and choose (s_{i-1}, t_{i-1}). In our practical examples, this strategy always succeeded in computing a sufficiently good basis for the lattice L.

3. Expansions of length greater than two

In this section, we extend the ideas of [2] to expansions of length greater than two. The drawback of the generalization is the fact that such expansions can only be determined for endomorphisms ψ with special properties. The problem of finding appropriate endomorphisms will be considered in Section 4.

Length-3 representations. We start with expansions of length three, i.e. a representation $k \equiv k_0 + k_1 \cdot \lambda + k_2 \cdot \lambda^2 \bmod n$, where each integer $|k_i| \approx n^{1/3}$. Let us assume that such a representation exists. With the characteristic polynomial of ψ, we know that $\lambda^2 \equiv \mathrm{Tr}(\psi) \cdot \lambda - N(\psi) \bmod n$. Thus we deduce

$$k \equiv \bigl(k_0 - k_2 \cdot N(\psi)\bigr) + \bigl(k_1 + k_2 \cdot \mathrm{Tr}(\psi)\bigr) \cdot \lambda \bmod n. \tag{1}$$

Since the trace of an endomorphism is at most twice the square root of the norm, equation (1) directly shows that k_0, k_1, k_2 cannot be computed from a balanced length-two representation. Let us therefore assume that we know coefficients $\tilde{k}_0, \tilde{k}_1$ of size $O(n^x), O(n^{1-x})$ for some $0 < x < 1$, respectively, such that $k \equiv \tilde{k}_0 + \tilde{k}_1 \cdot \lambda \bmod n$.

We can determine k_0, k_1, k_2 from $\tilde{k}_0, \tilde{k}_1$ as follows (define $x \,\mathtt{Div}\, t$ as $\lfloor x/t \rfloor$ and $x \,\mathtt{Mod}\, t$ as $x - (x \,\mathtt{Div}\, t)$):

- determine $k_2 = \tilde{k}_0 \,\mathtt{Div}\, (-N(\psi))$,
- determine $k_0 = \tilde{k}_0 \,\mathtt{Mod}\, (-N(\psi))$,
- determine $k_1 = \tilde{k}_1 - k_2 \cdot \mathrm{Tr}(\psi)$.

If the endomorphism ψ has very small trace, norm approximately $n^{1/3}$ and we have chosen $x = \frac{2}{3}$, then we directly see that all the coefficients k_i are of absolute size $O(n^{1/3})$. Thus for any endomorphism with norm approximately the cube root of n and very small trace, we can determine a length-3 representation of the desired form.

Length-4 representations. An expansion of length four can be determined with the same idea. Assume that $k \equiv k_0 + k_1 \cdot \lambda + k_2 \cdot \lambda^2 + k_3 \cdot \lambda^3 \bmod n$. Using the characteristic polynomial of ψ, we find that

$$\begin{aligned} k \equiv\ & \bigl(k_0 - N(\psi) \cdot (k_3 \cdot \mathrm{Tr}(\psi) + k_2)\bigr) \\ & + \bigl(-k_3 \cdot N(\psi) + k_1 + k_2 \,\mathrm{Tr}(\psi) + k_3 \,\mathrm{Tr}(\psi)^2\bigr) \cdot \lambda \bmod n. \end{aligned}$$

Assume that ψ has very small trace and norm $N(\psi) \approx n^{1/4}$. Let $\tilde{k}_0, \tilde{k}_1$ be the coefficients of a balanced length-two representation, each of absolute size $O(n^{1/2})$. Then we can find the coefficients k_i as

- $k_3 = \tilde{k}_1 \texttt{ Div } (-N(\psi))$,
- $k_2 = \big(\tilde{k}_0 \texttt{ Div } (-N(\psi))\big) - k_3 \operatorname{Tr}(\psi)$,
- $k_0 = \tilde{k}_0 \texttt{ Mod } (-N(\psi))$,
- $k_1 = \big(\tilde{k}_1 \texttt{ Mod } (-N(\psi))\big) - k_2 \operatorname{Tr}(\psi) - k_3 \operatorname{Tr}(\psi)^2$.

Obviously all the coefficients k_i are approximately of absolute size $O(n^{1/4})$.

Expansions of length greater than four. It is rather obvious that the same idea can also be used for determining expansions of length l greater than four. If the length is odd (say, $l = 2s+1$), then we have to start with a unbalanced length-two representation with coefficients of size $O(n^{(s+1)/l}), O(n^{s/l})$; for even lengths we can use a balanced length-two representation. If the trace of the endomorphism ψ is very small, and its norm is approximately $O(n^{1/l})$, then all the coefficients will also be of absolute size $O(n^{1/l})$. Since the following section will give concrete examples which use expansions of length 7, we shortly present the formulas for length 7.

Length-7 representations. Reducing a representation $k \equiv \sum_{i=0}^{6} k_i \cdot \lambda^i \bmod n$ of length 7 with the characteristic polynomial of ψ, we get the following result (define for shortness $t = \operatorname{Tr}(\psi)$ and $m = N(\psi)$):

$$\begin{aligned} k \equiv & -k_6 m^3 + (3k_6 t^2 + 2k_5 t + k_4) m^2 - (k_6 t^4 + k_5 t^3 + k_4 t^2 + k_3 t + k_2) m + k_0 \\ & + \lambda \cdot \big((3k_6 t + k_5) m^2 - (4k_6 t^3 + 3k_5 t^2 + 2k_4 t + k_3) m + k_6 t^5 + k_5 t^4 + k_4 t^3 \\ & \quad + k_3 t^2 + k_2 t + k_1 \big) \bmod n. \end{aligned}$$

It is straightforward how to compute the coefficients $k_6, \ldots, k_0$ from a unbalanced length-2 representation $\tilde{k}_0, \tilde{k}_1$ of size $O(n^{4/7}), O(n^{3/7})$, respectively, by using divisions with powers of $m = N(\psi)$. Therefore we do not present the formulas for all the coefficients here.

4. Finding appropriate endomorphisms

Obviously the ideas explained in the last section only have a practical significance if we can find appropriate endomorphisms. Evaluation of these endomorphisms should not be too expensive, and their norm and trace has to satisfy the given restrictions.

In [1], the authors describe cryptographic applications of finite fields $GF(p^r)$ for a prime p close to the computer word size. If the computer word size is 32 bit, then the extension degree r for fields usable in ECC applications is quite small. They therefore suggest to use so called *Optimal Extension Fields* (OEF) as basis for ECC. OEF are fields such that the characteristic p is chosen as $2^t - s$ with $t = 31, 32$, and $GF(p^r)$ can be generated by an irreducible polynomial of the form $X^r - m$ for small integers s, m, respectively. Then all the basic field operations and the evaluation of the Frobenius map in the field can be done very efficiently on processors with 32-bit arithmetic (for more details, see [1]).

This scenario is well-suited for the new algorithm. The fast evaluation of the Frobenius map in the OEF induces fast evaluation of the Frobenius endomorphism for an elliptic curve defined over an OEF. Frobenius expansions as in [12] however cannot directly be used, since the norm of the Frobenius endomorphism would enforce huge tables. But if we use cryptographically strong elliptic curves, defined over $GF(p)$, whose Frobenius endomorphism Φ_E has small trace, then we can use all the ideas of Section 3 with endomorphism $\psi = \Phi_E$. Table 1 presents a list of possible characteristics and cryptographically strong elliptic curves (order over the extension field has a big prime factor) for extension degree 7. All the curves given in these both tables are good candidates for fast software implementation of ECCs on computers with 32-bit arithmetic.

Char. p	Trace of Frobenius	a_4	a_6	Bit size of largest prime factor
$2^{31} - 225$	-2	64922124	2111117807	186
$2^{31} - 1765$	3	2091628385	2051548053	186
$2^{31} - 4075$	-1	2000110324	2119769106	186
$2^{31} - 4117$	2	1085363243	1624141799	181
$2^{32} - 17$	-2	2970079685	1823624568	177
$2^{32} - 185$	-1	3270452368	1606832491	182
$2^{32} - 1319$	3	3592814884	3004984892	192
$2^{32} - 1683$	2	418128152	3313617494	187

Table 1. Cryptographically strong elliptic curves defined over $GF(p)$ (in short Weierstrass normal form) for OEF of degree 7. For all examples, the OEF can be defined by the polynomial $x^7 - 2$. The group order of the group of points $E(GF(p^7))$ of the elliptic curve over the OEF has a prime factor of the given bit size.

In Table 2, we present statistical data for random scalars for both algorithms. The average bit size of the two coefficients in Gallant's algorithm as described in Section 2 is shown for all elliptic curves from Table 1. Moreover we show the average bit size of the coefficients in the new algorithm with length seven. The formulas for length seven contain the 5th power of the trace, such that the restriction on the trace is absolutely necessary (note the difference in the size of k_1 for curves with trace 3 compared to trace -1 curves). Since in all examples $n^{1/7}$ is smaller than the norm of the Frobenius algorithm, the coefficient k_6 becomes always null. We experimented also with several other sizes for the first and second component of the lattice basis, but $n^{4/7}$ and $n^{3/7}$ as explained above proved superior in all these examples.

s	c	Max. bit size scalars	Av. bit size $\lvert k_0\rvert, \lvert k_1\rvert$ in Algorithm [2]	Av. bit size $\lvert k_i\rvert$, $0 \le i \le 6$, with new idea
225	-2	186	(91, 91)	(30, 35, 33, 32, 32, 29, 0)
1765	3	186	(92, 92)	(31, 37, 35, 33, 32, 30, 0)
4075	-1	186	(92, 92)	(30, 32, 32, 31, 31, 30, 0)
4117	2	181	(89, 89)	(30, 32, 32, 30, 29, 27, 0)
17	-2	177	(85, 89)	(30, 33, 32, 32, 30, 14, 0)
185	-1	182	(90, 88)	(30, 32, 31, 31, 30, 19, 0)
1319	3	192	(95, 95)	(30, 37, 35, 33, 32, 30, 0)
1683	2	187	(92, 92)	(30, 32, 32, 30, 29, 27, 0)

Table 2. Practical data for elliptic curves over OEFs with characteristic $p = 2^{31} - s$ (first block), $p = 2^{32} - s$ (second block), and extension degree 7 (curves from Table 1). The Frobenius endomorphism has trace c. We compare the average bit size of k_0, k_1 from [2] and the average bit size of the coefficients with the method of Section 3 (for length 7). We tested 100000 randomly chosen scalars k of size at most the maximal prime dividing the group order of $E(GF(p^7))$.

Therefore we can evaluate the multiplication $k \cdot P$ for points $P \in E(GF(p^7))$ by first computing the coefficients k_i, $0 \le i \le 6$, determine all the points $\Phi_E^i(P)$ for $1 \le i \le 6$ with repeated evaluation of the Frobenius endomorphism (which is fast), and using the formula

$$k \cdot P = \sum_{i=0}^{6} k_i \cdot \Phi_E^i(P).$$

The latter formula leads us to the problem of simultaneous point multiplication of 7 points with 7 different scalars.

5. Simultaneous multiple point multiplication

The *simultaneous multiple point multiplication problem* is defined as follows: Given r points $P_0, \ldots, P_{r-1}$ and r scalars $k_0 \ge \cdots \ge k_{r-1} > 0$, compute the point $k_0 \cdot P_0 + \cdots + k_{r-1} \cdot P_{r-1}$. In [2], the authors use simultaneous point multiplication with $r = 2$. They suggest to use the signed sliding window-based method of [9]. This algorithm is a combination of the sliding window algorithm [3], the algorithm of Lim and Lee [6], and the fact that for elliptic curves binary expansions can be substituted by signed binary expansions.

The basic idea of the algorithm is the handling of the r scalars in parallel, starting from the most significant bit to the least significant bit (therefore window size of one; bigger and variable window sizes are also possible, see [9]). If we consider bits of the same "position", then all signed bit vectors $\underline{v} = (v_1, \ldots, v_r) \in \{0, 1, -1\}^r$ can occur. Therefore the computation for this bit position can be done with one addition, if all points $v_1 \cdot P_1 + \cdots + v_r \cdot P_r$ for all possible signed bit vectors $\underline{v}$ are precomputed and stored. Then the algorithm needs storage of $(3^r - 1)/2$ points and needs at most $\lceil \log_2(k_1) \rceil - 1$ point doublings and also point additions. A fortunate situation occurs if the bit vector of the bits at some position is the zero vector; in this case no addition is necessary. Since signed binary representations have on average more zero bits than standard binary representations, the change to signed representations gives an improvement on cost of a larger precomputation.

For simultaneous multiple point multiplications with 2 scalars as in [2] this algorithm is practical. In our situation however, we have 7 scalars, such that the size of the precomputed table becomes huge, which makes this algorithm impractical. Therefore we propose a different algorithm, a variant of an idea of de Rooij [11]. De Rooij proposed a multiple exponentiation algorithm based on the euclidean gcd algorithm (see also [8, Algorithm 14.113]). The idea of the algorithm is the following: assume that we want to compute $k_0 \cdot P_0 + k_1 \cdot P_1$. Then we use a division with remainder for the scalars k_i and get integers q, r with $k_0 = q \cdot k_1 + r$, $0 \le r < k_1$. Therefore

$$k_0 \cdot P_0 + k_1 \cdot P_1 = (q \cdot k_1 + r) \cdot P_0 + k_1 \cdot P_1 = r \cdot P_0 + k_1 \cdot (P_1 + q \cdot P_0).$$

Thus the initial parallel multiplication problem with scalars k_0, k_1 is reduced to a new problem with scalars r, k_1. If the quotient q is equal to one, this reduction can be achieved with only one point addition. De Rooij generalizes this idea to several scalars. He proposes to always reduce the biggest scalar in $k_0, \ldots, k_{r-1}$ with the second biggest one. Finally there will remain only one non-zero scalar with corresponding point, which will be computed with standard fast multiplication.

For elliptic curves, we see two small differences where special properties of elliptic curves can be used. Obviously the fast binary multiplications at the end and for quotients q can be substituted by signed binary multiplications. It should however be observed that signed binary expansions are only better than the standard binary version for $q \ge 7$, such that a mixed strategy seems appropriate.

Algorithm 1.

Input: Points $P_0, \ldots, P_{r-1}$ and positive scalars $k_0, \ldots, k_{r-1}$.

Output: $H = k_0 \cdot P_0 + \cdots + k_{r-1} \cdot P_{r-1}$.

Algorithm:

1. Determine the indices M, N of the largest and second largest scalar k_i.

2. While $k_N \neq 0$ do
   (a) set $q = \lfloor k_M/k_N \rfloor$ and $k_M = k_M - q \cdot k_N$,
   (b) compute $P_N = P_N + q \cdot P_M$ with signed binary multiplication for $q \geq 7$ and binary multiplication otherwise,
   (c) determine the indices M, N of the largest and second largest scalar k_i.
3. Compute $H = k_M \cdot P_M$ with signed binary multiplication (for $k_M \geq 7$) and binary multiplication otherwise.
4. Return H.

Practical test implementations showed that for two scalars this variant is slightly faster than the original de Rooij algorithm (on average, one to three point operations). Once the number of scalars gets bigger than 3, the two algorithms need essentially the same number of point operations, since almost all the integers q are one.

Table 3 presents practical results for Gallant's algorithm compared to the new idea described in Section 3 of this paper. We present two variants of Gallant's algorithm, one with the de Rooij variant described above and one using the sliding window algorithm of [9] with window size at most three. We list the average number of point operations for the elliptic curves listed in Table 1.

Char. p	Trace Φ_E	Gallant with Algorithm 1	Gallant with sliding window [9]	New algorithm with Algorithm 1
$2^{31} - 225$	-2	75.5 A, 69.6 D	45.5 A, 95 D	70 A, 11.3 D
$2^{31} - 1765$	3	75.5 A, 69.7 D	45.5 A, 95 D	72 A, 12.8 D
$2^{31} - 4075$	-1	75.5 A, 69.6 D	45.5 A, 95 D	67.3 A, 9.8 D
$2^{31} - 4117$	2	73.1 A, 67.5 A	44.4 A, 92.2 D	66.5 A, 10.5 D
$2^{32} - 17$	-2	71.5 A, 67.7 D	43.6 A, 89.6 D	65.7 A, 12.1 D
$2^{32} - 185$	-1	73.5 A, 68.5 D	44.6 A, 92.3 D	65.7 A, 10.7 D
$2^{32} - 1319$	3	77.9 A, 72 D	46.6 A, 98 D	74.1 A, 13.1 D
$2^{32} - 1683$	2	75.6 A, 69.8 D	45.5 A, 95.2 D	68.6 A, 10.8 D

Table 3. Average number of point operations (A = point addition/subtraction, D = point doubling) for the curves given in Table 1. 100000 tests with random scalars of size at most the largest prime divisor of the group order $E(GF(p^7))$ were done.

These results show that Algorithm 1 needs slightly more point operations than the sliding window method from [9]. It has however the advantage that no storage of points apart from the input points is needed, whereas the sliding window algorithm needs memory for a internal table. The information from Table 3 also proves that the total number of point operations in the new algorithm is significantly smaller than in Gallant's algorithm. In all examples, the new algorithm needed about 40% less point operations than the algorithm from [2] with simultaneous signed sliding window multiplication. Although the new algorithm needs 6 evaluations of the Frobenius endomorphism compared to one in Gallant's algorithm, we still expect the new algorithm to be superior. This prediction is based on the fact that optimal extension fields are used as basis for the chosen elliptic curves such that the Frobenius map can be efficiently computed.

Conclusion

We have presented a new fast multiplication algorithm for especially chosen elliptic curves defined over optimal extension fields. The new algorithm consists of two parts, computing a Frobenius representation of length 7, and using simultaneous point multiplications. We also gave a list of elliptic curves which can be used with the new multiplication algorithm. Since we are not aware of any better algorithm for the discrete logarithm problem for these special curves, we assume that the existence of a large prime factor in the group order $E(GF(p^7))$ ensures the security of these curves (up to current knowledge) for usage in elliptic curve cryptosystems. The new multiplication algorithm together with the efficient field arithmetic for OEFs makes these curves optimal candidates for very fast software implementations of ECC on 32-bit computers.

References

[1] Bailey, D., Paar, C., Efficient arithmetic in finite field extensions with application in elliptic curve cryptography. Preprint, submitted to J. Cryptology, 1999.

[2] Gallant, R., Lambert, R., Vanstone, S., Faster point multiplication on elliptic curves with efficient endomorphisms. Preprint, available from `http://www.cacr.math.uwaterloo.ca`, 2000.

[3] Gordon, D., A survey of fast exponentiation methods. J. Algorithms 27 (1998), 129–146.

[4] IEEE P1363 Draft: Standard Specifications for Public Key Cryptography. Available on `http://stdsbbs.ieee.org/groups/1363/index.html` .

[5] Koblitz, N., CM-curves with good cryptographic properties. In: Advances in Cryptology – CRYPTO '91 (ed. by J. Feigenbaum; Lecture Notes in Comput. Sci. 576), 279–287. Springer, Berlin 1992.

[6] Lim, C.H., Lee, P.J., More flexible exponentiation with precomputation. In: Advances in Cryptology – CRYPTO '94 (ed. by Y.G. Desmedt; Lecture Notes in Comput. Sci. 839), 95–107. Springer, Berlin 1994.

[7] Meier, W., Staffelbach, O., Efficient multiplication on certain non supersingular elliptic curves. In: Advances in Cryptology – CRYPTO '92 (ed. by E.F. Brickell; Lecture Notes in Comput. Sci. 740), 333–344. Springer, Berlin 1993.

[8] Menezes, A.J., van Oorschot, P.C., Vanstone, S.A., Handbook of Applied Cryptography. CRC Press, Boca Raton 1997.

[9] Miyaji, A., Ono, T., Cohen, H., Efficient elliptic curve exponentiation. In: Information and communications security (ICICS'97) (ed. by Y. Han, T. Okamoto, S. Qing; Lecture Notes in Comput. Sci. 1334), 282–290. Springer, Berlin 1997.

[10] Müller, V., Fast multiplication on elliptic curves over small fields of characteristic two. J. Cryptology 11 (1998), 219–234.

[11] de Rooij, P. Efficient exponentiation using precomputation and vector addition chains. In: Advances in Cryptology – EUROCRYPT'94 (ed. by A. De Santis; Lecture Notes in Comput. Sci. 950), 389–399. Springer, Berlin 1994.

[12] Smart, N., Elliptic curve cryptosystems over small fields of odd characteristic. J. Cryptology 12 (1999), 193–196.

Error-correcting codes and cryptography

Harald Niederreiter

Abstract. There have been fascinating applications of error-correcting codes to cryptography, starting with the construction of the well-known public-key cryptosystem of McEliece in 1978. This paper presents a survey of the uses of coding theory in the design of cryptographic schemes, and particularly of public-key cryptosystems. Furthermore, an account is given of recent developments in coding theory that can lead to better choices of suitable codes for cryptographic purposes.

1. Introduction

Coding theory and cryptography are two branches of information theory that have different goals (data correctness and data security, respectively), but nevertheless possess some links. Most of the relationships between coding theory and cryptography stem from the observation that coding theory is a source of hard problems that can form the basis of cryptosystems. Indeed, coding-theoretic ideas have a great potential in cryptography, and some, but certainly not all, of this potential has already been utilized. The most successful realizations of these ideas have led to cryptosystems that have more or less resisted cryptanalysis. In this paper we survey applications of coding theory to the design of cryptographic schemes, with an emphasis on public-key cryptosystems.

We first review some basic concepts of algebraic coding theory. As the signal alphabet for codes we always use $\mathbf{F}_q$, the finite field of order q. A *linear code* C over $\mathbf{F}_q$ is a nonzero linear subspace of the vector space $\mathbf{F}_q^n$ for some $n \geq 1$. The integer n is called the *length* of C and $k := \dim(C)$ is called the *dimension* of C. The (*Hamming*) *weight* $w(\mathbf{x})$ is the number of nonzero coordinates of $\mathbf{x} \in \mathbf{F}_q^n$, and the *minimum distance* d of C is the smallest weight of any nonzero vector in C. We collect these basic parameters of C by saying that C is a linear $[n, k, d]$ code over $\mathbf{F}_q$. Encoding is done by an $\mathbf{F}_q$-linear isomorphism $\psi : \mathbf{F}_q^k \to C$. This map can be represented by a $k \times n$ matrix G over $\mathbf{F}_q$, the *generator matrix* of C, in the sense that

$$\psi(\mathbf{a}) = \mathbf{a}G \in C \quad \text{for all} \quad \mathbf{a} \in \mathbf{F}_q^k.$$

Note that C is the row space of G and $\operatorname{rank}(G) = k$.

The linear code C can also be viewed as the set of all $\mathbf{c} \in \mathbf{F}_q^n$ such that $H\mathbf{c}^{\mathrm{T}} = \mathbf{0}$, where H is an $(n-k) \times n$ matrix over $\mathbf{F}_q$ with $\operatorname{rank}(H) = n - k$. In other words,

Public-Key Cryptography and Computational Number Theory
K. Alster, J. Urbanowicz and H. C. Williams (Eds.)

C is the null space of H. Any such matrix H is called a *parity-check matrix* of C. The matrices G and H are related by

$$GH^{\mathrm{T}} = 0.$$

If C is a linear $[n, k, d]$ code over $\mathbf{F}_q$, then C can correct up to $t := \lfloor (d-1)/2 \rfloor$ errors in each block of signals of length n. The vector space $\mathbf{F}_q^n$ can be partitioned into q^{n-k} cosets

$$\mathbf{a} + C = \{\mathbf{a} + \mathbf{c} : \mathbf{c} \in C\}.$$

If $\mathbf{c} \in C$ is transmitted and $\mathbf{y} \in \mathbf{F}_q^n$ is received, then we have an error vector

$$\mathbf{e} = \mathbf{y} - \mathbf{c} \in \mathbf{y} + C.$$

In an optimistic scenario, the most likely error vector is one with the minimum weight in the coset $\mathbf{y} + C$ of $\mathbf{y}$, i.e., a *coset leader* of $\mathbf{y} + C$. If a parity-check matrix H of C is known, then the coset of $\mathbf{y}$ is uniquely determined by the *syndrome* $H\mathbf{y}^{\mathrm{T}}$ of $\mathbf{y}$, since

$$H\mathbf{y}^{\mathrm{T}} = H\mathbf{z}^{\mathrm{T}} \Leftrightarrow \mathbf{y} + C = \mathbf{z} + C.$$

This leads to a general decoding algorithm for C: for a received vector $\mathbf{y} \in \mathbf{F}_q^n$, calculate the syndrome $H\mathbf{y}^{\mathrm{T}}$, find a coset leader $\mathbf{e}$ with the same syndrome, and then decode $\mathbf{y}$ as $\mathbf{x} = \mathbf{y} - \mathbf{e}$. Faster algorithms are known for special codes C. We refer to MacWilliams and Sloane [23] and van Lint [22] for further background on coding theory.

2. Public-key cryptosystems based on codes

The theoretical foundation for two public-key cryptosystems based on coding theory is the result of Berlekamp, McEliece, and van Tilborg [3] that the decoding problem for general linear codes is NP-complete.

The first proposal of a public-key cryptosystem using linear codes is due to McEliece [24]. Let C be a linear $[n, k, d]$ code over $\mathbf{F}_q$ and let the $k \times n$ matrix G be a generator matrix of C. Choose a random nonsingular $k \times k$ matrix S over $\mathbf{F}_q$ and a random $n \times n$ matrix P over $\mathbf{F}_q$ which is obtained from a nonsingular diagonal matrix by row permutations. Then form a new generator matrix $G' = SGP$ which may be viewed as a scrambled version of G. The McEliece public-key cryptosystem operates as follows.

Private key: S, G, P.

Public key: G'.

Encryption: The admissible plaintexts are row vectors $\mathbf{x} \in \mathbf{F}_q^k$. Given a plaintext $\mathbf{x}$, choose a random row vector $\mathbf{z} \in \mathbf{F}_q^n$ with weight $w(\mathbf{z}) \le t := \lfloor (d-1)/2 \rfloor$ and form the ciphertext $\mathbf{y} = \mathbf{x}G' + \mathbf{z} \in \mathbf{F}_q^n$.

Decryption: Given the ciphertext $\mathbf{y} \in \mathbf{F}_q^n$, compute $\mathbf{y}' = \mathbf{y}P^{-1} = \mathbf{x}SG + \mathbf{z}P^{-1}$. Note that $w(\mathbf{y}' - \mathbf{x}SG) = w(\mathbf{z}P^{-1}) = w(\mathbf{z}) \leq t$. Since $\mathbf{x}SG = (\mathbf{x}S)G$ belongs to the code C, an application of the decoding algorithm of C to $\mathbf{y}'$ yields $\mathbf{x}' := \mathbf{x}S$, from which we recover the plaintext $\mathbf{x} = \mathbf{x}'S^{-1}$.

A detailed treatment of the McEliece public-key cryptosystem can be found in the book of Horster and Sperling [15]. We defer the discussion of the security aspects of the McEliece public-key cryptosystem until after the presentation of the following cryptosystem.

The public-key cryptosystem introduced by the author [25] shares some features with the McEliece system. The main difference is that it is a public-key cryptosystem for low-weight plaintext vectors, like the well-known cryptosystem of Chor and Rivest [8]. The starting point in the design of the author's public-key cryptosystem is again a linear $[n, k, d]$ code C over $\mathbf{F}_q$, but now we consider a parity-check matrix H of C, which is an $(n-k) \times n$ matrix over $\mathbf{F}_q$. Choose a random nonsingular $(n-k) \times (n-k)$ matrix M over $\mathbf{F}_q$ and a random $n \times n$ matrix P over $\mathbf{F}_q$ which is obtained from a nonsingular diagonal matrix by row permutations. Then form a new parity-check matrix $H' = MHP$ which may be viewed as a scrambled version of H. Now the public-key cryptosystem works as follows.

Private key: M, H, P.

Public key: H'.

Encryption: The admissible plaintexts are column vectors $\mathbf{x} \in \mathbf{F}_q^n$ with weight $w(\mathbf{x}) \leq t := \lfloor (d-1)/2 \rfloor$. Given a plaintext $\mathbf{x}$, the corresponding ciphertext is $\mathbf{y} = H'\mathbf{x}$.

Decryption: Given the ciphertext $\mathbf{y}$, which is a column vector from $\mathbf{F}_q^{n-k}$, we first compute $\mathbf{y}' = M^{-1}\mathbf{y} = HP\mathbf{x}$. Now $\mathbf{x}' := P\mathbf{x}$ satisfies $w(\mathbf{x}') \leq t$, and so $\mathbf{x}'$ can be viewed as an error vector. Therefore, an application of the decoding algorithm of C to the syndrome $\mathbf{y}' = H\mathbf{x}'$ yields the error vector $\mathbf{x}'$. Finally, we recover the plaintext $\mathbf{x} = P^{-1}\mathbf{x}'$.

An important result for the cryptanalysis of the two public-key cryptosystems of McEliece and the author was shown by Li, Deng, and Wang [20]. They proved that with corresponding choices of code parameters, the two cryptosystems have equivalent security levels, in the sense that the problem of breaking either cryptosystem can be reduced to that of breaking the other cryptosystem by a simple polynomial-time algorithm. Therefore, in the following discussion of security aspects we will often consider the two cryptosystems simultaneously.

Attacks on the McEliece public-key cryptosystem, and thus on the author's public-key cryptosystem, were mounted for instance by Adams and Meijer [1], Chabaud [7], Gibson [10], Lee and Brickell [19], and van Tilburg [35]. These attacks used probabilistic algorithms and they have not been successful, except that greater care has to be taken in the choice of code parameters to thwart these attacks. In the case of the author's cryptosystem, it was already pointed out in

[26, Section 3.2] that the code parameters have to be chosen substantially larger than in the original paper [25]. The cryptanalytic attack announced by Korzhik and Turkin [18], who claimed to have broken the McEliece public-key cryptosystem, is not believed to be effective.

Canteaut and Sendrier [6] developed a new probabilistic algorithm for finding minimum-weight vectors in a large linear code. This yields another attack on the public-key cryptosystems of McEliece and the author. As a consequence, we need to choose larger code parameters in these cryptosystems than previously recommended. Canteaut and Sendrier [6] also carried out a comparative analysis of the McEliece cryptosystem, the author's cryptosystem, and RSA (with comparable security levels). They report that the author's cryptosystem allows a much faster encryption per information bit: it is faster than encryption in the McEliece cryptosystem by a factor 10 and faster than RSA encryption by a factor 48. Furthermore, they point out various other advantages of the author's cryptosystem over the McEliece cryptosystem. They suggest a key size of around 88000 bytes in the cryptosystems of McEliece and the author, according to the current state of the art in cryptanalysis. This large key size is still a drawback of these two cryptosystems, but progress in hardware technology may alleviate this problem. A comparison between the author's public-key cryptosystem and the Chor–Rivest public-key cryptosystem was undertaken in [25] and leads to the result that the first cryptosystem has a much better information rate than the second cryptosystem.

Berson [4] introduced cryptanalytic attacks against the McEliece cryptosystem that can be used if messages are resent or if messages satisfy a known linear relation. These attacks cannot be mounted against the author's cryptosystem since this cryptosystem is fully deterministic. Sun [34] carried out further work on Berson's attacks and proposed new variants of the McEliece cryptosystem which involve combinations with hash functions or block ciphers like DES and are immune against these attacks. Sendrier [31] designed an algorithm which, for a given linear code which is known to be a concatenation of an inner code and an outer code, recovers the inner code and the outer code. This has the implication that if a concatenated linear code is used in the public-key cryptosystems of McEliece and the author, then the security level is that of the inner code alone or of the outer code alone, whichever is greater. Hence we do not gain anything by employing concatenated linear codes in these cryptosystems. Gabidulin and Kjelsen [9] proposed a further scrambling of the public-key matrix in the cryptosystems of McEliece and the author to improve the security.

We now describe a third public-key cryptosystem based on error-correcting codes which was presented by Wu and Dawson at the Fourth International Conference on Finite Fields at Waterloo, Ontario, in August 1997. Let G be a generator matrix and H a parity-check matrix of a linear $[n, k, d]$ code over $\mathbf{F}_q$. Choose an $n \times (n-k)$ matrix J over $\mathbf{F}_q$ which is a pseudo-inverse of H, in the sense that $JHJ = J$. Then the Wu–Dawson public-key cryptosystem operates as follows.

Private key: JH.

Public key: G, J.

Encryption: The admissible plaintexts are row vectors $\mathbf{x} \in \mathbf{F}_q^k$. Given a plaintext $\mathbf{x}$, choose a random row vector $\mathbf{z} \in \mathbf{F}_q^{n-k}$ and form the ciphertext $\mathbf{y} = \mathbf{x}G + \mathbf{z}J^{\mathrm{T}}$.

Decryption: Given the ciphertext $\mathbf{y} \in \mathbf{F}_q^n$, we first compute $\mathbf{y} - \mathbf{y}(JH)^{\mathrm{T}}$. Using $GH^{\mathrm{T}} = 0$ and $JHJ = J$, we see that $\mathbf{y}(JH)^{\mathrm{T}} = \mathbf{z}J^{\mathrm{T}}$, and so $\mathbf{y} - \mathbf{y}(JH)^{\mathrm{T}} = \mathbf{x}G$. Knowing $\mathbf{x}G$, we can recover the plaintext $\mathbf{x}$ by linear algebra.

About one year after its introduction, the Wu–Dawson cryptosystem was broken by Roelse [30] who showed that a matrix R satisfying $\mathbf{y} - \mathbf{y}R^{\mathrm{T}} = \mathbf{x}G$ can be computed efficiently from the public key. Knowing such a matrix R is enough for decryption. The Wu–Dawson cryptosystem was independently broken by Blackburn [5]. The brief lifespan of the Wu–Dawson cryptosystem serves as a warning that designing a good public-key cryptosystem using coding theory also requires great care. The weakness of the Wu–Dawson cryptosystem is that it is not based on a hard problem of coding theory such as the decoding problem for general linear codes.

3. Further cryptographic schemes based on codes

We briefly survey some other types of cryptographic schemes that are based on coding theory. Rao and Nam [29] introduced a private-key (i.e., symmetric) cryptosystem which uses similar ingredients as the McEliece cryptosystem and in which the table of syndromes and coset leaders plays an important role. Li and Wang [21] modified the Rao–Nam cryptosystem and obtained a joint authentication and private-key encryption scheme which has the same security level and the same complexity of the encryption and decryption algorithms as the Rao–Nam cryptosystem. They noted that a similar modification can be made in the McEliece cryptosystem to get a public-key scheme for the purpose of both encryption and authentication.

Stern [32] designed a zero-knowledge protocol for identification which uses a parity-check matrix of a good binary linear code. The security of this scheme is based on the presumed intractability of the syndrome-decoding problem, i.e., the problem of computing the error vector from a given syndrome for a general linear code. Véron [38] proposed a dual version of this scheme which uses a generator matrix of the linear code instead of a parity-check matrix. This yields an improved transmission rate and reduces the complexity of computations.

Harari [11] proposed a secret-sharing scheme based on coding theory. Later, the same author [12] introduced an authentication scheme using binary linear codes. Some work was devoted to the construction of hash functions by means of coding theory. The inverse of the encoding function ψ in Section 1 is a candidate for a hash function, but it needs to be modified since it is a linear transformation.

Harari [13] suggested to combine decoding functions of binary linear codes with nonlinear functions on bit strings to obtain hash functions. Knudsen and Preneel [17] proposed a construction of hash functions in which block ciphers such as DES or IDEA are combined with linear codes over a finite extension field of $\mathbf{F}_2$.

Kabatianskii, Krouk, and Smeets [16] introduced a digital signature scheme which is based on the fact that for a linear code the set of correctable syndromes, although being in general not a vector space, contains as a subset a rather large vector space. This digital signature scheme improves on earlier such schemes proposed by Alabbadi and Wicker [2], Harn and Wang [14], and Wang [39]. As a curiosity we add that some authors, apparently not being familiar with the Chinese custom of listing the family name first, refer to Wang's digital signature scheme as the "Xinmei digital signature scheme" (Wang's full name is Wang Xinmei).

4. The choice of codes

We return to the public-key cryptosystems of McEliece and the author discussed in Section 2. In both cases, the criteria for the choice of suitable linear $[n, k, d]$ codes over $\mathbf{F}_q$ are the same. First of all, the numbers n, k, d, and q should be such that there exist many linear codes with these parameters (this prevents an attack by exhaustive search for the private key). Furthermore, the chosen linear code should allow a fast decoding algorithm (this guarantees fast decryption). Finally, the linear code should be selected in such a way that the *information rate* k/n is as large as possible for a given *relative minimum distance* d/n and a given q (this maximizes the information per bit of ciphertext). The last condition is exactly the same as that for the choice of good linear codes with regard to error correction. The asymptotic theory of linear codes, which will be sketched in the following, addresses the issue of this choice for long codes.

For a fixed prime power q, let U_q be the set ordered pairs $(\delta, R) \in [0,1]^2$ for which there exists a sequence $(C_i)_{i=1}^{\infty}$ of linear codes over $\mathbf{F}_q$ with the lengths of the C_i tending to ∞, with δ being the limit of the relative minimum distances of the C_i, and with R being the limit of the information rates of the C_i. It is known that there exists a continuous nonincreasing function α_q on $[0,1]$ such that

$$U_q = \{(\delta, R) : 0 \leq R \leq \alpha_q(\delta), 0 \leq \delta \leq 1\}.$$

In other words, U_q is the set in the first quadrant under the graph of α_q. Standard arguments of coding theory show that $\alpha_q(0)=1$ and $\alpha_q(\delta)=0$ for $(q-1)/q \leq \delta \leq 1$. The function α_q is unknown, but it is an important issue in coding theory to obtain good lower bounds for α_q on the open interval $(0, (q-1)/q)$. The asymptotic Gilbert–Varshamov bound, a classical result that goes back to the 1950s, says that

$$\alpha_q(\delta) \geq R_{GV}(q, \delta) := 1 - H_q(\delta) \quad \text{for all} \quad \delta \in \left(0, \frac{q-1}{q}\right),$$

where H_q is the q-ary entropy function

$$H_q(\delta) = \delta \log_q(q-1) - \delta \log_q \delta - (1-\delta)\log_q(1-\delta),$$

with $\log_q$ denoting the logarithm to the base q. For a long time, no improvements on the asymptotic Gilbert–Varshamov bound were obtained, and so it was believed by some that this bound is best possible. It came as a great surprise when in the early 1980s the asymptotic Gilbert–Varshamov bound was improved in a certain range by means of algebraic-geometry codes.

Let $\mathcal{A}$ be an (absolutely irreducible, smooth, projective) algebraic curve over $\mathbf{F}_q$ of genus g and let $P_1, \ldots, P_n$ be distinct $\mathbf{F}_q$-rational points on $\mathcal{A}$. For a divisor G of $\mathcal{A}$ with $P_1, \ldots, P_n$ not in the support of G, we consider the Riemann–Roch space

$$\mathcal{L}(G) = \{f \in F^* : \operatorname{div}(f) + G \geq 0\} \cup \{0\},$$

where F is the function field of $\mathcal{A}$ and $\operatorname{div}(f)$ the principal divisor corresponding to $f \in F^*$. Note that $\mathcal{L}(G)$ is a finite-dimensional vector space over $\mathbf{F}_q$. An *algebraic-geometry code* is defined as the image of the $\mathbf{F}_q$-linear map

$$f \in \mathcal{L}(G) \mapsto (f(P_1), \ldots, f(P_n)) \in \mathbf{F}_q^n.$$

If $\deg(G) < n$, then this construction yields a linear $[n, k, d]$ code over $\mathbf{F}_q$ with

$$k + d \geq n - g + 1.$$

In order to optimize this construction, we let $N_q(g)$ be the maximum number of $\mathbf{F}_q$-rational points that an algebraic curve over $\mathbf{F}_q$ of genus g can have and choose $\mathcal{A}$ to be an algebraic curve over $\mathbf{F}_q$ of genus g for which this maximum is attained. By letting g tend to ∞, we obtain the bound

$$\alpha_q(\delta) \geq R_{AG}(q,\delta) := 1 - \frac{1}{A(q)} - \delta \qquad \text{for} \quad 0 \leq \delta \leq 1,$$

where $A(q)$ is a positive number depending only on q which is defined by

$$A(q) := \overline{\lim_{g \to \infty}} \ \frac{N_q(g)}{g}.$$

For proofs of these results and for further details on algebraic-geometry codes, we refer to the books of Niederreiter and Xing [28], Stichtenoth [33], and Tsfasman and Vlădut [36].

In order to beat the asymptotic Gilbert–Varshamov bound, we need good lower bounds on $A(q)$. Based on the bound $A(q) \geq q^{1/2} - 1$ for squares q, Tsfasman, Vlădut, and Zink [37] showed that for squares $q \geq 49$ we have

$$R_{AG}(q,\delta) > R_{GV}(q,\delta)$$

if δ belongs to a suitable subinterval of $(0, (q-1)/q)$. This result disproved the conjecture that the asymptotic Gilbert–Varshamov bound is best possible. For non-squares q it took even longer to beat the asymptotic Gilbert–Varshamov bound.

Niederreiter and Xing [27] established the following lower bound on $A(q)$ for composite nonsquares q.

Theorem 1. *If $q = p^e$ with an odd prime p and an odd integer $e \geq 3$, then*

$$A(q) \geq \frac{q^{1/m} + 1}{(2q^{1/m} + 3)^{1/2} + 1},$$

where m is the least prime dividing e. A similar result holds for $p = 2$.

For any composite nonsquare q put

$$R_{NX}(q, \delta) = 1 - \frac{1}{b(q)} - \delta,$$

where $b(q)$ is the lower bound on $A(q)$ in Theorem 1. Then it follows from the above that

$$\alpha_q(\delta) \geq R_{NX}(q, \delta) \qquad \text{for } 0 \leq \delta \leq 1.$$

A comparison of the two functions $R_{NX}(q, \delta)$ and $R_{GV}(q, \delta)$ leads to the following result in [27].

Theorem 2. *Let $e \geq 3$ be an odd integer and let r be a prime power with $r \geq 100e^3$ if r is odd and $r \geq 576e^3$ if r is even. Then there exists an open interval $(\delta_1, \delta_2) \subseteq (0, 1)$ containing $(r^e - 1)/(2r^e - 1)$ such that*

$$R_{NX}(r^e, \delta) > R_{GV}(r^e, \delta) \qquad \textit{for all } \delta \in (\delta_1, \delta_2).$$

Algebraic-geometry codes are the only family of linear codes for which it has been possible so far to beat the asymptotic Gilbert–Varshamov bound.

5. Conclusions

The material presented in this paper demonstrates that error-correcting codes are a rich source for the construction of cryptographic schemes of various types, such as public-key cryptosystems, digital signature schemes, and hash functions. Well-designed public-key cryptosystems based on coding theory offer the following advantages:

- fast encryption;
- easy implementation;
- arbitrarily long linear codes of very good quality are available, for instance algebraic-geometry codes;
- breaking the systems is hard.

The only known drawback is the large key size that is required.

References

[1] Adams, C., Meijer, H., Security-related comments regarding McEliece's public-key cryptosystem. IEEE Trans. Inform. Theory 35 (1989), 454–455.

[2] Alabbadi, M., Wicker, S.B., Cryptanalysis of the Harn and Wang modification of the Xinmei digital signature scheme. Electronics Letters 28 (1992), 1756–1758.

[3] Berlekamp, E.R., McEliece, R.J., van Tilborg, H.C.A., On the inherent intractability of certain coding problems. IEEE Trans. Inform. Theory 24 (1978), 384–386.

[4] Berson, T.A., Failure of the McEliece public-key cryptosystem under message-resend and related-message attack. In: Advances in Cryptology – CRYPTO '97 (ed. by B.S. Kaliski; Lecture Notes in Comput. Sci. 1294), 213–220. Springer, Berlin 1997.

[5] Blackburn, S.R., Cryptanalysis of a public key cryptosystem due to Wu and Dawson. IEE Proc. Computers and Digital Techniques, to appear.

[6] Canteaut, A., Sendrier, N., Cryptoanalysis of the original McEliece cryptosystem. In: Advances in Cryptology – ASIACRYPT '98 (ed. by K. Ohta and D.Y. Pei; Lecture Notes in Comput. Sci. 1514), 187–199. Springer, Berlin 1998.

[7] Chabaud, F., On the security of some cryptosystems based on error-correcting codes. In: Advances in Cryptology – EUROCRYPT '94 (ed. by A. De Santis; Lecture Notes in Comput. Sci. 950), 131–139. Springer, Berlin 1995.

[8] Chor, B., Rivest, R.L., A knapsack-type public key cryptosystem based on arithmetic in finite fields. IEEE Trans. Inform. Theory 34 (1988), 901–909.

[9] Gabidulin, E.M., Kjelsen, O., Randomizing the public key for improving performances of public-key cryptosystems. In: Proc. Sixth Joint Swedish-Russian Internat. Workshop on Information Theory (Mölle, 1993), 297–301. Studentlitteratur, Lund 1993.

[10] Gibson, J.K., Equivalent Goppa codes and trapdoors to McEliece's public key cryptosystem. In: Advances in Cryptology – EUROCRYPT '91 (ed. by D.W. Davies; Lecture Notes in Comput. Sci. 547), 517–521. Springer, Berlin 1991.

[11] Harari, S., Application des codes correcteurs au partage du secret. Traitement du Signal 4 (1987), 353–356.

[12] —, A new authentication algorithm. In: Coding Theory and Applications (ed. by G. Cohen and J. Wolfmann; Lecture Notes in Comput. Sci. 388), 91–105. Springer, Berlin 1989.

[13] —, HCC: a hash function using error correcting codes. In: Cryptography and Coding (ed. by M. Darnell; Lecture Notes in Comput. Sci. 1355), 110–117. Springer, Berlin 1997.

[14] Harn, L., Wang, D.C., Cryptanalysis and modification of digital signature scheme based on error-correcting codes. Electronics Letters 28 (1992), 157–159.

[15] Horster, P., Sperling, R., Das Kryptosystem von McEliece. Hüthig Buch Verlag, Heidelberg 1990.

[16] Kabatianskii, G., Krouk, E., Smeets, B., A digital signature scheme based on random error-correcting codes. In: Cryptography and Coding (ed. by M. Darnell; Lecture Notes in Comput. Sci. 1355), 161–167. Springer, Berlin 1997.

[17] Knudsen, L., Preneel, B., Fast and secure hashing based on codes. In: Advances in Cryptology – CRYPTO '97 (ed. by B.S. Kaliski; Lecture Notes in Comput. Sci. 1294), 485–498. Springer, Berlin 1997.

[18] Korzhik, V.I., Turkin, A.I., Cryptanalysis of McEliece's public-key cryptosystem. In: Advances in Cryptology – EUROCRYPT '91 (ed. by D.W. Davies; Lecture Notes in Comput. Sci. 547), 68–70. Springer, Berlin 1991.

[19] Lee, P.J., Brickell, E.F., An observation on the security of McEliece's public-key cryptosystem. In: Advances in Cryptology – EUROCRYPT '88 (ed. by C.G. Günther; Lecture Notes in Comput. Sci. 330), 275–280. Springer, Berlin 1988.

[20] Li, Y.X., Deng, R.H., Wang, X.M., On the equivalence of McEliece's and Niederreiter's public-key cryptosystems. IEEE Trans. Inform. Theory 40 (1994), 271–273.

[21] Li, Y.X., Wang, X.M., A joint authentication and encryption scheme based on algebraic coding theory. In: Applied Algebra, Algebraic Algorithms and Error-Correcting Codes (ed. by H.F. Mattson, T. Mora, and T.R.N. Rao; Lecture Notes in Comput. Sci. 539), 241–245. Springer, Berlin 1991.

[22] van Lint, J.H., Introduction to Coding Theory, 3rd ed. Springer, Berlin 2000.

[23] MacWilliams, F.J., Sloane, N.J.A., The Theory of Error-Correcting Codes. North-Holland, Amsterdam 1977.

[24] McEliece, R.J., A public-key cryptosystem based on algebraic coding theory. DSN Progress Report 42–44, pp. 114–116. Jet Propulsion Lab., Pasadena, CA, 1978.

[25] Niederreiter, H., Knapsack-type cryptosystems and algebraic coding theory. Problems of Control and Inform. Theory 15 (1986), 159–166.

[26] —, Finite fields and cryptology. In: Finite Fields, Coding Theory, and Advances in Communications and Computing (ed. by G.L. Mullen and P.J.-S. Shiue), 359–373. M. Dekker, New York 1993.

[27] Niederreiter, H., Xing, C.P., Towers of global function fields with asymptotically many rational places and an improvement on the Gilbert-Varshamov bound. Math. Nachr. 195 (1998), 171–186.

[28] —, —, Rational Points on Curves over Finite Fields: Theory and Applications. Cambridge University Press, Cambridge, to appear.

[29] Rao, T.R.N., Nam, K.H., Private-key algebraic-code encryptions. IEEE Trans. Inform. Theory 35 (1989), 829–833.

[30] Roelse, P., Cryptanalysis of the Wu-Dawson public key cryptosystem. Finite Fields Appl. 5 (1999), 386–392.

[31] Sendrier, N., On the concatenated structure of a linear code. Applicable Algebra Engrg. Comm. Comput. 9 (1998), 221–242.

[32] Stern, J., A new identification scheme based on syndrome decoding. In: Advances in Cryptology – CRYPTO '93 (ed. by D.R. Stinson; Lecture Notes in Comput. Sci. 773), 13–21. Springer, Berlin 1994.

[33] Stichtenoth, H., Algebraic Function Fields and Codes. Springer, Berlin 1993.

[34] Sun, H.-M., Improving the security of the McEliece public-key cryptosystem. In: Advances in Cryptology – ASIACRYPT '98 (ed. by K. Ohta and D.Y. Pei; Lecture Notes in Comput. Sci. 1514), 200–213. Springer, Berlin 1998.

[35] van Tilburg, J., On the McEliece public-key cryptosystem. In: Advances in Cryptology – CRYPTO '88 (ed. by S. Goldwasser; Lecture Notes in Comput. Sci. 403), 119–131. Springer, Berlin 1990.

[36] Tsfasman, M.A., Vlădut, S.G., Algebraic-Geometric Codes. Kluwer, Dordrecht 1991.

[37] Tsfasman, M.A., Vlădut, S.G., Zink, T., Modular curves, Shimura curves, and Goppa codes, better than Varshamov-Gilbert bound. Math. Nachr. 109 (1982), 21–28.

[38] Véron, P., Improved identification schemes based on error-correcting codes. Appl. Algebra Engrg. Comm. Comput. 8 (1997), 57–69.

[39] Wang, X.M., Digital signature scheme based on error-correcting codes. Electronics Letters 26 (1990), 898–899.

Secret public key schemes

Jacques Patarin

Abstract. In this paper, we present an amazing concept concerning public-key cryptography: the concept of "secret" public-key schemes.

A "secret" public-key scheme, as its name implies, is a public-key algorithm, built so that the algorithm used in secret-key computations, to perform decryption, signature or authentication, remains secret and cannot be deduced from the public key.

This concept was introduced in [7] in the case of a very specific algorithm (HFEV), but we will see here that this concept is in fact very general.

We shall explain that it is indeed possible to build such "secret public-key schemes" in many ways, which are very different from each other. First, we will present some examples of "secret public-key schemes" by using multivariate polynomials. We will also present a new family of schemes, called "Chains" schemes. Then we will show that it is possible to build "secret public-key schemes" by using a theorem of logic, i.e. the Rice theorem, or by using a dissymetry on the amount of memory available for the prover and the verifier, or by signing the software itself that performs the public-key computation.

Finally, some examples of possible application of this strange and theoretical concept will be suggested.

1. Introduction

When looking at the history of cryptography, one realizes that the first algorithms used in cryptography were "secret secret-key" algorithms. By "secret-key algorithm", we mean, as usual, that Alice and Bob, the two persons that want to communicate with cryptography, had to agree on a common secret key K before using their algorithm. By "secret algorithm", we mean that the way Alice and Bob proceed, in particular the way they use K, is secret. Such "secret secret-key algorithms" (they are also called "proprietary" secret-key algorithms) are still massively used at present (in mobile phones, bank transactions, etc.) for encryption, authentication or digital signatures.

Nevertheless, in 1975, a new type of algorithm appeared with the publication of DES (cf. [5]). We will call it a "public secret-key scheme", because it is a secret-key scheme (a 56 bit secret key K has to be known by both Alice and Bob), but the DES algorithm is public: every one knows exactly how the key K is used to encrypt messages or to compute MACs. Since then, many other public secret-key

Public-Key Cryptography and Computational Number Theory
K. Alster, J. Urbanowicz and H. C. Williams (Eds.)

algorithms were intentionally published, for example FEAL, IDEA, BLOWFISH, etc., and the contest organized by the NIST to choose the AES algorithm was an important example that favored the study and the design of new public secret-key algorithms.

Moreover, at present, the cryptographic community agrees on the fact that a secret-key algorithm can be used for sensitive applications only if it is still secure when it is published, its secret key being the only part that remains secret. Reading the history of cryptography is sufficient in order to be convinced of this: it is far too dangerous to use weak algorithms, even if the users try to keep them secret. So a "secret secret-key" scheme must remain secure when it becomes a "public secret-key scheme". However, many companies still want to keep their secret-key algorithm secret as long as possible in order to give a strict minimum of information to potential enemies.

In 1976, Diffie and Hellman published their celebrated paper that described the concept of public-key cryptography. Then, in 1977, RSA, the most famous public-key algorithm, was published (cf. [17]). In fact, RSA is a "public public-key" algorithm, since all its details were described.

Now that we have seen "secret secret-key schemes", "public secret-key schemes" and "public public-key schemes", a natural question arises: do "secret public-key" schemes exist? The aim of the present paper is to study this question, and to give concrete examples of such algorithms. This concept was first presented in [7, p. 220] with an example. In the present paper, we will see that this concept is in fact very general.

In a "secret public-key" algorithm, what is secret is the algorithm A used for secret-key computations (not only the secret key used by A, but also A itself). The algorithm B, used for public-key computations, is public. Note that it must be impossible to deduce A from B. So RSA, or DSA, for example, could not be secret public-key schemes, because from their public key, it is easy to see that indeed they are RSA, or DSA, and to know what kind of secret computation are associated to them. Note also that, even if A becomes public (but the secret key K used by A remains secret), the algorithm has to remain secure. As a "secret secret-key" algorithm must remain secure when it becomes public and thus becomes a "public secret-key scheme", a "secret public-key scheme" must remain secure when it becomes public. This is conform with the so-called Kerckhoff's principle, and this principle is generally accepted in the modern cryptographic community.

We will present several examples of such algorithms in this paper. If some of them seem esoteric, certain others, on the contrary, are very natural. Moreover, several concrete applications of the theoretical principle will be quickly presented, in particular in the field of smartcards. As a matter of fact, some "secret public-key" algorithms have a flexibility and a simplicity of computations that make them very attractive in smartcard-based applications.

Part I
The standard notion of "secret public-key schemes"

2. Definitions

Let us detail the definition of a "secret public-key scheme" in the case of encryption, signature and authentication.

Definition in encryption. A "secret public-key scheme" for encryption is a scheme where no secret is required to encrypt (i.e. anybody can encrypt), but where the secret key to decipher <u>and</u> the exact algorithm to decipher are secret.

Note. By "secret" we mean here that, when a public key is given, it is not possible to make a finite list of possible schemes and to prove that this public key is a public key generated by one of the schemes of the list.

Definition in signature. A "secret public-key scheme" for signature is a scheme where no secret is required to verify the validity of a signature (when the message and its signature are given), but where the secret key to sign <u>and</u> the exact algorithm used to sign are secret.

Definition in authentication. A "secret public-key scheme" for authentication is a scheme where no secret is needed to check the validity of an authentication, but where the exact algorithm used by the prover to compute a valid authentication is secret.

Counter-example. If it is public that to encrypt a message M we compute $Y = M^3 \bmod n$, where n is an integer, then the fact that n is a product of 2, 3 or 4 primes can be maintained secret, but not the fact that the secret-key computations can be performed by computing $M = Y^d \bmod n$, where $3d = 1 \bmod \varphi(n)$, where φ is the Euler function of n. So, since we can find the secret-key function of this algorithm, it is not a secret public-key scheme. Similarly, ElGamal, Schnorr, Elliptic Curves algorithms, etc. are not secret public-key schemes (at least when their public key is given in the usual way).

	The secret-key computation algorithm is	
	Secret	Public
Symmetric Cryptography	Ex: proprietary algorithms	Ex: DES, AES, IDEA, etc.
Asymmetric Cryptography	The subject of this paper	Ex: RSA, ECC, etc.

Four kinds of cryptography

3. First examples: based on quadratic multivariate polynomials

Let us consider an algorithm whose public key is given by a set of m quadratic equations in n variables on a small finite field K:

$$y_i = \sum_{j=1}^{n}\sum_{k=1}^{n} \lambda_{ijk} x_j x_k + \sum_{j=1}^{n} \alpha_{ij} x_j + \beta_j, \qquad 1 \le i \le m.$$

The idea of designing some schemes with such a public key appeared in the 1980's in Japan with T. Matsumoto and H. Imai (cf. [8, 10]) and in the US with H. Fell and W. Diffie (cf. [6]). Since then, many cryptographic algorithms have been designed with such a public key. Some of them, as C^* (published in [8] and [10]) or $[C]$ (published in [8]) have been broken (see [2]), but some of them are still secure when their parameters are well chosen: for example UOV (cf. [7]), HFE (cf. [15, 9, 3]), HFEv$^-$ (cf. [7]), C^{*--}, HM (cf. [2]), etc.

The schemes which are not broken have generally the property that people who do not know the secret key cannot distinguish the public key (which has a trapdoor) from a random set of m quadratic equations in n variables (with no trapdoor).

As seen in [7, p. 220], instead of publishing those algorithms, one could have kept secret the way the computations are performed with the secret key, and publish only public keys. These algorithms (UOV, HFE, HM, etc.) would thus have been "secret public-key algorithms". Now, these algorithms have been published. However, they still have properties that more classical public public-key schemes (such as RSA, DSA, etc.) do not have: from the knowledge of their public key, we cannot be sure that it is indeed exactly these schemes that are used (because their public key is indistinguishable from a random set of quadratic equations). Moreover, it is possible to build many variations of these algorithms, and it is possible to build other algorithms whose public keys will also be indistinguishable from random quadratic equations. For instance, HFE uses the fact that we hide a monovariate equation (that we can solve) of relatively small degree d (in general $17 \le d \le 1000$) in a multivariate equation of degree 2. It is also possible to hide equations that we can solve (for instance by using Gröbner bases) in two (or three or four) relatively large variables in multivariate equations of degree 2.

Such generalizations of the basic HFE, suggested in [15], have not been intensively studied, and give an example of the many "secret" variations that can possibly be built. Instead of giving the details of this example in this paper, we will look for other types of public key (other than the systems of quadratic multivariate equations), where the concept of a secret public-key scheme appears.

Example of application: pro-active public keys. We will outline here an amazing application example of this "secret public-key scheme" concept. Let us suppose that a smartcard based application has been developed in the following way: the cards perform the secret-key computations, and give their public key to the terminals, which do not contain any secret.

The public key here is a system of quadratic multivariate equations as described above. Each card has its own public key (whose size is generally between 1 Kbyte and 3 Kbytes). These public keys are signed by the authority that makes the cards. The signature of this multivariate public key (certificate) is, for example, a static RSA signature (only the terminals compute RSA, to check the signature, the cards just send a fixed value). Suppose that a cryptographic weakness of the algorithm is discovered by the team that builds the cards, without this weakness being known by the users. It is then possible to change the algorithm into another algorithm which also uses, as the public key, a system of multivariate quadratic equations.

These new public keys are indistinguishable from the previous ones for the terminals, and thus we can change the algorithm at minimal cost: neither the terminals nor the users are aware of this change. (The new public keys will be signed by the authority.) After some time, all the cards will have been changed (smartcard are generally replaced every two years), and an algorithm will have replaced another one without discontinuity.

4. More general examples that use multivariate polynomials

Public keys, taking the form of a system of multivariate quadratic equations, are not the only ones that can be used to build "secret public-key schemes". More general systems of equations can indeed be used.

Equations of degree 3

For example, it is possible to use equations of degree 3 (or more). Most of the quadratic multivariate algorithms (HFE, UOV, etc.) have obvious generalizations in degree 3 (or more).

"Dragon" type equations

It is also possible to use "Dragon" type equations (see [16]), i.e. multivariate equations that "mix" the x_i and y_j variables (representing the cleartext or the ciphertext of the messages, or the signature and the message to be signed). For example, the public key of such a scheme can be a set of m equations of total degree 3 in n variables on a small finite field K:

$$\sum_{j=1}^{n}\sum_{k=1}^{n}\sum_{\ell=1}^{n}\gamma_{ijk\ell}x_jx_ky_\ell+\sum_{j=1}^{n}\sum_{k=1}^{n}\mu_{ijk}x_jy_k+\sum_{j=1}^{n}\sum_{k=1}^{n}\lambda_{ijk}x_jx_k$$
$$+\sum_{j=1}^{n}\alpha_{ij}x_j+\sum_{j=1}^{n}\beta_{ij}y_j+\delta_i=0 \qquad (1\le i\le m).$$

As a matter of fact, most of the multivariate algorithms have such generalizations (see [16]).

"Chains" of multivariate equations

Is it also possible to combine several systems of equations of degree 2 (or 3, or "Dragon"-type), by compositions and additions of the basic functions. We thus obtain much more general final functions: in particular, the equations that link the x_i variables of the cleartext (or the signature) to the y_i variables of the ciphertext (or the message to be signed) are not necessarily of low total degree any longer. We will call such algorithms "Chains".

In Figure 1, we give an example of a "Chains" algorithm. Here, s, t and u are three affine secret bijections from K^n to K^n, where $K = \mathbf{F}_q$ is a small finite field, and v, v', w, w' are four "intermediate variables": i.e. we will first compute v, v', w and w' from the cleartext x before computing the ciphertext y.

Public key. In this example of "Chains", the public key is made of the four following elements:

1. The equations that give v as a function of x. These equations are of the HFE type (cf. [15]), i.e. $v = t(f(s(x)))$, where f is a monovariate polynomial of small degree d, from $\mathbf{F}_{q^n}$ to $\mathbf{F}_{q^n}$, whose multivariate expression (in $(\mathbf{F}_q)^n$) is quadratic.
2. The equations that give v' in terms of x'. These equations are also multivariate quadratic equations (of the HFE type). We have: $v' = t(f'(s(x)))$, where f' is another polynomial of the same type as f, i.e. monovariate of degree d from $\mathbf{F}_{q^n}$ to $\mathbf{F}_{q^n}$, and quadratic multivariate from $(\mathbf{F}_q)^n$ to $(\mathbf{F}_q)^n$.
3. The equations that give w in terms of v. These equations are also quadratic multivariate equations (of the HFE type). We have: $w = u(g(t^{-1}(v)))$, where g is another polynomial of the same type as f.
4. The equations that give w' in terms of v'. These equations are also quadratic multivariate equations (of the HFE type). We have: $w' = u(g'(t^{-1}(v')))$, where g' is another polynomial of the same type as f.
5. Finally the fact that $y = w + w'$ is made public.

As a result, this public key is made of four public keys of the HFE type, together with the fact that $y = w + w'$.

Public key computations. From this public key, it is possible to compute the ciphertext y of a message x (by computing successively v, v', w, w' and finally $y = w + w'$).

(In signature, y represents the message, or the Hash of the message to be signed, and x represents the signature. It is possible to verify the validity of the signature x by computing v, v', w, w' and then $y = w + w'$ from x, and by verifying that we indeed obtain the Hash of the message.)

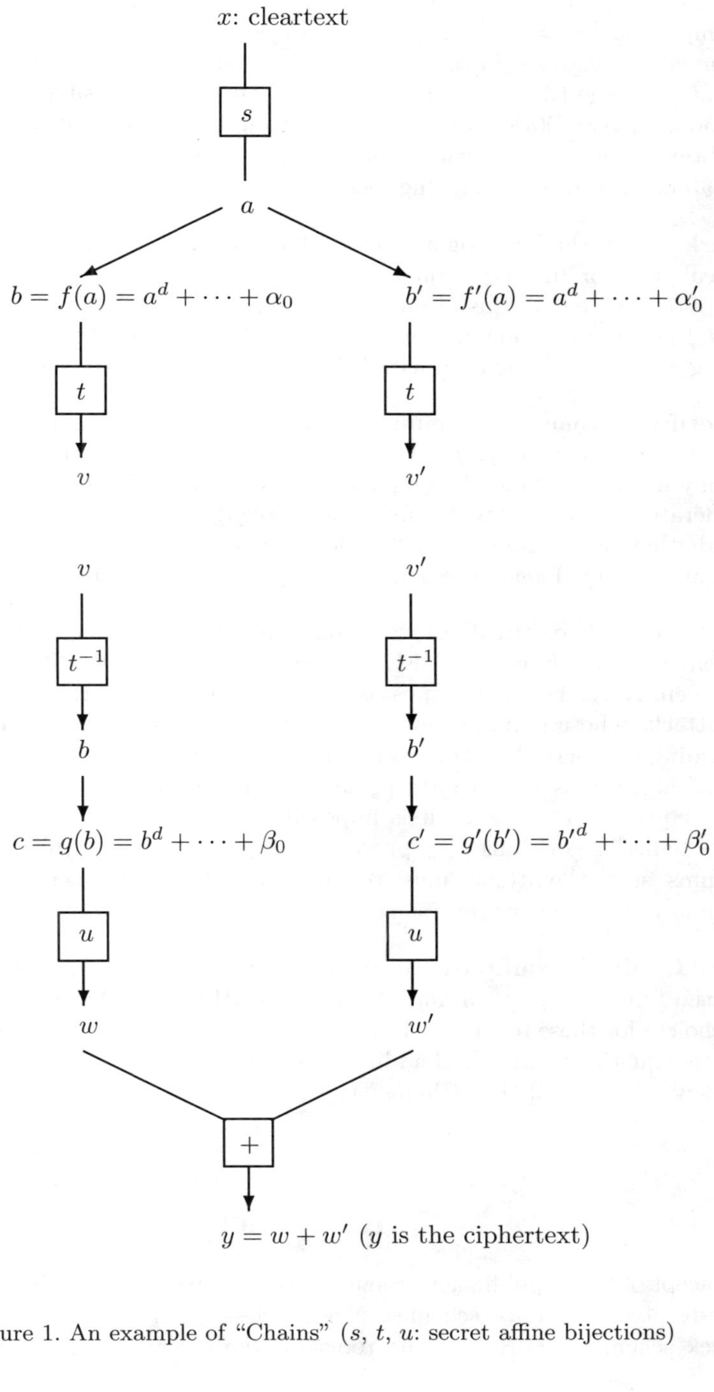

Figure 1. An example of "Chains" (s, t, u: secret affine bijections)

Secret-key computations. The secret key is given by the secret functions s, t, u, f, f', g and g'. When these secret functions are known, then from a value c, we can compute $u^{-1}(d) = u^{-1}(z) + u^{-1}(z') = c + c' = g(b) + g(b')$. Since b and b' are polynomials of degree d in a, $g(b) + g(b')$ is a polynomial F of degree d^2 in a. If d^2 is not too large (if $d^2 < 1000$ for example), it is then possible to solve in a the equation $u^{-1}(d) = F(a)$. (For instance, by using Berlekamp's algorithm.) We have thus obtained the possible values for a, hence the possible values for $x = s^{-1}(a)$, which allow decrypting or signing messages.

Remark. As for the HFE algorithm, the fact that F is not necessarily bijective is not a real problem. In encryption mode, we will easily detect the correct cleartext x, by using redundancy in this message x, or by sending SHA-1(x) together with the value d. In signature mode, we can sign x =SHA-1($M||R$), where M is the message to be signed, and R is random with a fixed size, and perform several tries for R.

Key setup. To compute the public key and secret key of this algorithm, first the secret functions s, t, u, f, f', g and g' are generated. s, t and u are generated randomly in the set of all affine bijective functions from K^n to K^n. f, f', g and g' are generated as random polynomials of degree d from $\mathbf{F}_{q^n}[X]$ (it is also possible to decide that the monomial $\alpha_d X^d$ of degree d of these polynomials will be chosen such that $\alpha_d = 1$). Then the corresponding public-key equations are generated.

Security for this example of "Chains". In [16, p. 406] it is mentioned that no really efficient attack against HFE is known at present, when its parameters are well chosen. Nevertheless, the most efficient attacks against HFE at present seem to be attacks whose aim is to invert the function given by the public key, rather than finding the secret key. However, this "inversion" attack does not work for the "Chains" algorithms we have just presented. As a matter of fact, from the value d and the equation $d = z + z'$, it is impossible to know the value of z (or z'). So these "Chains" algorithms may prove even harder to attack than the HFE-type algorithms, since the attacks must really try to retrieve the secrets (s, t, u, f, f', g, g'), and not simply try inversions of values.

Other "Chains" examples. We have presented here an example of "Chains", where the functions f, f', g and g' are of the HFE type. We can imagine many other choices for these functions. From similar public keys (for example four sets of quadratic equations and a final addition), it is then possible to build many "secret public-key schemes" of the "Chains" type.

5. Knapsack

The concept of secret public-key schemes was explicitly cited for the first time after the study of multivariate schemes. Nevertheless, after the publication of many knapsack schemes (cf. [13]), some tools for the discovery of this concept were already available. As a matter of fact, in knapsack-based schemes, the public key has always the same form (public values a_i, and unknown bits x_i to be found such that $\sum x_i a_i$ equals a given value). As a result, nothing compels the users to specify the way the trapdoor was introduced in such schemes. Nevertheless, almost all the knapsack-based schemes have now been broken (often due to the use of the LLL algorithm), so that the knapsacks do not constitute at present a very practical example of secret public-key schemes.

6. Error correcting codes

With public-key encryption schemes based on error correcting codes, it is also possible to design "secret public-key" schemes.

The most famous public-key encryption schemes based on error correcting codes are McEliece (cf. [11]) and Niederreiter ciphers (cf. [14]).

In a McEliece scheme, the ciphertext corresponding to the k bit message x is $y = xG' + e$, where e is a random n-bit error-vector of weight t.

This public transformation can also be given as a set of n equations of degree one of this type:

$$y_i = e_i + \sum_{j=1}^{k} \alpha_{ij} x_j, \qquad 1 \le i \le n.$$

Here, less than t bits of e_i can be different from 0. (Initially, it was suggested to have for example $n = 1024$, $k = 524$ and $t = 50$, however slightly larger values are now recommended for security, cf. [1].)

When such a public key is given, there is not a clear way to discover what exact corresponding secret-key algorithm is used.

The idea of McEliece scheme is to first select a particular linear code for which an efficient decoding algorithm is known, and then to disguise the code as a general linear code. For this, a Goppa code is generally used. However, different linear codes may be used with still the property that the public key looks as a general linear code. If such a code is used and kept secret, then we will have a "secret public" McEliece scheme.

In a Niederreiter cipher, a plaintext x is an n-bit vector of weight t and the associated ciphertext y corresponds to the syndrome of m relatively to the public code, $y = XH'^t$. This public transformation can also be given as a set of n equations of degree one of this type:

$$y_i = \sum_{j=1}^{n} \alpha_{ij} x_i, \qquad 1 \le i \le n,$$

where less than t bits of x_i can be different from 0.

As with McEliece scheme, it may be possible to choose a secret linear code with the property that the public key will look as a general linear code. If such a code is used and kept secret, then we will have a "secret public" Niederreiter scheme.

Part II
Extended notions of "secret public-key schemes"

7. Schemes whose algorithm is known at the last moment (and can be diversified)

The concept of "secret public-key scheme" can be generalized as we will see in this part. We will see that this extended notion is very general and is not restricted to schemes based on knapsacks, error correcting code, or multivariate polynomials. Let us start by using in an original way a standard public-key signature algorithm, for instance RSA, as we will see now.

Authentication

Let us consider the following authentication scheme:

Step 1. The prover sends the source code C of a computer program, together with its signature S. (S is the signature of C by an authority. For example, it is a static 1024 bits RSA signature.)

Remark. The authority will sign here only codes C of an authentication algorithm that this authority consider as a valid authentication scheme for the prover.

Step 2. The verifier sends a challenge R.

Step 3. The prover sends a response T to the challenge.

Step 4. The verifier checks on the one hand that S is indeed the signature of the source code C, and on the other hand that executing the source code C on the values R and T indeed gives the output $C(R,T) = 1$.

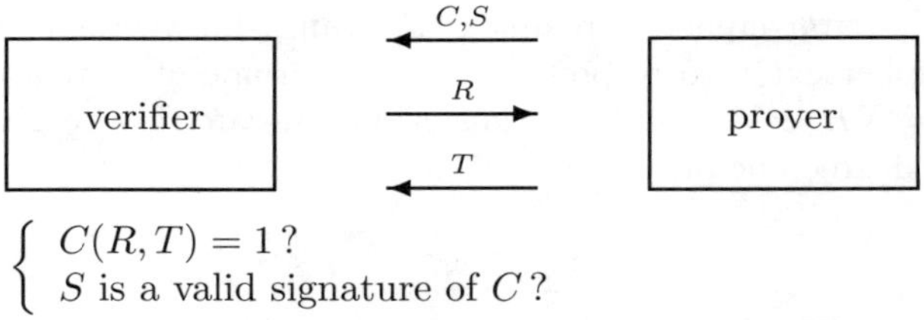

Figure 2. Schemes known at the last moment

This authentication scheme has two original characteristics. Firstly, the verifier does not know in advance whose authentication algorithm C is the code. For example, it can be RSA, or GQ, or HFE, or something else. Secondly, even when he obtains C, he does not need to understand whose exact algorithm C is the source code. He only has to know that C was correctly signed by the authority. (This particularity will be used in an even more dramatic way in Section 9 below.)

Such an authentication scheme can easily be transformed into a signature or an encryption scheme.

Signature

In signature mode, here is a possible protocol:

Computation of the signature. The signer signs the message M by using his customary signature algorithm A and his secret signature key: he then obtains a signature $s = A(M)$. The complete signature J of M is then made of three parts: s, C and S, where C is the source code of a fixed computer program (we will see below that the program C checks the validity of the signature s of M), and S is a (static) signature of C by the authority (for example, S is a static 1024 bits RSA signature of C).

Remark. Here again, the authority will sign here only codes C that this authority consider as a valid signature scheme for the signer.

Verification of the signature. The verifier thus has the message M and its signature s, C, S. The verification is performed in two steps. First, he checks that S is indeed a valid signature of C (by using the public key of the authority). Then, he checks that executing the source code C on the values M and s gives the correct output $C(M, s) = 1$.

If those two verifications are correct, the signature is a valid signature of M; otherwise, it is not.

This signature scheme shares the same two original characteristics as the previous authentication scheme. First, the verifier does not know in advance whose signature algorithm C is the source code. For instance, it can be RSA, GQ, HFE, or something else. Secondly, even if he obtains C, he does not need to understand whose exact algorithm C is the source code. He just has to know that C was correctly signed by the authority.

Encryption

In encryption mode, here is a possible protocol:

Encryption of a message. The public encryption key is made of two parts: the source code C of a computer program (we will see below that the program C encrypts in a public way the messages), and a signature S of C by the authority.

To encrypt a message M, we begin by checking that S is a valid signature of C (by using the public signature key of the authority). Then, we run the code C on the message M. The output $T = C(M)$ is the ciphertext.

Decryption of a message. To decrypt, we compute $M = A(k, T)$, where k is the secret key, and where A is the decryption algorithm. In such a scheme, C and its signature S can be given to the encryption people at the last moment, and they do not have to understand whose exact algorithm C is the source code. They just have to know that C was correctly signed by the authority.

8. Authentication based on the difference of the memory sizes

Let us speak now of another very different example of a "secret public-key" authentication scheme, whose security relies on the difference of the memory sizes available for the prover and the verifier.

Let us suppose that the prover is a small smartcard. At present, smartcards contain only several Kbytes of memory. We suppose that the smartcard is able to compute the function $\text{TDES}_K(x)$, where K is its secret key.

Let us now suppose that the verifier is a computer with a large memory (several Gigabytes for instance). The public key of the card consists in a look-up table containing, say, 100 million values $\text{TDES}_K(x)$, for x from 1 to 100 million. This public key is thus approximately 800 Megabytes in size, and is stored on the computer (this public key is intentionally much too large to be stored in the smartcard, but the smartcard is able to compute any of its values).

The authentication proceeds as follows:

Step 1. The prover (the computer) randomly chooses a number x between 1 and 100 million and sends it to the card.

Step 2. The card send back $Y = \text{TDES}_K(x)$. The prover checks that this value is correct, by examining his table.

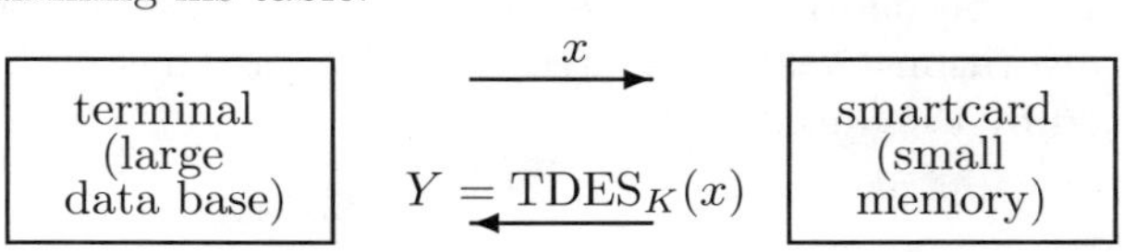

Figure 3. Schemes based on memory sizes

Advantages of this authentication scheme

1. Although this is a public-key authentication scheme, the computations performed by the card are extremely simple, fast, and require very little RAM.

2. It is easy to prove that, if the function $\mathrm{TDES}_K(x)$ is indistinguishable from a random permutation on 64 bits (and this is generally the classic hypothesis on the secret-key encryption functions such as TDES), then it is impossible to obtain sufficiently useful data from the public key to build fake cards (as soon as the memory of the cards remains very small). Moreover, we also obtain a proof that we have a "secret public-key scheme", since the observations of the public key does not allow to prove that it comes from a TDES.

Remark. The security level achieved here is the quotient of the memory size of the computer and the memory size of a false smartcard, i.e. about 10^6. This security level of 10^6 is a classical level of security in on-time authentication (for example similar security levels have been suggested for Fiat–Shamir or GQ authentication schemes). This comes from the fact that here, contrary to what appears in a signature scheme, where higher security level is mandatory, the attacker has to be in contact with the computer he wants to attack each time he tries an authentication.

Drawbacks of this authentication scheme

1. Obviously, this scheme is not very practical to implement and to use.
2. It is sometimes difficult to ensure that Y was computed by a smartcard (with a very small memory) and not by another computer that would have copied the public 80 Megabytes key and would have transmitted this value Y, either directly, or through the card.

9. The Rice theorem and schemes with a mysterious verifier

In the schemes mentioned in Section 7, it was pointless to understand the meaning of the source code C. It was sufficient to run it and to check its signature. When the source code C of a program is given, is it always easy to understand whose algorithm it is the code?

The answer to this question is no. For example, in Section 8 above, we saw that from the list of the values $\mathrm{TDES}_K(x)$, where x varies, it is impossible to know that we have such a TDES, if the key K is not revealed (if the TDES is indeed indistinguishable from a random permutation).

But more generally, it seems that the negative answer to this question is a consequence of a famous theorem of logic: the Rice theorem.

Definition 1. Let A be a subset of the set of integers. A is said to be *recursive* if a computer program exists such that, when it is given an integer n in input, tells, after a finite number of operations, whether $n \in A$ or $n \notin A$.

Definition 2. Let f be a function from a set $D \subset \mathbf{N}$ to $\mathbf{N}$. f is called a *recursive semi-function* if a computer program exists that computes $f(x)$ when it is given $x \in D$ in input, and never answers when $x \notin D$.

Theorem 1 (Rice, 1953). *Let F be a set of recursive semi-functions, non-empty and distinct from the set of all those functions. Let A be the set of the integers n such that n is the code of a semi-function of F. Then A is not recursive.*

(A proof of this theorem can be found for example in [12].)

(A "code" of a function f is, for example, an integer α such that when α is written in base 2, we obtain in machine language a program P that computes $f(n)$ when an integer n is given as input to this program P.)

Applications. The Rice theorem tells us that in practice no method exists to know:

1. whether two codes compute the same function (take for F a single element),
2. whether a code always answers 0 for any input (take for F the set containing only the zero function),
3. whether a code always give an answer for any input (take for F the set of recursive semi-functions defined on all $\mathbf{N}$),
4. whether a code always answers from input values less or equal to an integer a (take for F the set of semi-functions $\leq a$),

etc.

The Rice theorem is thus a justification of the fact that it is generally extremely difficult to understand what is performed by some computer programs from the study of their code.

This is a motivation to try to exploit this to build "secret public-key" schemes of a new type.

Example. Let C be the computer code of a program that takes in input a message M and a value S, and gives the output $C(M, S) = 1$ if and only if $\mathrm{TDES}_K(M) = S$. Is it possible to give such a code C without revealing K ? If yes, one could imagine a very simple signature algorithm: C would be the public key; to sign a message one simply compute $\mathrm{TDES}_K(M)$, and to verify a signature, one checks that $C(M, S) = 1$.

It seems difficult to prove the security of such a code C. However, if for instance C is written in assembly language and is 10 Megabytes large, who will try to understand what it does? To write such a code C, one could for instance build a program A that takes in input the code G of a computer program, and gives in output a code H of a computer program that does the same thing as G, but is longer and worse written. By re-using A several times, we will get a code C that does the same thing as G, but C is written in a crazy way, so one can hope that this code C will be very difficult to analyze...

Such an approach to cryptography really moves away from all the traditional cryptography (where the focus is put on the fact of using well-written programs, having security that relies on well-defined problems, etc.). We do not pretend that the use of such a scheme is very practical or recommended. However, it is amazing to realize that taking backwards all the principles, we can obtain something that could nevertheless work, and it is tempting to try to apply the great theorems of logic to cryptography. Unfortunately, the Rice theorem poses more questions than answers at present. For instance:

What can be precisely stated about the security of a scheme based on a code difficult to understand?

How to generate in practice codes that are difficult to understand?

What is the exact meaning of the Rice theorem in practice? Etc.

Remark. One can also object to another important point of these "secret public-key" algorithms based on the Rice theorem: if one does not want the verifier to have a specific code for each card, then one will use the same code in public key to verify several, and perhaps all the cards. (In the same way, with RSA, the same code is used to compute all the modular exponentiations, only the values of the modulus and of the inputs change, but not the algorithm that is implemented.) However, if this code computes secret protocols or values (that are not revealed by a careful study of the code), these secret protocols or values constitute a kind of "master secret" (because it is shared by several cards) inside the public key. It seems extremely dangerous to have such hidden master secrets in a public key, even if they are well hidden. In todays applications of cryptography, the advantage of a public key is, on the one hand to be public, but on the other hand to contain absolutely no secret, and to be linked to only one card. Thus, if all the secrets associated to a public key are revealed, this does not break the security of the other cards.

In brief, it is at present amazing to present the Rice theorem with the concept of public-key cryptography, but we do not recommend schemes with a security based on the Rice theorem for practical applications at present.

10. Conclusion

In this paper, we have presented the concept of a "secret public-key scheme". We have seen that very different "secret public-key schemes" can be built. Some of them use new types of public keys, such as the new "Chains" algorithms. An interesting property of these "Chains" algorithms is the fact that attacks that try to invert the function without finding the secret key seem to be pointless. This is due to a non-bijective operation: the addition of two unknown values in the schemes. Another interesting property of these "Chains" algorithms is the fact that the cleartext values x_i and the ciphertext values y_i are not necessarily linked with relations of small degree. This is because the y_i values are the results of additions and compositions of k quadratic transformations and k can be 2, 3, 4 or even larger. So the functions generated by the "Chains" algorithms can be more general than functions usually obtained with multivariate schemes. Some other schemes are based on error correcting codes. Others are connected to some general results of logic (the Rice theorem) and some secret public-key schemes are just unusual applications of well-known tools. This concept of "secret public-key schemes" does not only have a theoretical interest: from a practical point of view, it is possible to design more flexible schemes based on this concept. Some of these schemes are particularly interesting in certain smartcard applications, a good example being when the smartcard has no arithmetic coprocessor and can only compute simple functions.

References

[1] Canteaut, A., Sendrier, N., Cryptanalysis of the original McEliece cryptosystem. In: Advances in Cryptology – ASIACRYPT '98 (ed. by K. Ohta and D.Y. Pei; Lecture Notes in Comput. Sci. 1514), 187–199. Springer, Berlin 1998.

[2] Courtois, N., Goubin, L., Patarin, J., C^*_{-+} and HM, Variations around two schemes of T. Matsumoto and H. Imai. In: Advances in Cryptology – ASIACRYPT '98 (ed. by K. Ohta and D.Y. Pei; Lecture Notes in Comput. Sci. 1514), 35–49. Springer, Berlin 1998.

[3] Courtois, N., Klimov, A., Patarin, J., Shamir, A., Efficient algorithms for solving overdefined systems of multivariate polynomial equations. In: Advances in Cryptology – EUROCRYPT 2000 (ed. by B. Preneel; Lecture Notes in Comput. Sci. 1807), 392–407. Springer, Berlin 2000.

[4] Diffie, W., Hellman, M., New directions in cryptography. IEEE Trans. Inform. Theory 22 (1976), 644–654.

[5] Data Encryption Standard. Federal Information Processing Standards Publication 46. National Bureau of Standards, 15 January 1977.

[6] Fell, H., Diffie, W., Analysis of a public key approach based on polynomial substitution. In: Advances in Cryptology – CRYPTO '85 (ed. by H.C. Williams; Lecture Notes in Comput. Sci. 218), 340–349. Springer, Berlin 1986.

[7] Goubin, L., Kipnis, A., Patarin, J., Unbalanced oil and vinegar signature schemes. In: Advances in Cryptology – EUROCRYPT '99 (ed. by J. Stern; Lecture Notes in Comput. Sci. 1592), 206–222. Springer, Berlin 1999.

[8] Imai, H., Matsumoto, T., Algebraic methods for constructing asymmetric cryptosystems. In: Algebraic Algorithms and Error Correcting Codes (AAECC-3, Grenoble, 1985) (ed. by J. Calmet; Lecture Notes in Comput. Sci. 229), 108–119. Springer, Berlin 1986.

[9] Kipnis, A., Shamir, A., Cryptanalysis of the HFE public key cryptosystem. In: Advances in Cryptology – CRYPTO '99 (ed. by M. Wiener; Lecture Notes in Comput. Sci. 1666), 19–30. Springer, Berlin 1999.

[10] Matsumoto, T., Imai, H., Public quadratic polynomial-tuples for efficient signature-verification and message-encryption. In: Advances in Cryptology – EUROCRYPT '88 (ed. by C.G. Günther; Lecture Notes in Comput. Sci. 330), 419–453. Springer, Berlin 1988.

[11] McEliece, R.J., A public-key cryptosystem based on algebraic coding theory. DSN Progress Report 42–44, pp. 114–116. Jet Propulsion Lab., Pasadena, CA, 1978.

[12] Mendelson, E., Introduction to Mathematical Logic, Third Edition, p. 258. Wadsworth & Brooks/Cole Advanced Books & Software, Monterey 1987.

[13] Merckle, R., Hellman, M., Hiding information and signatures in trapdoor knapsacks. IEEE Trans. Inform. Theory 24 (1978), 525–530.

[14] Niederreiter, H., Knapsack-type cryptosystems and algebraic coding theory. Problems Control Inform. Theory 15 (1986), 159–166.

[15] Patarin, J., Hidden Field Equations (HFE) and Isomorphisms of Polynomials (IP): two new families of asymmetric algorithms. In: Advances in Cryptology – EUROCRYPT '96 (ed. by U. Maurer; Lecture Notes in Comput. Sci. 1070), 33–48. Springer, Berlin 1996.

[16] —, Asymmetric cryptography with a hidden monomial. In: Advances in Cryptology – CRYPTO '96 (ed. by N. Koblitz; Lecture Notes in Comput. Sci. 1109), 45–60. Springer, Berlin 1996.

[17] Rivest, R., Shamir, A., Adleman, L., A method for obtaining digital signatures and public-key cryptosystems. Comm. ACM 21 (1978), 120–126.

Index form surfaces and construction of elliptic curves over large finite fields

*Attila Pethő**

Abstract. Index form equations play an important role in algebraic number theory especially in computing all elements with given discriminant. Using geometric ideas Gaál, Pethő and Pohst [4] gave a practical method for the solution of index form equations over quartic number fields. Continuing their investigations we introduce the notion of index form surfaces associated to quartic polynomials and show that they are either empty or elliptic surfaces.

2000 Mathematics Subject Classification: 11D57, 14J27.

1. Introduction

Let $\mathbb{K}$ be an algebraic number field of degree n, denote by $\mathbb{Z}_{\mathbb{K}}$ its ring of integers and by $D_{\mathbb{K}}$ its discriminant. Let $1, \omega_1, \ldots, \omega_{n-1}$ be an integral basis of $\mathbb{Z}_{\mathbb{K}}$, $\underline{X} = (X_1, \ldots, X_{n-1})$,

$$L^{(i)}(\underline{X}) = L^{(i)}(X_0, \underline{X}) = X_0 + \omega_1^{(i)} X_1 + \cdots + \omega_{n-1}^{(i)} X_{n-1}, \quad i = 1, \ldots, n.$$

The *discriminant form* of $\mathbb{K}/\mathbb{Q}$ with respect to the basis $1, \omega_1, \ldots, \omega_{n-1}$ or simply the discriminant form of $\mathbb{K}/\mathbb{Q}$ is then defined by

$$D_{\mathbb{K}/\mathbb{Q}}(L(\underline{X})) = \prod_{1 \le i < j \le n} \left(L^{(i)}(\underline{X}) - L^{(j)}(\underline{X})\right)^2.$$

It is easy to see that

$$D_{\mathbb{K}/\mathbb{Q}}(L(\underline{X})) = D_{\mathbb{K}}\left(I_{\mathbb{K}/\mathbb{Q}}(\underline{X})\right)^2,$$

where $I_{\mathbb{K}/\mathbb{Q}}(\underline{X})$ is a homogeneous polynomial of degree $n(n-1)/2$ in $\mathbb{Z}[\underline{X}]$. The polynomial $I_{\mathbb{K}/\mathbb{Q}}(\underline{X})$ is called the *index form* of $\mathbb{K}/\mathbb{Q}$ with respect to the basis $1, \omega_1, \ldots, \omega_{n-1}$ or simply the index form of $\mathbb{K}/\mathbb{Q}$.

To find elements with given discriminant in algebraic number fields the index forms are playing an important role. Indeed, the element $\alpha = x_0 + x_1\omega_1 + \cdots +$

* Research supported in part by Centre de Recerca Matemática, Bellaterra and also by Hungarian National Foundation for Scientific Research Grant 25157/98.

$x_{n-1}\omega_{n-1} \in \mathbb{Z}_{\mathbb{K}}$ has discriminant d if and only if

$$I_{\mathbb{K}/\mathbb{Q}}(x_1, \dots, x_{n-1}) = \pm\sqrt{d/D_{\mathbb{K}}}. \tag{1}$$

T. Nagell [12] proved that if $\mathbb{K}$ is a quartic number field, then (1) has only finitely many solutions. His result was extended to arbitrary number fields by K. Győry [6]. It is important that Győry's result is effective, i.e. it implies an algorithm for the solution of (1). He generalized this result for relative extensions and even for finitely generated integral domains [7].

The solutions of index form equations for small degree fields are also tabulated. This is not too hard for $n = 3$, because then $I_{\mathbb{K}/\mathbb{Q}}(x_1, x_2)$ is a cubic form, thus (1) becomes a Thue equation of degree 3. Gaál and Schulte [5] developed an algorithm in this case and computed all power integral bases in cubic number fields with $|D_{\mathbb{K}}| < 50000$.

Forgetting the arithmetical meaning of equation (1) one can study the geometrical properties of the curve defined by that equation over an extension of $\mathbb{Q}$ or over a finite field. It is well known that this curve is of genus one, hence it is empty or is an elliptic curve. (See e.g. [16], Section I.2.)

Already for quartic number fields becomes the computation of integer solutions of (1) much harder. Győry's result implies an algorithm for the resolution of (1), but it is not practical. This is because he transforms (1) to finitely many S-unit equations in two unknowns over the normal closure of $\mathbb{K}$, and from the solutions of these unit equations derives the solutions of (1). In the worst, but most frequent, case the Galois group of $\mathbb{K}$ is isomorphic to S_4, the symmetric group of degree 4, and the normal closure of $\mathbb{K}$ has degree 24. Generally even a system of fundamental units of such a large degree field is very hard to compute. Therefore we must understand better the structure of the index form if we want to find the solutions of index form equations. If $\mathbb{K}$ has a quadratic subfield, then $I_{\mathbb{K}/\mathbb{Q}}(\underline{X})$ splits in $\mathbb{Q}[\underline{X}]$ and we have to deal with a system of equations. Based on this fact Gaál, Pethő and Pohst worked out methods for solving quartic index form equations if the Galois group of $\mathbb{K}/\mathbb{Q}$ is isomorphic to C_4, V_4 or D_8. Finally they succeeded to prove that one can get all integer points lying on $I_{\mathbb{K}/\mathbb{Q}}(\underline{X})$ by a simple transformation of the integer solutions of finitely many quartic Thue equations (see [4] and the references therein). You find a collection of recent computational results on index form equations of higher degree number fields in Gaál [1, 2].

We shall show in this note that if $\mathbb{K}$ is a quartic extension of $\mathbb{Q}$ then the surface defined by the equation

$$I_{\mathbb{K}/\mathbb{Q}}(\underline{X}) = m \neq 0$$

has a nice geometric structure: it is either empty or an elliptic surface, i.e. can be uniquely covered by elliptic curves. Exactly this structure makes relatively simple the computation of integer points on such a surface. This observation is behind the result of Gaál, Pethő and Pohst. We present the geometrical structure theorem in more general context.

The classification of index form surfaces associated to higher degree number fields should be the topics of further research.

The results of Sections 2 and 3 were proved several years ago and circulated in a manuscript. In Section 4 we summarize results of F. Leprévost et al. [8]. In those sections the index form surface is studied over a finitely generated integral domain over $\mathbb{Q}$. In the last section we collect some observation of P. Nagy [13], who performed computations concerning index form surfaces over finite fields.

I thank the referee for his comments, which helped to improve the clarity of the presentation.

2. Parametrization of the index forms

To formulate our results we generalize the notion of index form and introduce some notation.

Let k be a field (of arbitrary characteristic), a_1, a_2, a_3, a_4 be indeterminates over k and $K = k(a_1, a_2, a_3, a_4)$. Let $p(X) = X^4 + a_1X^3 + a_2X^2 + a_3X + a_4 \in K[X]$ of non-zero discriminant. Let $\alpha = \alpha_1, \dots, \alpha_4$ denote the zeros of $p(X)$ in some algebraic closure of K. Let $L = K(\alpha)$ and denote by $I_p(\underline{X})$ the index form associated to the basis 1, α, α^2, α^3 of L. Then

$$I_p(\underline{X}) = \prod_{1\le i<j\le 4} L_{i,j}(\underline{X}),$$

where

$$L_{i,j}(\underline{X}) = X_1 + (\alpha^{(i)} + \alpha^{(j)})X_2 + (\alpha^{(i)2} + \alpha^{(i)}\alpha^{(j)} + \alpha^{(j)2})X_3$$

for $1 \le i < j \le 4$.

By the fundamental theorem on symmetric polynomials, $I_p(\underline{X})$ is a form of degree 6 in $K[\underline{X}]$. We are given a parametrization of $I_p(\underline{X})$ in Theorem 1 below. Remark that if one specializes K to $\mathbb{Q}$ and a_1, a_2, a_3, a_4 to fixed integers, then L becomes a number field. If it is quartic and 1, α, α^2, α^3 is a power integral basis of the ring of integers of L, then $I_p(\underline{X})$ specializes to the classical notion of index forms.

Note that one can define analogously the index form of higher degree polynomials. The following Theorem 1 and its proof almost literally coincides with that of Gaál et al. [3]. However there is an essential difference; while in the cited paper a_1, a_2, a_3, a_4 denote integers such that the polynomial $p(X)$ is irreducible in $\mathbb{Q}[X]$, here we are dealing with much more general and abstract objects.

Theorem 1. *Let*

$$\begin{aligned} Q_1(\underline{X}) &= X_1^2 - a_1X_1X_2 + a_2X_2^2 + (a_1^2 - 2a_2)X_1X_3 + (a_3 - a_1a_2)X_2X_3 \\ &\qquad + (-a_1a_3 + a_2^2 + a_4)X_3^2 = \underline{X}Q_1\underline{X}^T, \\ Q_2(\underline{X}) &= X_2^2 - X_1X_3 - a_1X_2X_3 + a_2X_3^2 = \underline{X}Q_2\underline{X}^T, \end{aligned} \tag{2}$$

where Q_1 and Q_2 denote the matrix of the quadratic form with the same nameand $\underline{X}^T$ denotes the transpose of the vector $\underline{X}$. Let further

$$f_3(X,Y) = X^3 - a_2X^2Y + (a_1a_3 - 4a_4)XY^2 + (4a_2a_4 - a_3^2 - a_1^2a_4)Y^3. \tag{3}$$

Then we have

$$I_p(\underline{X}) = f_3(Q_1(\underline{X}), Q_2(\underline{X})). \tag{4}$$

Note that $f_3(X,1)$ is the cubic resolvent of $p(X)$.

Proof. Collect the factors in the definition of $I_p(\underline{X})$ as follows:

$$I_p(\underline{X}) = \big(L_{1,2}(\underline{X})L_{3,4}(\underline{X})\big)\big(L_{1,3}(\underline{X})L_{2,4}(\underline{X})\big)\big(L_{1,4}(\underline{X})L_{2,3}(\underline{X})\big).$$

Putting $\xi_1 = \alpha^{(1)}\alpha^{(2)} + \alpha^{(3)}\alpha^{(4)}$, $\xi_2 = \alpha^{(1)}\alpha^{(3)} + \alpha^{(2)}\alpha^{(4)}$ and $\xi_3 = \alpha^{(1)}\alpha^{(4)} + \alpha^{(2)}\alpha^{(3)}$ we obtain

$$I_p(\underline{X}) = \big(Q_1(\underline{X}) - \xi_1 Q_2(\underline{X})\big)\big(Q_1(\underline{X}) - \xi_2 Q_2(\underline{X})\big)\big(Q_1(\underline{X}) - \xi_3 Q_2(\underline{X})\big)$$

after a simple, but lengthy computation. A similar simple and lengthy computation proves (4). We use in both parts of the proof only the well known Vieta's formulae. □

3. Structure of the index form surfaces

Let M be an extension of K and let $0 \neq \mu \in M$. We shall now study the structure of the set

$$\mathcal{F}(M,\mu) = \{P = (x_1, x_2, x_3) \in M^3 : I_p(P) = \mu\},$$

i.e. the set of M-rational points on the *index form surface* $I_p(\underline{X}) - \mu$.

Let us call $P_1, P_2 \in \mathcal{F}(M,\mu)$ *equivalent* if

$$Q_1(P_1) = Q_1(P_2) \quad \text{and} \quad Q_2(P_1) = Q_2(P_2).$$

This is obviously an equivalence relation on the set $\mathcal{F}(M,\mu)$. Moreover, there exists by Theorem 1 a bijective correspondence between the equivalence classes of $\mathcal{F}(M,\mu)$ and the M-rational points on the curve $f_3(X,Y) = \mu$. It can be defined formally as follows: For $u, v \in M$ such that $f_3(u,v) = \mu$ let

$$\mathcal{F}_{(u,v)} = \mathcal{F}_{(u,v)}(M,\mu) = \{P \in \mathcal{F}(M,\mu) : Q_1(P) = u,\, Q_2(P) = v\}.$$

Then Theorem 1 implies

$$\mathcal{F}(M,\mu) = \bigcup_{\substack{u,v,\in M\\ f(u,v)=\mu}} \mathcal{F}_{(u,v)}.$$

It can happen that $\mathcal{F}(M,\mu) = \emptyset$ and even if $\mathcal{F}(M,\mu) \neq \emptyset$ the set $\mathcal{F}_{(u,v)}$ can be empty for some $(u,v) \in M^2$. Indeed, consider for example the polynomial $p(x) = x^4 + 4a^2$, with $a \in \mathbb{Q}$. The polynomials f_3, Q_1, Q_2 corresponding to the index form $I_p(\underline{X})$ by Theorem 1 are $f_3(X,Y) = X^3 - 16a^2XY^2$, $Q_1(\underline{X}) = X_1^2 + 4a^2X_3^2$, $Q_2(\underline{X}) = X_2^2 - X_1X_3$.

The rank of the elliptic curve $X^3 - 16a^2XY^2 = 1$ over $\mathbb{Q}$ is zero, and the only rational points on it are $(1,0)$, $(-1/2, 3/8a)$, $(-1/2, -3/8a)$. This implies

$$\mathcal{F} = \mathcal{F}_{(1,0)} \cup \mathcal{F}_{(-1/2,3/8a)} \cup \mathcal{F}_{(-1/2,-3/8a)}$$

for

$$\mathcal{F} = \mathcal{F}(\mathbb{Q}, 1) = \{P \in \mathbb{Q}^3 \,:\, I_p(P) = 1\}.$$

Obviously $(1,0,0) \in \mathcal{F}_{(1,0)}$, hence $\mathcal{F}$ is not empty. On the other hand, the sets $\mathcal{F}_{(-1/2,\pm 3/8a)}$ are empty because the equation $X_1^2 + 4a^2X_3^3 = -1/2$ is not solvable in $\mathbb{Q}$.

The above example is due to Susanne Schmitt. She also proved that $\mathcal{F}$ from the example is infinite if and only if $a = c_1c_2(c_1+c_2)(c_1-c_2)z^2$, where $c_1, c_2 \in \mathbb{Z}$, $\gcd(c_1, c_2) = 1$ and $z \in \mathbb{Q}$.

If M is algebraically closed then $\mathcal{F}_{(u,v)} \neq \emptyset$ for every $u, v \in M$ with $f_3(u,v) = \mu$. Indeed, choosing $x_3 = 0$ we obtain the system of equations $x_1^2 - a_1x_1x_2 + a_2x_2^2 = u$, $x_2^2 = v$, which is always solvable in $x_1, x_2 \in M$, thus $(x_1, x_2, 0) \in \mathcal{F}_{(u,v)}$.

As the intersection of two conics is an elliptic curve, $\mathcal{F}_{(u,v)}$ is an elliptic curve for every $u, v \in M$ with $f_3(u,v) = \mu$. We will compute a model for these curves more explicitly.

In the proof of the following theorem we are using the method of the proofs of Lemmata 3 and 4 of [4] (cf. the remark before Theorem 1).

Theorem 2. *Let M be an extension of K, $0 \neq \mu \in M$. Let $u, v \in M$ be such that $f_3(u,v) = \mu$ and $P = (x_1, x_2, x_3) \in \mathcal{F}_{(u,v)}$. Put $Q_3 = vQ_1 - uQ_2$. Then there exists for every $R \in \mathcal{F}_{(u,v)}$ a vector $S \in M^3$ such that one of the coordinates of S is 0,*

$$\ell(S)R = S\ell(S) - PSQ_3S^T, \quad \text{with} \quad \ell(S) = 2PQ_3S^T$$

and S satisfies the equation

$$v\ell(S)^2 = (S\ell(S) - PSQ_3S^T)Q_2(S^T\ell(S) - SQ_3S^TP^T) \tag{5}$$

if $v \neq 0$, and

$$4u(PQ_3S^T)^2 = ((SQ_2S^T)P - 2(PQ_2S^T)S)Q_1((SQ_2S^T)P^T - 2(PQ_2S^T)S^T)$$

if $v = 0$.

Proof. Assume that $v \neq 0$ and let $P = (x_1, x_2, x_3) \in \mathcal{F}_{(u,v)}$. Then one of the coordinates of P is non-zero. Assume that this is x_3. Let $R \in \mathcal{F}_{(u,v)}$. Then there exist $r, s_1, s_2 \in M$ such that

$$R = rP + S,$$

with $S = (s_1, s_2, 0)$. Using the fact that $R, P \in \mathcal{F}_{(u,v)}$ we obtain the following chain of identities:

$$\begin{aligned} 0 &= vQ_1(R) - uQ_2(R) = vRQ_1R^T - uRQ_2R^T \\ &= v(rP+S)Q_1(rP^T + S^T) - u(rP+S)Q_2(rP^T+S^T) \end{aligned}$$

$$= r^2(vPQ_1P^T - uPQ_2P^T) + r\big(P(vQ_1 - uQ_2)S^T + S(vQ_1 - uQ_2)P^T\big)$$
$$+ vSQ_1S^T - uSQ_2S^T.$$

We have $vPQ_1P^T - uPQ_2P^T = vQ_1(P) - uQ_2(P) = 0$. As Q_1 and Q_2 are symmetric matrices, Q_3 is symmetric too. Thus $(PQ_3S^T)^T = SQ_3P^T$. Further, as PQ_3S^T is a constant, we obtain $PQ_3S^T + SQ_3P^T = 2PQ_3S^T = \ell(S)$ and

$$r\ell(S) = -SQ_3S^T.$$

This implies the assertion on R at once.

We have further $Q_2(R) = v$. If $v \neq 0$ then by multiplying this equation by $\ell(S)^2$ and inserting the formula for $\ell(S)R$ we obtain (5) immediately.

Finally, if $v = 0$, then as $f_3(u, 0) = u^3 = \mu \neq 0$ we have $u \neq 0$. Working now with the relation $Q_1(R) = u$ we obtain the asserted equation for S. □

One of the coordinates of S is zero by Theorem 2. Denote by s_1 and s_2 the other two coordinates of S. Then the expression on the right-hand side of (5) is a quartic form in s_1 and s_2. We denote it by $f_4(s_1, s_2)$. The coefficients of $f_4(s_1, s_2)$ belong obviously to M. Redoing the computation of Sections 3.2 and 3.3 of [4] one can prove the following theorem.

Theorem 3. *We use the notation of Theorem 2 and denote by α_i, $i = 1, \dots, 4$, the zeros of $p(X)$ and by $f_4(s_1, s_2)$ the quartic form staying on the right-hand side of (5). Then we have*

$$f_4(s_1, s_2) = v^2 \prod_{i=1}^{4} \big((x_3\alpha_i - x_2 + a_1x_3)s_1 - (x_3\alpha_i^2 + x_3a_1\alpha_i + x_3a_2 - x_1)s_2\big).$$

4. Elliptic surfaces associated to cubic and quartic polynomials

We give in this section an overview on results of the paper of F. Leprévost et al. [8].

We proved in the previous section that in the parametrization of index form surfaces associated to quartic polynomials the curve

$$C: \quad f_3(X, Y) = 1, \tag{6}$$

where $f_3(X, Y)$ is defined by (3), plays an important role. Observe that the coefficients of f_3 depend only on the coefficients of the quartic polynomial $p(x) = x^4 + a_1x^3 + a_2x^2 + a_3x + a_4$. The point $P = (1, 0)$ lies obviously on the curve (6). In the sequel we assume that $\mathrm{char}(k) \neq 2, 3$.

Let

$$H(X, Y) = -\frac{1}{4}\left(\frac{\partial^2 f_3}{\partial X^2}\frac{\partial^2 f_3}{\partial Y^2} - \left(\frac{\partial^2 f_3}{\partial X \partial Y}\right)^2\right)$$

denote the quadratic and

$$G(X,Y) = \frac{\partial f_3}{\partial X}\frac{\partial H}{\partial Y} - \frac{\partial f_3}{\partial Y}\frac{\partial H}{\partial X}$$

the cubic covariants of f_3. They satisfy the classical identity

$$4H^3 = G^2 + 27D_{f_3}f_3^2, \tag{7}$$

where D_{f_3} denotes the discriminant of $f_3(X,1)$. Remark that D_{f_3} is equal to the discriminant of $p(x)$.

We have $f_3(1,0) = 1$. If we define

$$A(4) := H(1,0) = 12a_4 + a_2^2 - 3a_1a_3 \quad \text{and}$$
$$B(4) := G(1,0) = 27a_1^2a_4 - 9a_1a_2a_3 + 2a_2^3 - 72a_2a_4 + 27a_3^2,$$

the point $P' = (A(4), B(4))$ lies on the curve

$$E: \quad y^2 = 4x^3 - 27D_{f_3}.$$

The curve E is 3-isogenous to the curve

$$E_t: \quad Y^2 = X^3 - 16D_{f_3},$$

where the 3-isogeny is given by the map

$$\varphi: \quad (x,y) \longrightarrow (X,Y) = \left(\frac{4(x^3 - 27D_{f_3})}{9x^2}, \frac{4y(x^3 + 54D_{f_3})}{27x^3}\right).$$

Hence the point

$$P_0 := \varphi(P') = \left(\frac{4(A^3(4) - 27D_{f_3})}{9A^2(4)}, \frac{4B(4)(A^3(4) + 54D_{f_3})}{27A^3(4)}\right)$$

lies on the elliptic curve E_t, provided $A(4) \neq 0$.

The main result of F. Leprévost et al. [8] is the following.

Theorem 4. *Let $k = \mathbb{Q}$ and assume that the transcendence degree of $K = \mathbb{Q}(a_1, a_2, a_3, a_4)$ is 4. Then the rank of $E_t(K)$ is at least one, P_0 is of infinite order and the group of torsion points of $E_t(K(\sqrt{D_{f_3}}))$ is isomorphic to $\mathbb{Z}/3\mathbb{Z}$.*

Using the specialization $(a_1, a_2, a_3, a_4) = (-1, -6, 2, 14)$ and the point $P = (1,0)$ they also proved:

Theorem 5. *The surface $I_p(X_1, X_2, X_3) = 1$ is an elliptic surface. Its Kodaira dimension and the dimension of its Albanése variety is one.*

5. Index form surfaces over finite fields

Let $q > 3$ be a rational prime and $k = \mathbb{F}_q$ be the finite field of q elements. If a_1, a_2, a_3, a_4 denote indeterminates over k and $K = k(a_1, a_2, a_3, a_4)$ then the point P_0, defined in the last section, belongs to $E_t(K)$.If we specialize a_1, a_2, a_3, a_4 to

$\tilde{a}_1, \tilde{a}_2, \tilde{a}_3, \tilde{a}_4 \in \mathbb{F}_q$, such that $D_{\tilde{f}_3}\tilde{A}(4) \neq 0$ the curve $\tilde{E}_t(\mathbb{F}_q)$ becomes an elliptic curve and the image $\tilde{P}_0$ of P_0 is a rational point of $\tilde{E}_t(\mathbb{F}_q)$.

If $q \equiv 2 \pmod 3$, then $|\tilde{E}_t(\mathbb{F}_q)| = q + 1$, i.e. the curve is supersingular (see A. Menezes [9], pp. 26, 27). The situation is much more interesting if $q \equiv 1 \pmod 3$, which we shall assume in the sequel.

In his diploma work P. Nagy [13] studied for randomly chosen primes $q \equiv 1 \pmod 3$ and $p(x) \in \mathbb{F}_q[x]$ how often $\tilde{P}_0$ generates a subgroup of $\tilde{E}_t(\mathbb{F}_q)$ of small index. He used the following algorithm:

1. $q \leftarrow$ random odd prime with $q \equiv 1 \pmod 3$,
   $p(x) \leftarrow x^4 + a_1x^3 + a_2x^2 + a_3x + a_4 \in \mathbb{F}_q[x]$ random polynomial,
   $t \leftarrow$ discriminant of $p(x)$,
2. if $t = 0$ or $A(4) = 12a_4 + a_2^2 - 3a_1a_3 = 0$ in $\mathbb{F}_q$ then goto 1,
3. $N \leftarrow$ the order of $\tilde{E}_t(\mathbb{F}_q)$,
4. if $N = q + 1$ then goto 1,
5. factorize N. If failed goto 1,
6. compute $\tilde{P}_0$ and its order M,
7. output $q, \tilde{P}_0, N, N/M$.

Remarks.

1. The algorithm was tested for 80, 100, 120 and 200 decimal digit primes.

2. To compute $N = |\tilde{E}_t(\mathbb{F}_q)|$ Cornacchia's algorithm was used, its complexity is $O(\log^2 q)$ (cf. R. Schoof [15]).

3. To factorize N only trial division with the first 100 000 primes was performed. If N or one of its divisors passed this test and the Miller–Rabin test [10, 14] then it was declared to be prime. We did not used deterministic primality tests.

4. Assume that the prime factorization of N is $N = p_1^{\alpha_1} \cdot \dots \cdot p_r^{\alpha_r}$ with $p_1 < \dots < p_r$, $\alpha_i \geq 1$ is computed. Then the order of $\tilde{P}_0$ is the smallest integer M such that M divides N and $M\tilde{P}_0 = \mathcal{O}$.

The algorithm was implemented in SIMATH 4.3 [17]. The computation were done on a PC with 166 MHz PENTIUM-MMX processor. Our experiences are the following:

- For 100 decimal digit primes the algorithm was performed 4000 times. We were able to compute the order of $\tilde{P}_0$ 440 times. The largest index was 1350. Curves with a point of small index were found in 1.3 minutes.
- For 120 decimal digit primes the algorithm was performed 2200 times. We were able to compute the order of $\tilde{P}_0$ 186 times. The largest index was 1158. Curves with a point of small index were found in 3.2 minutes.
- For 200 decimal digit primes the algorithm was performed 1632 times. We were able to compute the order of $\tilde{P}_0$ 89 times. The largest index was 223. Curves with a point of small index were found in about 30 minutes.

References

[1] Gaál, I., Power integral bases in algebraic number fields. In: Number Theory (ed. by K. Győry, A. Pethő and V.T. Sós), 243–254. Walter de Gruyter, Berlin–New York 1998.

[2] —, Power integral bases in algebraic number fields, II. In: Algebraic Number Theory and Diophantine Analysis (ed. by F. Halter-Koch and R.F. Tichy), 153–161. Walter de Gruyter, Berlin–New York 2000.

[3] Gaál, I., Pethő, A., Pohst, M., On the resolution of index form equations in quartic number fields. J. Symbolic Comput. 16 (1993), 563–584.

[4] —, —, —, Simultaneous representation of integers by a pair of ternary quadratic forms—with an application to index form equations in quartic number fields. J. Number Theory 57 (1996), 90–104.

[5] Gaál, I., Schulte, N., Computing all power integral bases of cubic fields. Math. Comp. 53 (1989), 689–696.

[6] Győry, K., Sur les polynomes á coefficients entiers et de discriminant donné, III. Publ. Math. Debrecen 23 (1976), 141–165.

[7] —, Effective finiteness theorems for polynomials with given discriminant and integer elements with given discriminant over finitely generated domains. J. Reine Angew. Math. 346 (1984), 54–100.

[8] Leprévost, F., Fermigier, S., Fieker, C., Sur certaines surfaces elliptiques et courbes elliptiques de Mordell de rang ≥ 1 associées á des discriminants de polynômes cubiques ou quartiques. J. Number Theory 78 (1999), 149–165.

[9] Menezes, A., Elliptic Curve Public Key Cryptosystems. Kluwer Academic Publ., Boston–Dordrecht–London 1997.

[10] Miller, G., Riemann's hypothesis and test for primality. J. Comput. System Sci. 13 (1976), 300–317.

[11] Mordell, L.J., Diophantine Equations. Academic Press, New York–London 1969.

[12] Nagell, T., Sur les discriminants des nombres algébriques. Ark. Mat. 7 (1967), 265–282.

[13] Nagy, P., Nyilvános kulcsú titkosítások (Public Key Cryptosystems). Diploma work, University of Debrecen, 2000.

[14] Rabin, M., Probabilistic algorithms for testing primality. J. Number Theory 12 (1980), 128–138.

[15] Schoof, R., Counting points on elliptic curves over finite fields. J. Théor. Nombres Bordeaux 7 (1995), 219–254.

[16] Silverman, J.H., Tate, J., Rational Points on Elliptic Curves. Springer, New York 1992.

[17] `http://emmy.math.uni-sb.de/~simath`, homepage of the Simath group.

On the size of solutions of the inequality $\phi(ax+b) < \phi(ax)$

*Herman te Riele**

Abstract. An estimate is given of the size of a solution $n \in \mathbb{N}$ of the inequality $\phi(an+b) < \phi(an)$, $\gcd(a,b) = 1$. Experiments indicate that this gives a useful indication of the size of the *minimal* solution.

1991 Mathematics Subject Classification: 11A25, 11Y70.

1. Introduction

Let $\phi(m)$ be the Euler totient function. Recently, D.J. Newman [5] has shown that for any nonnegative integers a, b, c, and d with $ad \neq bc$, there exist infinitely many positive integers n for which

$$\phi(an+b) < \phi(cn+d). \tag{1}$$

For the case $a = c = 30$, $b = 1$, $d = 0$, Newman stated that there are no solutions n with $n < 20\,000\,000$ and that a solution may be beyond the reach of any possible computers. Two years later, Greg Martin [3] found the smallest solution for this case, which turned out to be a number as large as 1116 decimal digits.

In this paper, we will analyse Newman and Martin's approach to this problem which enables us, for the case $a = c$, $\gcd(a,b) = 1$, $d = 0$, to give an estimate of the size of an n satisfying (1). Experiments indicate that this estimate also gives a useful indication of where the *minimal* solution of (1) can be expected.

Notation. By p_k we mean the k-th prime and by P_k the product $p_1 p_2 \cdots p_k$.

Acknowledgements. I like to thank Greg Martin and two anonymous referees for their constructive criticism which led to an improved presentation of this paper.

* Part of this research was carried out while the author was visiting the Mathematical Sciences Research Institute (Berkeley, CA) in September 2000.

Public-Key Cryptography and Computational Number Theory
K. Alster, J. Urbanowicz and H. C. Williams (Eds.)

2. A solution of $\phi(30n+1) < \phi(30n)$

We first consider the special case $a = c = 30$, $b = 1$, $d = 0$. As Martin showed, if n satisfies $\phi(30n+1) < \phi(30n)$, then

$$\frac{\phi(30n+1)}{30n+1} < \frac{\phi(30n)}{30n+1} < \frac{\phi(30)n}{30n} = \frac{4}{15} = 0.26666\ldots, \qquad (2)$$

(using $\phi(ab) \leq \phi(a)b \ \forall a, b \in \mathbb{N}$). Since ϕ is multiplicative and since $\phi(p^e)/p^e = \phi(p)/p$ for any prime p and any $e \geq 2$, the smallest m for which $\phi(m)/m$ has a given value, is squarefree. Therefore, we look for solutions of the inequality $\phi(30n+1) < \phi(30n)$ among the numbers

$$m_k := \prod_{i=4}^{k} p_i, \quad k = 4, 5, \ldots,$$

which satisfy

$$m_k \equiv 1 \bmod 30 \quad \text{and} \quad \frac{\phi(m_k)}{m_k} < \frac{4}{15}. \qquad (3)$$

Such m_k exist with high probability because the numbers

$$\frac{\phi(m_k)}{m_k} = \prod_{i=4}^{k} \left(1 - p_i^{-1}\right), \quad k = 4, 5, \ldots$$

decrease monotonically to zero, and because the residues $m_k \bmod 30$, $k = 4, 5, \ldots,$ seem to be uniformly distributed. For example, in the first 800 terms, the $\phi(30) = 8$ possible values

$$1, 7, 11, 13, 17, 19, 23, 29$$

occur with frequencies

$$100, 99, 107, 104, 110, 100, 85, 95,$$

respectively.

With help of the GP/Pari package [1], we have found that

$$m_{388} \equiv 1 \bmod 30 \quad \text{and} \quad \frac{\phi(m_{388})}{m_{388}} = 0.26631\ldots < \frac{4}{15}, \qquad (4)$$

and that there is no m_k with $4 \leq k < 388$ which satisfies these conditions. Now we check whether the number $n_{388} := (m_{388} - 1)/30$ actually is a solution of the inequality $\phi(30n+1) < \phi(30n)$. It turns out that $n_{388} = 2^3 n'$ where $n' = 5.502175051\ldots \times 10^{1124}$ has no prime divisors $\leq p_{50000} = 611953$. Using the well-known result that if n' has no prime divisors $\leq B$ then

$$\frac{\phi(n')}{n'} > \left(1 - \frac{1}{B}\right)^{\log n' / \log B},$$

we find

$$\frac{\phi(30n_{388})}{30n_{388}} = \frac{\phi(240n')}{240n'} = \frac{4}{15}\frac{\phi(n')}{n'}$$

$$> \frac{4}{15}\left(1 - \frac{1}{611953}\right)^{\log n'/\log 611953} = 0.26658\ldots.$$

Since

$$\frac{30n_{388}}{30n_{388}+1} = 1 - 7.57... \times 10^{-1128},$$

we conclude that

$$\frac{\phi(30n_{388})}{30n_{388}+1} > 0.26657.$$

Combining this with (4) we have

$$\frac{\phi(30n_{388}+1)}{30n_{388}+1} = 0.26631... < 0.26657 < \frac{\phi(30n_{388})}{30n_{388}+1}$$

which implies that $\phi(30n_{388}+1) < \phi(30n_{388})$.

So $n_{388} = 4.401740040... \times 10^{1125}$ is a solution of the inequality $\phi(30n+1) < \phi(30n)$, but it is *not* the smallest one. Martin [3] found this by computing the minimum number of distinct prime factors of such an n, viz., 382, by explicitly giving a solution with 382 distinct prime factors, and by showing that there are no smaller ones. Martin's minimum solution is given by

$$n = (z-1)/30, \quad \text{where} \quad z = \left(\prod_{i=4}^{383} p_i\right) p_{385}p_{388},$$

and

$$n = 2.329098101... \times 10^{1115}.$$

3. An estimate of the size of a solution of $\phi(an+b) < \phi(an)$, $\gcd(a,b) = 1$

In this section we will mimic and analyse the step described in Section 2 to find an $m_k \equiv 1 \bmod 30$ for which $\phi(m_k)/m_k < \phi(30)/30$, for the more general case $a = c$, $\gcd(a,b) = 1$, $d = 0$ in (1). So we consider the inequality

$$\phi(an+b) < \phi(an), \quad \gcd(a,b) = 1, \tag{5}$$

and look for a number $m_k \equiv b \bmod a$ for which $\phi(m_k)/m_k < \phi(a)/a$. We expect this m_k to be a solution of (5) and, also, that its size is not too far from the size of the *smallest* solution of (5) as we have seen in Section 2 for the case $a = 30$, $b = 1$.

As in Section 2, consider the products of the small primes which are not in a:

$$m_k := \frac{P_k}{\gcd(P_k, a)} \quad \text{for} \quad k = 1, 2, \ldots, \tag{6}$$

which satisfy

$$m_k \equiv b \bmod a \quad \text{and} \quad \frac{\phi(m_k)}{m_k} < \frac{\phi(a)}{a}. \tag{7}$$

Write $m_k = an_k + b$. We derive an estimate of the expected size of the smallest m_k satisfying (7) as follows. This m_k must satisfy

$$\phi(an_k + b) \approx \phi(an_k). \tag{8}$$

We assume that $b \ll an_k$ so that $an_k + b \approx an_k$. Dividing gives

$$\frac{\phi(an_k + b)}{an_k + b} \approx \frac{\phi(an_k)}{an_k}. \tag{9}$$

For the left hand side of (9) we have, using (6) [1)]

$$\frac{\phi(an_k + b)}{an_k + b} = \frac{\phi(m_k)}{m_k} = \frac{a}{\phi(a)} \frac{\phi(P_k)}{P_k} = \frac{a}{\phi(a)} \prod_{p \le p_k} \left(1 - \frac{1}{p}\right).$$

For the right hand side of (9) we assume that

$$\frac{\phi(an_k)}{an_k} \approx \frac{\phi(a)}{a}.$$

This requires that the prime divisors of n_k which are *not* in a are not too small. Substitution in (9) gives

$$\prod_{p \le p_k} \left(1 - \frac{1}{p}\right) \approx \left(\frac{\phi(a)}{a}\right)^2.$$

With Mertens's Theorem [2, §22.8]:

$$\prod_{p \le x} \left(1 - \frac{1}{p}\right) \sim \frac{e^{-\gamma}}{\log x} \quad \text{as} \quad x \to \infty,$$

where γ is Euler's constant ($= 0.5772...$), it follows that

$$\log p_k \approx e^{-\gamma} \left(\frac{a}{\phi(a)}\right)^2. \tag{10}$$

We estimate the corresponding size of n_k as follows. We have

$$an_k + b = m_k = \frac{P_k}{\gcd(P_k, a)},$$

1) with k such that $p_k \ge$ the largest prime in a.

so that

$$\log n_k \approx \log P_k - \log a - \log(\gcd(P_k, a)).$$

By the Prime Number Theorem [2, Chapter 22],

$$\log P_k = \sum_{p \le p_k} \log p = \theta(p_k) \sim p_k, \quad \text{as} \quad p_k \to \infty,$$

where $\theta(\cdot)$ is Chebyshev's function. So we could simplify our estimate of $\log n_k$ by replacing $\log P_k$ by p_k, but this introduces an undesirable error. Summarizing, we have the following

Estimate. *An estimate of the size of a solution of the inequality*

$$\phi(an+b) < \phi(an), \quad \text{with} \quad \gcd(a,b) = 1,$$

is given by $\log n \approx \log P_k - \log a - \log(\gcd(P_k, a))$, *where* k *is such that* $\log p_k \approx e^{-\gamma}(a/\phi(a))^2$.

For $a = 30$, $b = 1$ this gives: $p_k \approx 2685$, $\log n \approx 2600$, $\log_{10} n \approx 1129$ while in Section 2 we found $k = 388$, $p_{388} = 2677$ and $\log_{10} n_{388} = 1125.643\ldots$.

Remark. Greg Martin [4] pointed out that when a is the product of several primes, $a/\phi(a)$ has order of magnitude $\log\log a$ and if such an a has D digits, then it follows from the analysis given above that the smallest solution to $\phi(an+b) < \phi(an)$ will have about $\exp(c(\log D)^2)$ digits, for some constant c. In particular, there is in general no polynomial-time algorithm for finding the least solution to this inequality, for the simple reason that just writing down the answer takes longer than any polynomial function of D!

4. A program for finding a solution of $\phi(an+b) < \phi(an)$, $\gcd(a,b) = 1$

We have written a GP/Pari program[2)] which finds a solution of (5), for given a and b, in the same way as we found the solution of $\phi(30n+1) < \phi(30n)$ in Section 2. This program has two steps:

Step 1. Find the smallest $k \in \mathbb{N}$ for which m_k as defined in (6) satisfies (7).

Step 2. For this m_k define $n_k := (m_k - b)/a$. Find a lower bound for the quotient $\phi(an_k)/(an_k)$ by dividing out all the prime factors of n_k up to some fixed bound B. Let

$$n_k := n'n''n''',$$

2) This program is available from the author upon request.

where
— n' consists of the prime factors of n_k which are in a,
— n'' consists of the (known) prime factors of n_k which are *not* in a, and which are not greater than B, and
— n''' consists of the (unknown) prime factors of n_k which are greater than B.

Then

$$\frac{\phi(an_k)}{an_k} = \frac{\phi(a)}{a}\frac{\phi(n'')}{n''}\frac{\phi(n''')}{n'''} > \frac{\phi(a)}{a}\frac{\phi(n'')}{n''}\left(1-\frac{1}{B}\right)^{\log n'''/\log B} =: R.$$

Now check whether $\phi(m_k)/m_k$, as computed in Step 1, satisfies

$$\frac{\phi(m_k)}{m_k} < R\,\frac{an_k}{an_k+b}\,.$$

If so, it follows that

$$\frac{\phi(an_k+b)}{m_k} < \frac{\phi(an_k)}{m_k}\,,$$

so that n_k is a solution of (5). If not, continue with Step 1 to find the next smallest solution of (7). □

We have run this program for $b = 1$ and $a = 6, 30, 42$ with $B = p_{15000} = 163841$ and for $b = 1$, $a = 210$ with $B = p_{100000} = 1299709$, and compared the values of p_k and $\log_{10} n$, as estimated using Section 3, with the values of p_k and $\log_{10} n$ computed with this program. The results are given in Table 1.

	estimated		computed			
a $(b = 1)$	p_k	$\log_{10} n$	k	p_k	$\log_{10} n$	$\tilde{k}$
$6 = 2\cdot 3$	157	57.796...	36	151	57.796...	35
$30 = 2\cdot 3\cdot 5$	2685	1129.072...	388	2677	1125.643...	385
$42 = 2\cdot 3\cdot 7$	971	397.081...	171	1019	421.063...	161
$210 = 2\cdot 3\cdot 5\cdot 7$	46476	20048.160...	4981	48413	20880.507...	4789

Table 1. Comparison of estimated (according to Section 3) and computed values of p_k and $\log_{10} n$, where the computed value of $n = (m_k - b)/a$, with $m_k = P_k/\gcd(P_k, a)$, satisfies $\phi(an+b) < \phi(an)$, $\gcd(a, b) = 1$. The last column lists the minimal value $\tilde{k}$ of k for which $\phi(m_k)/m_k < \phi(a)/a$.

The main reason for the difference between the estimated and computed values of p_k and $\log_{10} n$ is that the condition $m_k \equiv 1 \bmod a$ is only satisfied in about 1 in every $\phi(a)$ cases (on the assumption of the uniform distribution of the residues $m_k \bmod a$).

The last column of Table 1 lists the minimal value $\tilde{k}$ of k for which $\phi(m_k)/m_k < \phi(a)/a$, where $m_k = P_k/\gcd(P_k, a)$. Since this inequality is a *necessary condition* for any solution, we can use our computed solution and this $\tilde{k}$ to find the minimal solution. For example, for $a = 6$, $b = 1$, we have $\tilde{k} = 35$, so

$$m = p_3 p_4 \cdots p_{35} = 5 \cdot 7 \cdots 149$$

is the smallest product of consecutive primes ≥ 5 which satisfies the inequality $\phi(m)/m < 1/3$. In addition, for this m we have $m \equiv 1 \bmod 6$, $\phi(m) = 8.2531... \times 10^{55}$ and

$$\phi(m-1) = \phi(2 \cdot 3 \cdot 1381 \cdot 70140112179047 \cdot p39) = 8.2838... \times 10^{55},$$

where $p39$ is a prime of 39 decimal digits, easily computable from $m-1$ and the other given factors of $m-1$. So this m is also the *minimal* solution $\equiv 1 \bmod 6$ of the inequality $\phi(m) < \phi(m-1)$.

Table 1 lists sizes of estimated and computed solutions for various values of a, with $b = 1$. In fact, our program finds solutions for *all* those values of b for which $\gcd(a, b) = 1$, and since we have no indications that the residues $m_k \bmod a$ are *not* uniformly distributed, we expect the solutions for $b \neq 1$ to have about the same size as those given for $b = 1$ in Table 1.

References

[1] Batut, C., Bernardi, D., Cohen, H., Olivier, M., User's Guide to PARI-GP. `http://www.parigp-home.de`. PARI-GP was developed at Bordeaux by a team led by Henri Cohen. It is maintained now by Karim Belabas at the Université Paris-Sud Orsay with the help of many volunteer contributors.

[2] Hardy, G.H., Wright, E.M., An Introduction to the Theory of Numbers, fifth edition. Clarendon Press, Oxford 1995.

[3] Martin, G., The smallest solution of $\phi(30n+1) < \phi(30n)$ is .... Amer. Math. Monthly 106 (1999), 449–451.

[4] —, Private communication, January 24, 2001.

[5] Newman, D.J., Euler's ϕ function on arithmetic progressions. Amer. Math. Monthly 104 (1997), 256–257.

Security of DL-encryption and signatures against generic attacks—a survey

Claus Peter Schnorr

Abstract. We survey recent results on the security of DL-cryptosystems and DL-signatures against generic attacks [Sh97, SJ99, SJ00] assuming the random oracle model (ROM) and the generic group model (GM). We comment on the relevance of these results towards applications.

Key words and phrases: Generic algorithm, generic group model, random oracle model, signed ElGamal encryption, generalized signed ElGamal encryption, one-more decryption attack, chosen ciphertext attack, digital signature, DL-signature, one-more signature forgery, blind Schnorr signature, interleaved, parallel attack.

1. Introduction and summary

While there are no security proofs without unproven assumptions the most important practical attacks against DL-signatures and DL-cryptosystems are generic in that either the hash function or the cyclic group or both are modelled by black-box oracles. The random oracle model (ROM) assumes an ideal random hash function while the generic group model (GM) assumes an ideal group G of prime order q with group operations done by a group oracle. The ROM is a stronger assumption than assuming the hash function to be collision resistant—collisions of random functions have negligibly small probability. The GM is a stronger assumption than assuming that the discrete logarithm is intractable—it has been shown that the discrete logarithm has exponential complexity in the GM [Ne94, Sh97]. The GM is appropriate for elliptic and hyperelliptic curves as the known attacks are all generic. The GM is less appropriate for groups of units as it does not cover attacks using sieving or index calculus. The ROM goes back to Fiat and Shamir [FS86] and has been further enhanced by Bellare and Rogaway [BR93], while the generic group model goes back to Nechaev [Ne94] and Shoup [Sh97].

Assuming a combination of the ROM and the GM we study security against attacks that do not exploit particular weaknesses of the hash function H or of the group G or of a combination of G and H. While we do not have to rely on any unproven assumption, it is the case that our security guarantee hinges on the existence of strong hash functions H and groups G for which the combination

Public-Key Cryptography and Computational Number Theory
K. Alster, J. Urbanowicz and H. C. Williams (Eds.)

(G, H) has no weaknesses. The strong assumption of the ROM+GM makes sense in cases where traditional security proofs fail. Prominent examples are the security of Schnorr identification against active adversaries shown in [Sh97], security of signed ElGamal encryption [SJ00] and of blind Schnorr signatures [SJ99]. A security proof in the ROM+GM encourages further analysis under relaxed assumptions. For instance it is reasonable to ask whether the random function H can be replaced by Canetti's "oracle hashing"—replacing $H(z)$ by $(h(z, r), r)$ for a one-way h and a random one-time key r. It would be nice to show security under this replacement.

Security of signed ElGamal encryption. It has long been believed that a suitably tagged ElGamal ciphertext is secure against the strong adaptive chosen ciphertext attack (CCA) of Rackoff and Simon [RS92]. CCA-security means indistinguishability of the plaintext m against any other message when given a ciphertext of m—the target ciphertext—and free access to a decryption oracle except for the target ciphertext. The idea of tagging an ElGamal ciphertext appears first in [ZS92], and one of the proposed schemes can be proven to be CCA-secure. CCA-secure schemes together with a security proof have been proposed in [SG98, CS98, ABR98, FO99, Sh00]. The schemes in [ZS92, SG98, CS98, ABR98, Sh00] either use an involved tag construction or key generation to simplify the reduction to the discrete log or to the Diffie–Hellman problem, the tag in [ABR98] uses a private key encryption scheme.

Signed ElGamal encryption is a very practical encryption scheme where an ElGamal encryption [E85] is tagged by a Schnorr signature [Sc91]—the public signature key is part of the ElGamal ciphertext. Signed ElGamal encryption was independently proposed by Tsiounis and Yung [TY98] and Jakobsson [J98]. Herein, an ElGamal ciphertext (g^r, mh^r) is authenticated by a *Schnorr signature* [Sc91] providing a proof of knowledge of the plaintext m and of the secret r—the public signature key g^r is part of the ciphertext. CCA-security of *signed ElGamal encryption* has been shown in [TY98] under the assumption that the signer really "knows" the secret signature key r. That assumption holds in the ROM if there is only a logarithmic number of interactions with the decryption oracle[1)].

While CCA-security of signed ElGamal encryption is still open under traditional assumptions Schnorr and Jakobsson [SJ00] prove CCA-security of signed ElGamal encryption in the ROM+GM. Moreover, it is shown in [SJ00] that signed ElGamal encryption is secure against the novel *one more decryption attack.* That proof is based on a novel and strong concept of generic plaintext awareness that is inspired by the notion of plaintext awareness set forth in [BR94]. It is possible to extract from a generic algorithm that generates a valid signed ElGamal ciphertext the corresponding plaintext m along with the one-time keys r, s used for encryption and for the tagged signature. There is a generic extractor that controls the

1) The FFS-extractor of Feige–Fiat–Shamir, in the oracle replay mode of Pointcheval and Stern [PS96a], extracts the secret signature key from signed ElGamal encryptions. The FFS-extractor has a constant delay factor, and thus can in polynomial time at most be iterated a logarithmical number of times.

generic steps of the generic attacker. The explicit description of the attacker is given to the extractor not just the black-box of the attacker.

The CCA-security of signed ElGamal encryption in the ROM+GM favorably contrasts with the partially unsettled situation of OAEP proposed in [BR94]. Recently, it was shown in [FOPS00] that the RSA-OAEP is CCA-secure in the ROM under the RSA-assumption filling a proof gap pointed out in [Sh00b]. But the exact CCA-security of RSA-OAEP remains open—the reduction to the RSA-assumption in [FOPS00] is not tight—also CCA-security for general one-way trapdoors is still unsettled. The CCA-security of signed ElGamal encryption in the ROM+GM indicates that signed ElGamal encryption is CCA-secure for all strong groups G and strong H.

The one-more decryption attack. In the ROM+GM it is possible to prove security against the one-more decryption attack in which an attacker interacts some l times with a decryption oracle and gets partial information about $l+1$ plaintexts corresponding to given ciphertexts. Various applications of e-commerce are possible provided that a decryption oracle is accessible that decrypts arbitrary ciphertexts against a corresponding fee [SJ00]. For such an application it is important that an attacker cannot get even partial information about more than l plaintexts by asking the decryption oracle some l times.

Security of blind Schnorr signatures. We study the security of blind Schnorr signatures against the one-more signature forgery in which an attacker interacts some l times with the legitimate signer and produces from these l interactions $l+1$ signatures. We first present a novel, parallel attack that succeeds in a one-more signature forgery against blind Schnorr and blind Okamoto–Schnorr signatures. The attack merely requires a solution of the ROS-problem, a possibly intractable problem: find an overdetermined, solvable system of linear equations modulo q with random inhomogeneities (right sides). Specifically, given a system of $t \gg l$ linear equations modulo q in l unknowns with random right sides, find a solvable subsystem of $l+1$ equations—a solvable subsystem corresponds to an $(l+1)\times(l+1)$-submatrix of rank l.

The generic parallel attack has the interesting feature not to depend on the public key. Traditional security proofs do not seem to work in the presence of such an attack. Usually, traditional security proofs use the attacker to solve a DL-problem or a decisional Diffie–Hellman-problem associated with the public key. As the generic parallel attack uses a solution of the ROS-problem that is not related to the public key, the attack cannot help to solve a DL- or a DDH-problem. How then could [PS00, P00, PS96b] prove security? The known security results only cover cases where solutions of the ROS-problem exist with negligible probability. The generic parallel attack points to an inherent weakness of the security result of Pointcheval and Stern.

The main result Theorem 26 of [PS00] shows[2)] that an attacker mounting a one-more signature forgery with a probability of success $\varepsilon > 4Q^{l+1}/q$ can be used

2) The security proofs of Pointcheval, Stern show that blind Okamoto–Schnorr signatures are secure against parallel interactive attacks [PS96b, PS00] provided that

to compute a discrete logarithm. Here Q is the number of hash queries, l is the number of interactions with the signer and q is the prime order of G. For an elliptic curve G of order $\leq 2^{200}$ and $Q = 2^{50}$ we must have $l \leq 3$ since $\varepsilon \leq 1$. For a subgroup G of units of order $\leq 2^{1000}$ we must have $l \leq 20$.

The generic parallel attack shows that any improved security guarantee, covering larger values of l, requires that the ROS-problem is intractable. The security of blind DL-signatures against the one-more signature forgery hinges on the intractability of the ROS-problem. The ROS-problem deserves further study because restricting the signer to merely a few parallel interactions is impractical for applications in e-commerce. It is important that the server issuing anonymous digital cash can operate in an unrestricted parallel way.

Practically meaningful security guarantees against the one-more signature forgery require in addition to the ROM and the intractability of the DL-problem also the intractability of the ROS-problem. Conversely, assuming the intractability of the ROS-problem a practical security guarantee for blind Schnorr signatures holds in the ROM+GM, see Theorem 5. A generic attacker performing t generic steps, including some l interactions with the signer, cannot produce $l+1$ signatures with a better probability than $\binom{t}{2}/q$. For elliptic curves G of order $q \approx 2^{200}$ this guarantee covers up to $t = 2^{100}$ generic steps including up to 2^{100} parallel signer interactions that can be interleaved in an arbitrary way. Moreover, this shows that blind Schnorr signatures have the same security level in the ROM+GM as the double-keyed blind Okamoto–Schnorr signatures, thus blind Schnorr signatures can save a considerable overhead. In conclusion our result suggests to use blind Schnorr signatures in connection with strong elliptic curves rather than double-keyed blind Okamoto–Schnorr signatures using subgroups of units.

The structure of the paper. In Section 2 we introduce the generic model for interactive algorithms that use a hash oracle and an oracle for decryption/signatures. In Section 3 we survey and comment on the security of signed ElGamal encryption presented in [SJ00]. In Section 4 we consider the security of blind Schnorr signatures against interactive attacks based on [SJ99].

2. The random oracle and the generic model

The random oracle model (ROM). Let H be an *ideal* hash function, modelled as an oracle, that given an input (query) outputs a random number in the range of H. Hash functions for signatures are of the type $H : G \times M \to \mathbf{Z}_q$, where G

the number of interactions with the signer is polylogarithmic—polylog($|q|$) for the binary length $|q|$ of q. The polylog bound on the number of signer interactions has not been explicitly mentioned in [P00] but it is required as the proof is based on the results of [PS00]. In [P98] a third party—the *checker*—has been introduced, and it is shown that the resulting three-party signature protocol is secure for a polynomial number of *synchronized* signer interactions, where the synchronization forces the completion of each step for all the different protocol invocations before the next step of any other invocation is started.

is a group of prime order q, M is a range of messages and $\mathbf{Z}_q$ denotes the field of integers modulo q. Formally, H is a random function $H : G \times M \to \mathbf{Z}_q$ chosen at random over all functions of that type with uniform probability distribution. There is an ongoing debate on whether the assumption of a random hash function is realistic or too generous. The problem is that random functions can in principle not be implemented by public algorithms. Canetti, Goldreich, Halevi [CGH98] present an artificial "counter-example" that is provably secure in the ROM but which cannot be implemented in a secure way replacing the random oracle by a computable function family. In [CGH98] a mechanism for the implementation of random hash functions has been added to the ROM. The artificial "counter-example" is defined relative to that mechanism using the function ensemble that implements the random oracle. Nevertheless, the security achievable in the ROM seems to in practice eliminate all attacks at hand.

The Generic Model (GM). Let G be an *ideal* group of prime order q with generator g. Generic algorithms for G do not use the binary encodings of the group elements, as they access group elements only for group operations and equality tests. Nechaev [Ne94] proves that the discrete logarithm problem is hard in such a model. The generic model of algorithms was further elaborated on by Shoup [Sh97]. We present the Shoup model in a slightly different setup[3)] and we extend it to algorithms that interact with a decryption oracle. We do not assume a random binary encoding of group elements. We exemplify the difference of the two setups for the baby-step-giant-step DL-algorithm. While our generic algorithms do not allow for efficient sorting of group elements this does not affect the number of generic steps as equality tests of group elements are free of charge.

Encryptions are for the private/public key pair (x, h), where x is random in $\mathbf{Z}_q$ and $h = g^x$. We describe the extended generic model in detail, first focusing on non-interactive algorithms and thereafter on algorithms interacting with oracles for hashing and decryption.

The *data of a generic algorithm* is partitioned into group elements in G and non-group data. The *generic steps* for group elements are multivariate exponentiations:

- mex : $\mathbf{Z}_q^d \times G^d \to G$, $(a_1, \ldots, a_d, g_1, \ldots, g_d) \mapsto \prod_i g_i^{a_i}$ with $d \geq 0$.

The cases $d = 2$, $a_1 = 1$, $a_2 = \pm 1$ present multiplication/division. The case $d = 0$ presents *inputs* in G—e.g., g, h are inputs for the DL-computation.

3) We count the same generic steps as in [Sh97]; however, we allow arbitrary multivariate exponentiations while Shoup merely uses multiplication and division. The technical setup in [Sh97] looks different as groups G are *additive* and associated with a random injective encoding $\sigma : G \to S$ of the group G into a set S of bit strings—the generic algorithm performs arbitrary computations on these bit strings. Addition/subtraction is done by an oracle that computes $\sigma(f_i \pm f_j)$ when given $\sigma(f_i), \sigma(f_j)$ and the specified sign bit. As the encoding σ is random it contains only the information about which group elements coincide—this is what we call the set of *collisions*.

Definition. A (non-interactive) *generic algorithm* is a sequence of t generic steps [4)]

- $f_1, \dots, f_{t'} \in G$ (inputs) $1 \le t' < t$,
- $f_i = \prod_{j=1}^{i-1} f_j^{a_j}$ for $i = t'+1, \dots, t$, where $(a_1, \dots, a_{i-1}) \in \mathbf{Z}_q^{i-1}$ depends arbitrarily on i, the non-group input and the set $\mathcal{CO}_{i-1} := \{(j,k) \mid f_j = f_k, 1 \le j < k \le i-1\}$ of previous *collisions* of group elements.

Typical non-group inputs are given elements in $\mathbf{Z}_q$ contained in given ciphertexts or signatures. $\mathcal{CO}_t$ is the set of all collisions of the algorithm.

Some group inputs f_i depend on random coin flips, e.g., the random public key $h = g^x$ depends on the random secret key $x \in_R \mathbf{Z}_q$. The *probability space* consists of the random group elements of the input. The logarithms $\log_g f_i$ of the random inputs f_i play the role of *secret parameters.* Information about the secret parameters can only be revealed by collisions. E.g., $g^a = f_i^b$ implies $\log_g f_i = a/b$. We let the non-group input and the generator g not depend on random bits.

The *output* of a generic algorithm consists of

- non-group data that depend arbitrarily on the non-group input and on the set $\mathcal{CO}_t$ of all collisions,
- group elements $f_{\sigma_1}, \dots, f_{\sigma_d}$ where the integers $\sigma_1, \dots, \sigma_d \in \{1, \dots, t\}$ depend arbitrarily on the non-group input and on $\mathcal{CO}_t$.

For the sake of clarifying the GM, we give an example of a generic algorithm:

The baby-step-giant-step DL-algorithm. This algorithm is given q and $g, h \in G$ and computes $\log_g h \in \mathbf{Z}_q$ in $2\sqrt{q}$ generic steps.

1. Compute $k := \lceil \sqrt{q} \rceil$, $l := \lceil q/k \rceil$ so that $lk - k < q \le lk$. The computation of the non-group data k, l is for free.
2. Form the lists $L_1 := \{g^i \mid 0 \le i < k\}$ in $k-1$ multiplications and $L_2 := \{hg^{jk} \mid 0 \le j < l\}$ in l multiplications. Clearly, $L_1 \cap L_2 \ne \emptyset$.
3. Find a collision by testing all equalities $g^i = hg^{jk}$. Note that the detection of the collision is for free. An equality implies $\log_g h = i - jk \bmod q$.

While this algorithm performs $\#L_1 \times \#L_2$ "free" equality tests, the corresponding Turing machine—in the [Sh97]-setup—constructs a collision differently, using only $O(\sqrt{q}\log_2 q)$ equality tests. It sorts the binary encodings of the g^i and inserts the encodings of hg^{jk} into the sorted list.

Going back to the description of the model we work in, we now elaborate on *interactive, generic algorithms.* We count the following generic steps:

- group operations, $\text{mex} : \mathbf{Z}_q^d \times G^d \to G$, $(a_1, \dots, a_d, g_1, \dots, g_d) \mapsto \prod_i g_i^{a_i}$,
- queries to the hash oracle H,

4) We can allow a generic algorithm to perform a number t of generic steps, where t varies with the input. We can let the algorithm decide after each step whether to terminate depending arbitrarily on the given non-group data. Then the number t of generic steps depends on the computed non-group data.

- interactions with a decryption/signature oracle—see 3.2, 4.2 [5)].

A *generic adversary* $\mathcal{A}$—attacking an encryption scheme—is an interactive algorithm that interacts with a decryptor. It performs some t generic steps resulting in $t' \leq t$ group elements $f_1, \ldots, f_{t'}$. $\mathcal{A}$ iteratively selects the next generic step—a group operation, a query to H, an interaction with the decryptor—depending arbitrarily on the non-group input and on previous collisions of group elements.

The *input* consists of the generator g, the public key $h \in G$, the group order q, a collection of messages and ciphertexts and so on, all of which can be broken down into group elements and non-group data.

The computed *group elements* $f_1, \ldots, f_{t'} \in G$ are the group elements contained in the input, such as g, h. When counting the number of group operations, we count each input as one operation. As a decryptor interaction is counted as a generic step the number t' of group elements is bounded by the number t of generic steps, $t' \leq t$. We have $t = t'$ for a non-interactive $\mathcal{A}$.

The given *non-group data* consists of the non-group data contained in the input, the previous hash replies $H(Q)$ of queries Q, and the set of previous collisions of group elements.

An interaction with the *decryptor/signer* (defined in subsection 3.2/4.2) is a two round deterministic protocol. A claimed ciphertext is sent to the decryptor, which performs a generic group operation using the secret decryption key x, verifies the Schnorr signature using the public key g^r contained in the ciphertext, and—in case that this signature is correct—outputs the decrypted message. If the signature is invalid the decryptor outputs a random element of G. $\mathcal{A}$'s interactions with the decryptor are sequential as the interleaving of these two-round interactions is necessarily trivial.

$\mathcal{A}$'s *output* and *transmission* to the decryptor consists of non-group data NG and previously computed group elements f_σ, where NG and σ, $1 \leq \sigma \leq t'$, depend arbitrarily on given non-group data.

$\mathcal{A}$'s *transmission* to the hash oracle H depends arbitrarily on given group elements and given non-group data. The *probability space* consists of the random H and the random input group elements.

The *restriction of the generic model* is that $\mathcal{A}$ can use group elements only for generic group operations, equality tests and for queries to the hash oracle, whereas non-group data can be arbitrarily used without charge. The computed group elements $f_1, \ldots, f_{t'}$ are given as explicit multiplicative combinations of group elements in the input and from decryptor interactions. Let the group elements in the input and from decryptor interactions be $g_1, \ldots, g_l$. By induction on j, a computed $f_j \in G$ is of the form $f_j = g_1^{a_{j,1}} \cdots g_l^{a_{j,l}}$, where the exponents $a_{j,1}, \ldots, a_{j,l} \in \mathbf{Z}_q$ depend arbitrarily on given non-group data. $\mathcal{A}$ can arbitrarily use the coefficients

5) Other types of interactions are possible for other signature/encryption schemes, other cryptographic protocols using groups of non-prime order, groups of unknown order or using several distinct groups.

$a_{j,1}, \ldots, a_{j,l}$ from this explicit representation of f_j. A generic adversary is deterministic, which is not a restriction as its coin flips would be useless[6)].

Trivial collisions. We call a collision $(i,j) \in \mathcal{CO}_t$ *trivial* if $f_i = f_j$ holds with probability 1, i.e., if it holds for all choices of the secret data such as the secret key x and the random bits r of the encipherer. We write $f_i \equiv f_j$ for a trivial collision[7)]. Trivial collisions do not release any information about the secret data while non-trivial collisions can completely release some secret data. Trivial collisions can be ignored, and so, we can exclude them from $\mathcal{CO}_t$ so that $\mathcal{CO}_t$ consists only of non-trivial collisions.

2.1. Results for the GM

This subsection refers to a generic, non-interactive adversary $\mathcal{A}$ that performs some t generic steps in attacking the indistinguishability of ElGamal encryption. Given q, the public key $h = g^x$, two messages $m_0, m_1 \in G$ and an ElGamal ciphertext $\mathrm{cip}_b = (g^r, m_b h^r)$ for random $r, x \in_R \mathbf{Z}_q$ and $b \in_R \{0,1\}$, $\mathcal{A}$ guesses b. There is no hash function involved, we assume the GM but not the ROM. We show that $\mathcal{A}$ does not succeed better than with probability $\frac{1}{2} + 2\binom{t}{2}/q$.

The probability space consists of the random group elements $g^r, g^x, m_b g^{rx}$, or equivalently of the random $r, x \in_R \mathbf{Z}_q$ and $b \in_R \{0,1\}$. Let $\mathcal{A}$ compute the group elements $f_1, \ldots, f_t$. We let the *Main Case* be the part of the probability space where there are no non-trivial collisions among $f_1, \ldots, f_t$, i.e., $\mathcal{CO}_t = \emptyset$.

Lemma 1 ([SJ00]). *Non-trivial collisions among $f_1, \ldots, f_t$ occur with no better probability than $2\binom{t}{2}/q$. The probability refers to the random b, r, x.*

The factor 2 is the degree of the polynomial $\log_g m_b h^r = rx \log_g m_b \in \mathbf{Z}_q[r,x]$. If $m_b h^r$ is removed from the input then all polynomials $\log_g f_j \in \mathbf{Z}_q[r,x]$ are linear, and the factor 2 disappears. The degree of the polynomials $\log_g f_j \in \mathbf{Z}_q[r,x]$ enters into Lemma 1 via Lemma 2 that is attributed to Schwartz [Sch80].

Lemma 2. *A multivariate polynomial $F \in \mathbf{Z}_q[X_1, \ldots, X_k]$ of total degree d satisfies for random $x_1, \ldots, x_k \in \mathbf{Z}_q$ that $\Pr_{x_1,\ldots,x_k}[F(x_1, \ldots, x_k) = 0] \le d/q$.*

The leakage of secret information through the absence of collisions. Here we pay attention to the fact that b, r, x are not perfectly random if $\mathcal{CO}_t = \emptyset$. By

6) $\mathcal{A}$ could select interior coin flips that maximize the probability of success—there is always a choice for the internal coin flips that does not decrease $\mathcal{A}$'s probability of success. It is useless for $\mathcal{A}$ to generate random group elements—in particular ones with unknown DL. Using one generic step, $\mathcal{A}$ could replace random elements in G by some deterministic g^a where $a \in \mathbf{Z}_q$ is chosen as to maximize the probability of success.

7) Trivial collisions occur in testing correctness of an ElGamal ciphertext (g^r, mh^r) and its message m. In case of a correct message-ciphertext pair the test results in a trivial collision. Also, identical repetitions of a group operation yield a trivial collision.

Lemma 1 a $2\binom{t}{2}/q$-fraction of the probability space is excluded in the Main Case. The Shannon entropy of the secret parameters b, r, x decreases accordingly. We can neglect this minor leakage of secret information through the absence of collisions. Thus, for a "collision-free" attacker the secret data are statistically independent of the computed non-group data:

Lemma 3. *If there are no non-trivial collisions of group elements then the random b, r, x are statistically independent of the computed non-group data except that the b, r, x leading to collisions are excluded.*

Proof. The random b, r, x, enter into the generic computation only via the group elements $g^r, g^x, m_b g^{rx}$. Therefore, b, r, x enter into non-group data only via non-trivial collisions of group elements. □

Proposition 1 (Generic DL-complexity lower bound [Ne94, Sh97]). *Let $\mathcal{A}$, upon input g and $h = g^x \in_R G$, output $y \in \mathbf{Z}_q$. Then $\Pr_x[y = \log_g h] \leq \binom{t}{2}/q + \frac{1}{q}$.*

Proof. We use Lemma 1 and 3 for a generic $\mathcal{A}$ with input g, h—without inputs $g^r, m_b h^r$. The factor 2 in Lemma 1 disappears as the polynomials $\log_g f_j$ have total degree ≤ 1. For a collision-free $\mathcal{A}$, x is statistically independent of the non-group output y, and thus $\Pr_h[y = \log_g h] = \frac{1}{q}$. By Lemma 1, non-trivial collisions occur at most with probability $\binom{t}{2}/q$. □

Proposition 2 (Indistinguishability). *Let a non-interactive generic algorithm $\mathcal{A}$ be given g, h, two messages $m_0, m_1 \in G$ and a ciphertext $(g^r, m_b h^r)$ for random $r \in_R \mathbf{Z}_q$ and $b \in_R \{0, 1\}$. Let $\mathcal{A}$ output a guess b' for b. Then $\Pr_{b,x,r}[b' = b] \leq \frac{1}{2} + 2\binom{t}{2}/q$.*

Proof. Suppose that there is no non-trivial collision of group elements in $\mathcal{A}$'s computation. Then, b, r, x are, by Lemma 2, statistically independent of the non-group output b', thus $\Pr_{b,x,r}[b' = b] = \frac{1}{2}$. Non-trivial collisions occur with probability $\leq 2\binom{t}{2}/q$. □

The decisional Diffie–Hellman problem (DDH-problem). This problem is to distinguish g^{xy} and g^z when given g, g^x, g^y for random $x, y, z \in_R \mathbf{Z}_q$. Proposition 2 implies the hardness of the DDH-problem in GM which was first shown by Shoup [Sh97].

Proposition 3 ([Sh97]). *Let a generic algorithm $\mathcal{A}$ be given g, g^x, g^y and $\{g^{xy}, g^z\}$ in random order for random $x, y, z \in_R \mathbf{Z}_q$ and output a guess for the order of g^{xy}, g^z. Then $\mathcal{A}$ succeeds with no better probability than $\frac{1}{2} + 2\binom{t}{2}/q$.*

2.2. Security of Schnorr identification, following [Sh97]

An identification scheme is an interactive protocol that allows one party $\mathcal{P}$ to prove its identity to another party $\mathcal{V}$. To do this $\mathcal{P}$ has a private key and a corresponding public key. Such a scheme is considered secure if an attacker cannot feasibly convince $\mathcal{V}$ it is conducting the protocol with $\mathcal{P}$. One can allow the adversary to first interact with $\mathcal{P}$, pretending to be $\mathcal{V}$ (but not necessarily following $\mathcal{V}$'s protocol), in order to gain some useful information about $\mathcal{P}$'s secret key. Such an attack is called "active". An attack where no prior interaction with $\mathcal{P}$ is allowed is called "passive".

The Schnorr identification scheme [Sc91] runs as follows. Let G be a group of prime order q, with a publicly known generator g. $\mathcal{P}$'s private key is a random element $x \in_R \mathbf{Z}_q$, the corresponding public key is $h = g^x$. In the first step of the protocol $\mathcal{P}$ picks a random $r \in_R \mathbf{Z}_q$, computes $h' = g^r \in G$ and sends h' to $\mathcal{V}$. Upon receiving h', $\mathcal{V}$ picks a random challenge $c \in_R \mathbf{Z}_q$ and sends c to $\mathcal{P}$. Upon receiving c, $\mathcal{P}$ computes $y = r + xc \in \mathbf{Z}_q$ and sends y to $\mathcal{V}$. $\mathcal{V}$ accepts if $g^y = h'h^c$ and otherwise rejects.

It has been shown in [Sc91] that this identification scheme is secure against passive attacks provided that the DL-problem is hard. Shoup shows security against active attacks in the GM. An interaction with $\mathcal{P}$ is counted as generic step.

Proposition 4 ([Sh97]). *Let a generic adversary $\mathcal{A}$ be given $g, h = g^x$ and q and performing some t generic steps, including interactions with $\mathcal{P}$. Then $\mathcal{A}$ cannot make $\mathcal{V}$ accept with a better probability than t^2/q.*

2.3. Zeroknowledge versus generic verifiers

Let $\mathcal{L}$ be a language in the class NP of non-deterministic polynomial time languages. An interactive proof of membership for $\mathcal{L}$ is an interactive protocol between a prover $\mathcal{P}$ and a verifier $\mathcal{V}$. For a common input y, $\mathcal{P}$ wants to convince $\mathcal{V}$ of a proof of membership for "$y \in \mathcal{L}$" known to $\mathcal{P}$. The zeroknowledge property of the interactive protocol $(\mathcal{P}, \mathcal{V})$ shows that $\mathcal{V}$ cannot learn anything about $\mathcal{P}$'s proof of membership by executing the protocol with $\mathcal{P}$. The protocol $(\mathcal{P}, \mathcal{V})$ is called *zeroknowledge* if there is a simulator $\mathcal{S}$ which given y generates the same distribution of data as do $\mathcal{P}$ and $\mathcal{V}$, without using $\mathcal{P}$'s proof of membership.

A zeroknowledge simulator $\mathcal{S}$ for an interactive proof of membership needs some control over the verifier $\mathcal{V}$. For many protocols it is sufficient that $\mathcal{S}$ uses the verifier as a black-box—we call $\mathcal{S}$ a *black-box simulator*. A simulator $\mathcal{S}$ that is given the Turing machine of $\mathcal{V}$ is more powerful as $\mathcal{S}$ completely controls all steps of $\mathcal{V}$. For generic verifiers in the generic group model it makes sense to let the simulator control the generic group steps of $\mathcal{V}$—we call $\mathcal{S}$ a *generic verifier simulator*.

M. Fischlin [F00] demonstrates the power of generic verifier simulators. He presents a three round negligible-error proof of membership for arbitrary NP-languages together with a polynomial time zeroknowledge simulator $\mathcal{S}$. Fischlin's generic verifier simulator $\mathcal{S}$ achieves something that is impossible for black-box simulators and arbitrary Turing machine verifiers. Goldreich and Krawczyk [GK96] show that polynomial time black-box simulators for three round negligible-error proofs of membership merely exist for languages in BPP—which is conjectured to be a proper subclass of NP.

Fischlin's example shows that generic verifier zeroknowledge does not imply black-box TM verifier zeroknowledge. This is no surprise as there are two reasons: generic verifiers are in their generic steps more restricted than TM-verifiers, and black-box simulators are less powerful than generic verifier simulators that control the generic group steps of the verifier.

3. Signed ElGamal encryption

3.1. Introduction

We analyse a very practical public key cryptosystem in terms of its security against the strong *adaptive chosen ciphertext attack* (CCA) of [RS92], in which an attacker can access a decryption oracle on arbitrary ciphertexts (except for the target ciphertext).

Notions of security. Let G be a cyclic group of prime order q with generator g, and let $\mathbf{Z}_q$ be the field of integers modulo q. A Diffie–Hellman key pair consists of a random secret key $x \in \mathbf{Z}_q$ and the corresponding public key $h = g^x \in G$. Diffie–Hellman keys give rise to many cryptographic schemes, for example *ElGamal encryption* [E85]. An ElGamal ciphertext of message $m \in G$ is a pair $(g^r, mh^r) \in G^2$ for random $r \in \mathbf{Z}_q$. Assuming the intractability of the DDH-problem, ElGamal encryption is *indistinguishable*, and equivalently *semantically secure* [GM84]—it is secure against a passive, merely eavesdropping adversary. Formally, an attacker, given distinct messages m_0, m_1 and a corresponding target ciphertext cip_b for random $b \in \{0, 1\}$, cannot guess b better than with probability $\frac{1}{2}$. However, ElGamal encryption is completely insecure against various active attacks, where a decryption oracle can be used under appropriate conditions.

A powerful active attack is the CCA-attack of Rackoff and Simon [RS92]. CCA-security means indistinguishability against an adversary that can freely use a decryption oracle except for the target ciphertext. Dolev, Dwork and Naor [DDN91,00] propose another notion of security against active attacks, called *non-malleability*. Here the adversary—which is given a decryption oracle—tries to create another ciphertext that is related in an interesting way to the target ciphertext. Non-malleability and CCA-security have been shown to be equivalent [DDN00].

Security in the ROM+GM. The [TY98]-assumption that the signer of a Schnorr signature really "knows" the secret key can be proved in the ROM+GM. The core of the security proof in [SJ00] is a generic extractor that extracts from a generic CCA-attacker the plaintext and the secret one-time keys of a ciphertext constructed by the attacker, see Theorem 1. Theorem 2 shows security against a generic CCA-attacker performing some $t = o(\sqrt{q})$ interactions and generic group steps. A CCA-attacker can freely use a decryption oracle except for the target ciphertext. We show that a generic CCA-attacker using t generic steps, and given distinct messages m_0, m_1, a target ciphertext cip_b for random $b \in_R \{0, 1\}$, cannot predict b with probability better than $\frac{1}{2}+t^2/q$. This probability is over the random hash function H, the random public encryption key h, the coin tosses of the encipherer, and the random bit b. This bound is almost tight, as a generic attacker, given the public key h, can compute the secret decryption key with probability $\binom{t}{2}/q$ in t generic steps. This result improves the known security guarantees for signed ElGamal encryption. Moreover, our security proofs extend to a straightforward distributed threshold version of signed ElGamal encryption, see [SG98] for the threshold setting.

Furthermore, we introduce the *one-more decryption attack* and we show that signed ElGamal encryption is secure against this attack. In the one-more decryption attack the adversary attempts to partially decrypt $l+1$ ciphertexts by asking a decryption oracle some l times. For motivation of the one-more decryption attack, we propose a practical scheme for buying digital information anonymously. It is based on blind decryption and security against the *random one-more attack*, a weak version of the one-more decryption attack, but does not require CCA-security.

3.2. Definition of signed ElGamal encryption

We first define Schnorr signatures, based on an ideal hash function $H : G \times M \to \mathbf{Z}_q$, where M is the set of messages. Hereafter we define signed ElGamal encryption as well as the generalized concepts of the original and of signed ElGamal encryption.

Private/public key for signatures. The *private key* r is random in $\mathbf{Z}_q$. The corresponding *public key* $\bar{h} = g^r \in G$ is random in G, $r = \log_g h$.

A *Schnorr signature* on a message m is a triple $(m, c, z) \in M \times \mathbf{Z}_q^2$ such that $H(g^z \bar{h}^{-c}, m) = c$. In order to *sign* a message $m \in M$, pick a random $s \in_R \mathbf{Z}_q$, compute g^s, $c := H(g^s, m)$ and $z := s + cr$. Output the *signature* (m, c, z).

In order to *verify* a signature (m, c, z) check that $H(g^z \bar{h}^{-c}, m) = c$. The signature protocol produces a correct signature since $g^z \bar{h}^{-c} = g^{s+cr} \bar{h}^{-c} = g^s$ [8)].

The private/public key pair for encryption is x, $h = g^x$ where x is random in $\mathbf{Z}_q$. The basic encryption scheme is for messages in $M = G$, ElGamal ciphertexts are in $G \times M$, the tagged Schnorr signature signs pairs in $G \times M$ and uses a random hash function $H : G^2 \times M \to \mathbf{Z}_q$.

In order to *encipher* a message $m \in G$, pick random $r, s \in_R \mathbf{Z}_q$, compute g^r, $m\,h^r$, $c := H(g^s, g^r, mh^r)$ and $z := s+cr$ and output the *ciphertext* $(g^r, mh^r, c, z) \in G^2 \times \mathbf{Z}_q^2$.

A decryption oracle (*decryptor*) is a function that decrypts valid ciphertexts: The user sends a claimed ciphertext $(\bar{h}, \bar{f}, c, z)$ to the decryptor. The decryptor checks that $H(g^z \bar{h}^{-c}, \bar{h}, \bar{f}) = c$ and sends, if that test succeeds, $m := \bar{f}/\bar{h}^x$ to the

8) Signatures (c, z) with $c = 0$ have the undesirable property that they hold for all public keys h. It makes sense to exclude such signatures even though they appear only with negligible probability.

user. If the test fails the decryptor sends a random message in G. For simplicity, we disregard the impact of that random message to the probability.

The decryption is correct as $\bar{h} = g^r$, $\bar{f} = m\,h^r$ yields $\bar{f}/\bar{h}^x = m\,g^{rx}g^{-rx} = m$.

A signed ciphertext (g^r, mh^r, c, z) consists of an ElGamal ciphertext (g^r, mh^r) and a Schnorr signature (c, z) of the "message" (g^r, mh^r) for the public signature key g^r. The secret signature key r is the one-time key for encryption. In the ROM the signature (c, z) does not contain any information about m as (c, z) depends on m exclusively via some hash value that is statistically independent of m.

Threshold distributed version. The validity of the ciphertext $(\bar{h}, \bar{f}, c, z)$ is tested prior to and separate from decryption. Hence, the security properties of the scheme are preserved in the more general setting of threshold cryptography, see [SG98]. It is possible for a distributed entity to perform the decryption in a controlled manner after each server first having verified that indeed the decryption is allowed i.e., that the signature in the ciphertext is valid. If this were not locally verifiable, it would make a threshold decryption severely more complex.

Comparison with other secure DL-cryptosystems. We count the number of exponentiations[9)] per encryption/decryption and the number of on-line exponentiations per encryption (exponentiations not depending on the message).

	exp./enc.	on-line/enc.	exp./dec.
Signed ElGamal enc.	3	0	2
[FO99]	2	2	2
[ABR98]	2	0	1
[CS98], [Sh00]	4	1	2
[SG98], TDH1, TDH2	5	2	5

The relative efficiency of [FO99], [ABR98] is due to the usage of further cryptographic primitives. [FO99] uses private encryption, [ABR98] uses private encryption and message authentication code. Signed ElGamal encryption and TDH1, TDH2 of [SG98] are amenable to a secure distributed threshold decryption. Signed EG-encryption is plaintext aware in a strong sense. Signed ElGamal encryption virtually combines all the good properties.

Generalized (signed) ElGamal encryption. Our security results extend to the variant of ElGamal encryption proposed by Naor, Reingold [NR97]. This variant has two major advantages. *Firstly*, for long messages our generalized encryption is very fast and its data expansion rate approaches 1. *Secondly*, the generalized encryption does not require messages to be encoded into the group generated by

9) We count an expression $g^z\bar{h}^{-c}$ as 1 exponentiation even though it is slightly more expensive than a full exponentiation.

the public key h [10]. The generalized encryption solves the security flaw in textbook ElGamal encryption pointed out in [BJN00].

Let the message space M be an arbitrary additive group, e.g., $M = \mathbf{Z}_q^n$. Let $H : G^2 \times M \to \mathbf{Z}_q$ be a random hash function and let $H_M : G \to M$ be a second random hash function that is statistically independent of H. Then replace in the basic encryption scheme $mh^r \in G$ by $m + H_M(h^r) \in M$.

The generalized ElGamal ciphertext is $(g^r, \bar{f})$, where $\bar{f} = m + H_M(h^r)$, the generalized signed ElGamal ciphertext is $(g^r, \bar{f}, c, z)$, and $c = H(g^s, g^r, \bar{f})$, $z = s + cr$. Decrypt a signed ciphertext $(\bar{h}, \bar{f}, c, z)$ into $\bar{f} - H_M(\bar{h}^x)$ provided that the signature (c, z) of $(\bar{h}, \bar{f})$ is correct, i.e., $H(g^z \bar{h}^{-c}, \bar{h}, \bar{f}) = c$.

For $M = \mathbf{Z}_q^n$ the bit length of the ciphertext is $\log_2 \|G\| + (n+2)\log_2 q$, the message is $n \log_2 q$ bits long and $\|G\|$ is the bit length of the group elements. The data expansion rate is $1 + \frac{2}{n} + \frac{\log_2 \|G\|}{n \log_2 q}$ which is near to 1 for large n.

The short generalized ciphertexts are as secure as the original ones. Encryption requires only a long [11] and a short hash as well as a long and a short addition. The three exponentiations g^r, h^r, g^s can be done beforehand.

First extension of Proposition 2. Obviously Proposition 2 extends to generalized ElGamal ciphertexts $(g^r, m + H_M(h^r))$ provided that $H_M : G \to M$ is a random function. Whereas $\mathcal{A}$ can freely use the hash values $H_M(f_1), \ldots, H_M(f_t)$ of the computed group elements these hash values are statistically independent random numbers except for collisions $f_i = f_j$.

Second extension of Proposition 2. Assuming the ROM, Proposition 2 extends to signed ElGamal encryption and to generalized signed ElGamal encryption. This is because the added Schnorr signature does not contain any information about the plaintext.

3.3. Security against interactive attacks

Consider the security of signed ElGamal encryption in ROM+GM. Theorem 1 shows that signed ElGamal encryption is plaintext aware in a strong sense. The proof of Theorem 1 is implicit in [SJ00], it is the core for proving Theorems 2 and 3.

Theorem 1 (Plaintext awareness [SJ00]). *Let a generic adversary $\mathcal{A}$ be given for input g, h and signed ElGamal ciphertexts* $\mathrm{cip}_1, \ldots, \mathrm{cip}_d$. *Let $\mathcal{A}$ produce within t*

---

10) Encoding of arbitrary bit sequences into sequences of group elements is easy for particular groups such as $\mathbf{Z}_q^*$ that correspond to an interval of integers. For general groups, even for subgroups of $\mathbf{Z}_N^*$ or subgroups of elliptic curves, an encoding into group elements is impractical. Known extensions of ElGamal encryption—see e.g., [MOV97] section 8.26—do not solve this encoding problem.

11) Long hash values in $\mathbf{Z}_q^n$ can be generated using a random hash function $H_M : G \to \mathbf{Z}_q^n$ according to the following, or some related, approach: $(H_M(f, 1), \ldots, H_M(f, n))$.

generic steps a signed ElGamal ciphertext cip_{d+1}. *Then there is a generic algorithm* $\mathcal{A}'$ *which given the same input produces within t generic steps the plaintext and the one-time keys corresponding to* cip_{d+1}. $\mathcal{A}'$ *succeeds except for an event of probability* $3/q$. *The probability refers to the random* h, H *and the random one-time keys of the encipherer.*

$\mathcal{A}'$ succeeds in extracting the plaintext corresponding to cip_{d+1} whenever $\mathcal{A}$ has produced a correct ciphertext cip_{d+1}—except for an event of probability $\frac{3}{q}$. Moreover, $\mathcal{A}'$ extract the one-time keys $r, s \in \mathbf{Z}_q$ from the construction of $\mathrm{cip}_{d+1} = (g^r, mh^r, c, z), g^s = g^z h^{-c}$.

The generic extractor is more efficient than known black-box extractors for DL-signatures and DL-identification. The black-box extractor of [FFS88] has a constant delay factor compared to the attacker. The generic extractor does not induce any delay, the number of generic steps does not increase compared to the attacker. Only the attacker's probability of success decreases by at most $\frac{3}{q}$. The power of the generic extractor relies on the fact that it can be iterated some T number of times decreasing the attacker's probability of success by at most $3T/q$. Theorem 1 can be extended from given ciphertexts $\mathrm{cip}_1, \ldots, \mathrm{cip}_d$ to ciphertexts that are produced in an adaptive way by querying an enciphering oracle about ciphertexts of chosen messages.

Theorem 2 shows indistinguishability against a CCA-adversary $\mathcal{A}$. This is equivalent to non-malleability against CCA [DDN00]. We refer to non-malleability as defined in [DDN00] and to the strong chosen ciphertext attack proposed by Rackoff and Simon [RS92]. The adversary has access to a decryption oracle which can be used arbitrarily except for the target ciphertext. The adversary is given a target ciphertext cip_b and a decryption oracle for the decryption of arbitrary ciphertexts except for cip_b. The attack is called adaptive because the queries to the decryption oracle may depend on the challenges and their corresponding answers. We let the generic adversary $\mathcal{A}$ perform some t generic steps: group operations, inputs in G, queries to the oracle H, and queries to the decryption oracle not including the target ciphertext.

Theorem 2 ([SJ00]). *Let the attacker* $\mathcal{A}$ *be given* g, h, *distinct messages* m_0, m_1, *a target ciphertext* cip_b *corresponding to* m_b *for a random bit* $b \in_R \{0, 1\}$, *and oracles for H and for decryption. Then a generic* $\mathcal{A}$ *using t generic steps cannot predict b with a better probability than* $\frac{1}{2} + t^2/q$. *The probability space consists of the random* x, H, b *and the one-time keys of the encipherer.*

Theorem 3 shows that signed ElGamal encryption is secure against the one more decryption attack. An adversary can—after some l interactions with the decryption oracle—not decrypt more than l ciphertexts. More precisely, he gets non-negligible information about at most l encrypted plaintexts.

Theorem 3 ([SJ00]). *Let the attacker* $\mathcal{A}$ *be given* g, h, *ciphertexts* $\mathrm{cip}_1, \ldots, \mathrm{cip}_d$, *the corresponding messages* $m_1, \ldots, m_d$ *in random order and oracles for H and for decryption. Let the generic* $\mathcal{A}$ *perform t generic steps including some* $l < d$ *arbitrary queries to the decryption oracle. Then* $\mathcal{A}$ *cannot produce* $l + 1$ *message-ciphertext pairs with a probability better than* $\frac{1}{d-l} + t^2/q$. *The probability space* *consists of the random x, H, the one-time keys of the encipherer and the random ordering of the messages.*

The notions of CCA-security and of security against the one-more decryption attack are of independent interest. Consider the case of blind decryption where the decryption oracle is queried about a blinded ciphertext that is statistically independent of the ciphertext that is actually decrypted. In this case—which is important for acquiring digital information anonymously—CCA-security is meaningless as there is no well defined target ciphertext.

CCA-security implies a weak form of security against the one-more decryption attack. A one-more decryption attack can be transformed into a CCA-attack by guessing the ciphertext for which the attacker forms a correct message-ciphertext pair without querying the decryption oracle. However, this transform induces a big delay factor as large as the number of given ciphertexts in the data bank.

Trading encrypted information. Suppose a user wants to buy sensitive digital information, e.g., digital music, videos, pictures, stock market analysis, etc. Let the digital information be freely accessible in encrypted form via a public data bank. For simplicity, let each encrypted package cost \$1. Let the users have access to a public decryption oracle that charges \$1 per decryption. For the security of such trade of encrypted information the encryption scheme must be secure against the one-more decryption attack.

This type of service does not require CCA-security. However, it would be nice to have an encryption that allows for blind decryption so that no information is revealed in a decryptor interaction. Blind decryption guarantees anonymity of the buyer of digital information. It is well known that the original ElGamal ciphertexts allow for blind decryption[12)]. Even though, ElGamal encryption is insecure against the one-more decryption attack we show below that it is secure against the weaker *random one-more attack*, where the enciphered plaintexts are statistically independent messages—e.g. secret keys that are unknown to the attacker.

Efficient scheme for anonymously buying digital information (ABDI). Let the information packages m_i of the public data bank be each encrypted under a private key k_i of a secure symmetric encryption scheme. Let m_i contain a content description descr_i of m_i and a signed ElGamal ciphertext $\mathrm{cip}(k_i) = (g^r, k_i h^r, c, z)$ of the key $k_i \in G$. Let (c, z) be a signature of $(g^r, k_i h^r, \mathrm{descr}_i)$ with public key g^r. Suppose a user wants to anonymously buy l packages m_i of his choice. He checks the Schnorr signature (c, z) of $\mathrm{cip}(k_i) = (g^r, k_i h^r, c, z)$ in package m_i and stops if the signature is invalid. Otherwise, he blinds the ElGamal ciphertext $(g^r, k_i h^r)$ into $(g^{r+s}, u k_i h^{r+s})$ for random $s \in \mathbf{Z}_q$, $u \in G$, and asks the decryption oracle to decrypt $(g^{r+s}, u k_i h^{r+s})$. As the blinded ciphertext is statistically independent of $(g^r, k_i h^r)$ no information is revealed about which k_i he gets. As the user pays for l decryptions it is important that he cannot get $l + 1$ keys k_i.

12) Blind decryption of the ElGamal ciphertext (g^r, mh^r): The user picks random $u \in G$ and $s \in \mathbf{Z}_q$ and asks for decryption of (g^{r+s}, umh^{r+s}). He gets m from the plaintext um transmitted by the decryptor by multiplication with u^{-1}.

Security against the random one-more attack. Consider the above ABDI for random $k_i \in G$. Clearly, $l+1$ keys $k_i \in_R G$ have Shannon *entropy* $(l+1)\log_2 q$. But each decryption reveals no more than $\log_2 q$ bits of a plaintext in G, $|G| = q$. Thus, l decryptions cannot reveal $l+1$ statistically independent keys k_i.

Of course, the private key encryption scheme must be secure. Specifically, ciphertexts are supposed not to reveal any information about the secrete encryption keys. The security argument does not require CCA-security of the public-key encryption scheme, but merely semantical security. Therefore, ElGamal encryption can be replaced by other self-reducible encryption schemes [AFK89]. Our proposal can be considered as an implementation of a $\binom{T}{l}$-oblivious transfer, where T is the number of encrypted packages in the data bank and l is the number of decryption requests.

Another application would be an electronic service for delivering sensitive, possibly unpleasant messages like court orders, summons, admonitions and so on. Such messages can be sent in encrypted form, given access to a decryption oracle that combines the decryption with an acknowledgement of the receipt of the decrypted message. This makes sure that a recipient can only read the message by acknowledging receipt. For such a service it would be important that the encryption is CCA-secure, so that the receipt correctly specifies the revealed message. However, we also need security against the one-more decryption attack as users may want to decrypt several ciphertexts. Signed ElGamal encryption can be used for such a service.

Security with short hash values. Let the hash values of H be random in an interval $[0, 2^k[\subset [0, q[\cong \mathbf{Z}_q$. In the situation of Theorem 2 a CCA-attacker does not succeed better than with probability $\frac{1}{2}+t^2/q+l(2^{-k}-\frac{1}{q})$, where l is the number of decryptor interactions. This shows that random hash values can securely range over a set of $\sqrt{q}$ values.

4. Security of blind Schnorr signatures

4.1. Introduction and survey

We are interested in blind signatures as required for anonymous digital cash. Blind signatures are generated by an interaction with the signer who controls the private signature key. We study security against the *one-more signature forgery* introduced in [PS96b], where security means that an attacker cannot obtain $l+1$ valid signatures from l interactions with the signer. The most general case are parallel attacks, where the parallel interactions are non-synchronous and arbitrarily interleaved. Security against this attack is important in e-commerce, where it translates into that an adversary cannot "create additional money". We first present the generic parallel attack which escaped in all previous works.

The generic parallel attack uses a solution of the ROS-problem: find a random, overdetermined, solvable system of $l+1$ linear equations in l unknowns modulo q

with random inhomogeneities (right sides). This may well be an intractible problem. Given a solution of the ROS-problem, the attacker easily succeeds in a parallel one-more signature forgery. The generic, parallel attack works in the same way for Schnorr and for Okamoto–Schnorr signatures, it does not even use the public key. Therefore, the attacker cannot help to solve a DL-, DH- or a DDH-problem. Theorem 4 evaluates the average number of solutions of the ROS-problem. A solution exists and the parallel attack is possible with 4 signer interactions and 2^{50} hash queries for an elliptic curve of order $q \approx 2^{200}$.

Assuming the intractability of the ROS-problem, Theorem 5 gives a practical security guarantee against the one-more signature forgery in the ROM+GM. A generic adversary using t generic steps cannot succeed better than with probability $\binom{t}{2}/q$.

4.2. Interactive generation of blind Schnorr signatures

We describe signer interactions, an interactive protocol that enables a user to generate Schnorr signatures of messages of its choice. We first describe the setting and the structure of the signatures, after which we review the protocol for generation of signatures. We later show how this can be used to generate blind signatures of the same type. Signatures will be based on an ideal hash function $H : G \times M \to \mathbf{Z}_q$, where M is the set of messages.

Private/public key pairs. The *private key* x of the signer is random in $\mathbf{Z}_q$. The corresponding *public key* is $h = g^x \in G$, a random group element. We have $x = \log_g h$.

Recall the definition of Schnorr signatures from Subsection 3.2. A triple (m, c, z) is a Schnorr signature for the public key $h = g^x$ if $H(g^z h^{-c}, m) = c$.

A signer interaction is an interactive protocol between the signer and a user consisting of three rounds. The signer picks a random $r \in_R \mathbf{Z}_q$ computes g^r and sends the commitment g^r. The user selects $c \in \mathbf{Z}_q$ and sends the challenge c. The signer responses by sending $z := r + cx$.

The user can generate from this protocol the standard signature (m, c, z) by selecting $c := H(g^r, m)$, but he has more options than that. We will study all possibilities to produce signatures by a sequence of arbitrary interactions with the signer. We let $(r, c, z) \in \mathbf{Z}_q^3$ denote the signer interaction consisting of the signer's random coin r, the user's *challenge* c and the signer's *response* z.

Non-interactive proof of knowledge. Schnorr signatures provide in the ROM a non-interactive proof of knowledge of the secret key $x = \log_g h$ [PS96a].

Blind signature protocol. A signer interaction (r, c, z) can be used to generate the *standard signature* (m, c, z) or a transformation (m, c', z') of this signature. We call the signature protocol *blind* if it generates a signature (m, c', z') that is statistically independent of the interaction corresponding to the triple (r, c, z) providing the view of the signer. The user can generate such an independent signature (m, c', z') from random numbers $\alpha, \beta \in_R \mathbf{Z}_q$.

signer		*user*
picks $r \in_R \mathbf{Z}_q$		picks $\alpha, \beta \in_R \mathbf{Z}_q$
computes g^r	$\xrightarrow{g^r}$	
	$\xleftarrow{c}$	$c := H(g^{r+\alpha}h^{\beta}, m) + \beta$
$z := r + cx$	$\xrightarrow{z}$	$z' := z + \alpha,\ c' := c - \beta$

Validity. For the output of the interaction $(m, c', z') = (m, c - \beta, z + \alpha)$ we have $g^{z'} h^{-c'} = g^{r+cx+\alpha} h^{-c+\beta} = g^{r+\alpha} h^{\beta}$. Hence $H(g^{z'} h^{-c'}, m) = c - \beta = c'$, and thus (m, c', z') is a valid signature.

Blindness property. The generated signature $(m, c-\beta, z+\alpha)$ is—for a constant interaction (r, c, z)—uniformly distributed over all signatures on message m due to the random $\alpha, \beta \in_R \mathbf{Z}_q$. Each signature (m, c', z') is produced for a unique pair α, β, namely $\alpha = z' - z$, $\beta = c - c'$.

4.3. The generic parallel attack

We present a variant of the attack that does not use the generator g and the public key h. We first present the attack for Schnorr signatures. Thereafter, we extend it to Okamoto–Schnorr signatures. This shows that Okamoto–Schnorr signatures do not protect better against the generic parallel than plain Schnorr signatures.

The signer sends commitments $g_1 = g^{r_1}, \dots, g_l = g^{r_l}$. The attacker $\mathcal{A}$ computes $f_k = g_1^{a_{k,1}} \cdot \dots \cdot g_l^{a_{k,l}}$ and $H(f_k, m_k)$ for $k = 1, \dots, t''$. Then $\mathcal{A}$ solves $l+1$ of the t'' equations (1) in the unknowns $c_1, \dots, c_l$ over $\mathbf{Z}_q$.

$$H(f_k, m_k) = \sum_{\ell=1}^{l} a_{k,\ell}\, c_\ell \text{ for } k = 1, \dots, t''. \tag{1}$$

$\mathcal{A}$ sends the solutions $c_1, \dots, c_l$ as challenges to the signer. The signer sends back $z_\ell := r_\ell + c_\ell x \in \mathbf{Z}_q$ for $\ell = 1, \dots, l$. For each solved equation (1), the attacker gets a valid signature (m_k, c'_k, z'_k) by setting

$$c'_k := \sum_{\ell=1}^{l} a_{k,\ell}\, c_\ell = H(f_k, m_k) \text{ and } z'_k := \sum_{\ell=1}^{l} a_{k,\ell} z_\ell .$$

In the ROM the values $H(f_k, m_k)$ are random. The coefficients $a_{k,\ell}$ selected by the attacker are arbitrary values. The solution $(c_1, \dots, c_l)$ of $l+1$ of the t equations (1) does not depend on g, h. As $\mathcal{A}$ does not use g, h it is impossible that $\mathcal{A}$ helps in black-box mode to solve a DL-, DH- or DDH-problem for G.

The generic parallel attack is intrinsic parallel. By Theorem 4 the number l of parallel interactions with the signer must be at least logarithmic in q. Otherwise, the probability $\binom{t''}{l+1}/q$ for the existence of an ROS-solution is negligible.

The generic parallel attack against Okamoto–Schnorr signatures. We follow the notation of [PS00]. There are two public keys h and $y = g^{-r}h^{-s}$ for random secret keys $r, s \in_R \mathbf{Z}_q$ while $\log_g h$ is unknown. A signature of message m is a tuple $(m, \varepsilon, \sigma, \rho) \in M \times \mathbf{Z}_q^3$ satisfying $H(g^\rho h^\sigma y^\varepsilon, m) = \varepsilon$.

The signer picks random $t_\ell, u_\ell \in_R \mathbf{Z}_q$ and sends commitments $g_\ell = g^{t_\ell} h^{u_\ell}$ for $\ell = 1, \ldots, l$ to the attacker $\mathcal{A}$. $\mathcal{A}$ selects coefficients $a_{k,\ell} \in \mathbf{Z}_q$, computes $f_k = g_1^{a_{k,1}} \cdots g_l^{a_{k,l}}$ and $H(f_k, m_k)$ for $k = 1, \ldots, t''$. $\mathcal{A}$ solves $l+1$ of the t'' linear equations (1) modulo q in the unknowns $c_1, \ldots, c_l$. $\mathcal{A}$ sends the solutions $c_1, \ldots, c_l$ as challenges to the signer. The signer sends back $R_\ell := t_\ell + c_\ell r$, $S_\ell := u_\ell + c_\ell s \in \mathbf{Z}_q$ for $\ell = 1, \ldots, l$. For each solved equation (1) $\mathcal{A}$ gets a valid signature $(m_k, \varepsilon, \rho_k, \sigma_k)$ by setting

$$\varepsilon_k = H(f_k, m_k) = \sum_{\ell=1}^{l} a_{k,\ell}\, c_\ell, \;\; \rho_k = \sum_{\ell=1}^{l} a_{k,\ell}\, R_\ell, \;\; \sigma_k = \sum_{\ell=1}^{l} a_{k,\ell}\, S_\ell.$$

Correctness. From the equations (1) we get that

$$g^{\rho_k} h^{\sigma_k} y^{\varepsilon_k} = \prod_{\ell=1}^{l} g_\ell^{a_{k,\ell}} = f_k \text{ and } H(g^{\rho_k} h^{\sigma_k} y^{\varepsilon_k}, m_k) = \varepsilon_k.$$

Conclusion. The attacker $\mathcal{A}$ does not use the public g, h, y. Thus, it is impossible to use a successful $\mathcal{A}$ to solve a DL- DH- or DDH-problem. The generic parallel attack has been excluded in Theorem 26 [PS00] by assuming that the attacker has a probability of success $4t''^{(l+1)}/q$ which is greater than the probability $\binom{t''}{l+1}/q$ for the existence of a solution of the ROS-problem. By Theorem 4 a solution of a subsystem of $l+1$ equations (1) exists and the generic parallel attack is possible with $l = 4$ parallel interactions and 2^{50} hash queries for a 200 bit order q of an elliptic curve. Thus it is impossible to get a practical security guarantee for elliptic curves of order $\leq 2^{200}$ by the proof methods of [PS00].

The average number of solutions for the ROS-problem. Consider distinct equations

$$a_{k,1}\, c_1 + \cdots + a_{k,l}\, c_l = F(a_{k,1}, \ldots, a_{k,l}) \text{ for } k = 1, \ldots, t'' \tag{2}$$

in the unknowns $c_1, \ldots, c_l$ over $\mathbf{Z}_q$ for a random function $F : \mathbf{Z}_q^l \mapsto \mathbf{Z}_q$. We evaluate the expected number of solvable subsystems consisting of $l+1$ equations.

Theorem 4 ([SJ99]). *For arbitrary coefficients $a_{k,\ell}$ the average number of solvable subsystems of $l+1$ out of the t'' equations (2) is at most $\binom{t''}{l+1}/q$. For statistically independent coefficients $a_{k,\ell} \in_R \mathbf{Z}_q$ the average number of solvable subsystems with $l+1$ equations (2) is $\binom{t''}{l+1} q^{-1}(1 - q^{-1} + O(q^{-2}))$.*

How to solve the ROS*-problem in the Turing machine model.* A simple method to find all solvable subsystems of $l+1$ out of t'' equations (2) tries all $\binom{t''}{l}$ choices for $1 \leq \sigma_1 < \cdots < \sigma_l < t''$—with overwhelming probability each subsystem has at most one solution—sorts these solutions and checks for a collision. A collision yields a solution of $l+1$ equations. Conversely, a solution of $l+1$ equations coincides with a collision. This method requires at least $\binom{t''}{l}$ arithmetic steps over $\mathbf{Z}_q$ and succeeds with probability $\leq \binom{t''}{l+1}/q$. That method is *optimal* for $l = 1$ [13)].

In a variant of the above method one forms for a subfamily of some T—from the $\binom{t''}{l}$ choices of l out of t'' equations—sorts the solutions of the corresponding l-tuples of equations, and searches for two colliding solutions. Two solutions collide with probability at most $\frac{1}{q}$. The restricted search does not succeed with a better probability than $\binom{T}{2}/q$ using T arithmetic steps over $\mathbf{Z}_q$—for an arbitrary selection of the coefficients $a_{k,\ell}$ and the T subsystems. We pose it as an open problem to beat that $\binom{T}{2}/q$ bound.

From the work of Håstad [H97] we know that beating the naive method—in maximizing a subset of solvable linear equations of an overdetermined system—is NP-hard. This associated NP-hard problem seems to indicate that the ROS-problem is not trivial. It seems to be irrelevant that the adversary can choose himself the coefficients $a_{k,\ell}$ of the equations (2) as the inhomogeneities $F(a_{k,0}, \ldots, a_{k,\ell})$ are independent random numbers. Moreover, the complexity of the ROS-problem hardly depends on the selected coefficients $a_{k,\ell}$. The ROS-problem for particular coefficients seems to be as hard as in worst case.

4.4. Security of blind signatures against interactive attacks

This subsection refers to a generic adversary $\mathcal{A}$ performing some t generic steps—including some l interactions $(r_1, c_1, z_1), \ldots, (r_l, c_l, z_l)$ with the signer—some t' group elements and some t'' queries to the hash oracle. We let $\mathbf{r} = (r_1, \ldots, r_l)$ denote the signers random coins. Let $f_1 = g$, $f_2 = h = g^x$, $f_3, \ldots f_{t'} \in G$ denote the group elements of $\mathcal{A}$'s computation. The generic $\mathcal{A}$ computes $f_j = g^{a_{j,-1}} h^{a_{j,0}} g_1^{a_{j,1}} \cdots g_l^{a_{j,l}}$, where $g_1 = g^{r_1}, \ldots, g_l = g^{r_l}$ are the signer's commitments and the exponents $a_{j,l} \in \mathbf{Z}_q$ depend arbitrarily on the previously computed non-group data. As each signer interaction yields one group element g^{r_ℓ} we have that $t'' = t - t' \geq 0$ is the number of interactions with the hash oracle. We first present the basic Lemma $1'$ and $2'$ that extend Lemma 1 and 2 from a non-interactive attacker to an adversary using an hash oracle and a signature oracle.

Interactions with the signature oracle, (signer *for short*). The generic algorithm sends to the signer a message $m \in M$ and a public key $h \in G$, and receives from the signer a random signature $(m, e, y) \in M \times \mathbf{Z}_q^2$ with public key h. The signer picks a random $r \in_R \mathbf{Z}_q$ forms $e := H(g^r, m)$ by asking the hash oracle, and sets $y := r + ex$, where $x = \log_g h$ is the secret key.

---

13) For $l = 1$ there is a solution of two of the t'' equations (2) if and only if there is a collision among the $\mathbf{Z}_q$-numbers $F(a_{k,1}, \ldots, a_{k,l})/a_{k,1}$ for $k = 1, \ldots, t''$. These $\mathbf{Z}_q$-numbers are pairwise statistically independent due to the random F. A collision occurs with probability $\binom{t''}{2}/q$, it can be found in $O(t'' \log t'')$ arithmetic steps by sorting.

Lemma 1′ ([SJ99]). Collisions among $f_1, \ldots, f_{t'}$ occur at most with probability $\binom{t'}{2}/q$. The probability refers to the random h, H and the random coins $\mathbf{r}$ of the signer.

Lemma 2′ ([SJ99]). If there are no collisions among $f_1, \ldots, f_{t'}$ the random x is statistically independent of the computed non-group data except that the random coins $(\mathbf{r}, x)$ leading to collisions are excluded.

Theorems 4 and 5 show that Schnorr signatures are secure against the one-more signature forgery in the ROM+GM. These theorems cover blind signatures as required for anonymous electronic cash. This is the first sharp security result for simple DL-signatures in the interactive setting. We characterize the different power of sequential and of parallel attacks. Parallel attacks that beat the success rate $\binom{t}{2}/q$ of sequential attacks must solve the ROS-problem: find an overdetermined solvable system of linear equations modulo q with random inhomogeneities.

Theorem 5 ([SJ99]). *Let a generic adversary $\mathcal{A}$ be given the generator g, the public key h, an oracle for H and let $\mathcal{A}$ interact with the signer some l times. Let $\mathcal{A}$ perform t generic steps including l parallel signer interactions. If a generic adversary $\mathcal{A}$ succeeds in a one-more signature forgery with a better probability of success than $\binom{t}{2}/q$ then $\mathcal{A}$ must solve the* ROS*-problem: given an oracle for a random function F find $l+1$ distinct vectors $(a_{k,1}, \ldots, a_{k,l}) \in \mathbf{Z}_q^l$ and a solution $c_1, \ldots, c_l \in \mathbf{Z}_q$ of the equations $\sum_{l=1}^{l} a_{k,l}\, c_l = F(a_{k,1}, \ldots, a_{k,l})$ for $k = 1, \ldots, l+1$.*

Security of Schnorr signatures against the chosen message attack (CMA). Theorem 5 implies for $l = 0$ that Schnorr signatures are secure in ROM+GM against existential forgery, where an attacker is given the public signature key and tries to produce some correct signature (m, c, z). This security result extends to the case that the adversary $\mathcal{A}$ has a signature oracle (signer) and can ask the oracle for signatures of messages of its choice. An interaction with the signer is counted as generic step. The goal of the attack is to generate a new signature which is not produced by the signer.

Theorem 6 (Security against CMA). *Let $\mathcal{A}$ be a generic algorithm that is given g, the public signature key $h \in_R G$ and oracles for H and signatures. Using t generic steps—group operations and hash queries and queries to the signer—$\mathcal{A}$ cannot produce a new Schnorr signature with a probability better than $\frac{3}{q} + \binom{t}{2}/q$. The probability space consists of the random h, H.*

Theorem 6 is closely related to the security in GM of Schnorr identification against active attacks proved by Shoup [Sh97]. From that proof the CMA-security of Schnorr signatures follows by a very close argument replacing the random queries of the verifier in the Schnorr identification by random hash values and using the oracle replay mode proposed in [PS96a].

However, the security of Schnorr identification in GM does not bear any security against the one-more signature forgery. The latter security means that an active attacker against the Schnorr identification scheme cannot transform l parallel interactions with an identification oracle into $l+1$ identifications. This would be a form of the man-in-the-middle-attack which has been explicitly excluded in [Sh97]. Moreover, such a security result requires the intractability of the ROS-problem.

Acknowledgement. I thank Marc Fischlin for useful comments while reading the manuscript.

References

[AFK89] Abadi, M., Feigenbaum, J., Kilian, J., On hiding information from an oracle. J. Computer System Sci. 39 (1989), 21–50.

[ABR98] Abdalla, M., Bellare, M., Rogaway, P., DHES: An encryption scheme based on the Diffie-Hellman problem. Contributions to P1363, ftp://stdgbbs.ieee.org/pub/p1363/contributions/aes-uhf.ps

[BDPR98] Bellare, M., Desai, A., Pointcheval, D., Rogaway, P., Relations among notions of security for public-key encryption schemes. In: Advances in Cryptology – CRYPTO '98 (ed. by H. Krawczyk; Lecture Notes in Comput. Sci. 1462), 26–45. Springer, Berlin 1998.

[BR93] Bellare, M., Rogaway, P., Random oracles are practical: a paradigm for designing efficient protocols. In: Proceedings of the 1st ACM Conference on Computer and Communication Security (CCS '93), 62–73. ACM Press, New York 1993.

[BR94] —, —, Optimal asymmetric encryption – how to encrypt with RSA. In: Advances in Cryptology – EUROCRYPT '94 (ed. by A. De Santis; Lecture Notes in Comput. Sci. 950), 92–111. Springer, Berlin 1995.

[BJN00] Boneh, D., Joux, A., Nguyen, P.Q., Why textbook ElGamal and RSA encryption are insecure. In: Advances in Cryptology – ASIACRYPT 2000 (ed. by T. Okamoto; Lecture Notes in Comput. Sci. 1976), 30–43. Springer, Berlin 2000.

[BL96] Boneh, D., Lipton, R.J., Algorithms for black-box fields and their application in cryptography. In: Advances in Cryptology – CRYPTO '96 (ed. by N. Koblitz; Lecture Notes in Comput. Sci. 1109), 283–297. Springer, Berlin 1996.

[C97] Canetti, R., Towards realizing random oracles: hash functions that hide all partial information. In: Advances in Cryptology – CRYPTO '97 (ed. by B.S. Kaliski; Lecture Notes in Comput. Sci. 1294), 455–469. Springer, Berlin 1997.

[CGH98] Canetti, R., Goldreich, O., Halevi, S., The random oracle methodology, revisited. In: STOC '98, 209–218. ACM Press, New York 1999. http://www.acm.org/pubs/contents/proceedings/stoc/276698/

[CS98] Cramer, R., Shoup, V., A practical public key cryptosystem provably secure against adaptive chosen ciphertext attack. In: Advances in Cryptology – CRYPTO '98 (ed. by H. Krawczyk; Lecture Notes in Comput. Sci. 1462), 13–25. Springer, Berlin 1998.

[DH76] Diffie, W., Hellman, M.E., New directions in cryptography. IEEE Trans. Inform. Theory 22 (1976), 644–654.

[DDN91] Dolev, D., Dwork, C., Naor, M., Non-malleable cryptography. In: STOC '91, 542–552. ACM Press, New York 1991.

[DDN00] —, —, —, Non-malleable cryptography. SIAM J. Comput. 30 (2000), 391–437.

[E85] ElGamal, T., A public key cryptosystem and a signature scheme based on discrete logarithms. IEEE Trans. Inform. Theory 31 (1985), 469–472.

[FFS88] Feige, U., Fiat, A., Shamir, A., Zero-knowledge proofs of identity. J. Cryptology 1 (1988), 77–94.

[FS87] Fiat A., Shamir, A., How to prove yourself: practical solutions to identification and signature problems. In: Advances in Cryptology – CRYPTO '86 (ed. by A.M. Odlyzko; Lecture Notes in Comput. Sci. 263), 186–194. Springer, Berlin 1987.

[F00] Fischlin, M., A note on security proofs in the generic model. In: Advances in Cryptology – ASIACRYPT 2000 (ed. by T. Okamoto; Lecture Notes in Comput. Sci. 1976), 458–469. Springer, Berlin 2000.

[FO99] Fujisaki, E., Okamoto, T., Secure integration of asymmetric and symmetric encryption schemes. In: Advances in Cryptology – CRYPTO '99 (ed. by M. Wiener; Lecture Notes in Comput. Sci. 1666), 537–554. Springer, Berlin 1999.

[FOPS00] Fujisaki, E., Okamoto, T., Pointcheval, D., Stern, J., RSA-OAEP is still alive!. Cryptology ePrint Archive 2000/061.

[GK96] Goldreich, O., Krawczyk, H., On the composition of zero-knowledge proof systems. SIAM J. Comput. 25 (1996), 169–192.

[GM84] Goldwasser, S., Micali, S., Probabilistic encryption. J. Comput. System Sci. 28 (1984), 270–299.

[H97] Håstad, J., Some optimal inapproximability results. STOC '97, 1–10. ACM Press, New York 1999.
`http://www.acm.org/pubs/contents/proceedings/stoc/258533/`

[J98] Jakobsson, M., A practical mix. In: Advances in Cryptology – EUROCRYPT '98 (ed. by K. Nyberg; Lecture Notes in Comput. Sci. 1403), 448–461. Springer, Berlin 1998.

[MOV97] Menezes, A., van Oorschot, P., Vanstone, S., Handbook of Applied Cryptography. CRC Press, Boca Raton 1997.

[NR97] Naor, M., Reingold, M., Number-theoretic constructions of pseudo-random functions. In: 38th Symposium on Foundations of Computer Science, 458–467. IEEE, Washington 1997.

[Ne94] Nechaev, V.I., On the complexity of a deterministic algorithm for a discrete logarithm (Russian). Mat. Zametki 55.2 (1994), 91–101. English transl.: Math. Notes 55 (1994), 165–172.

[O92] Okamoto, T., Provably secure identification schemes and corresponding signature schemes. In: Advances in Cryptology – CRYPTO '92 (ed. by E.F. Brickell; Lecture Notes in Comput. Sci. 740), 31–53. Springer, Berlin 1993.

[P98] Pointcheval, D., Strengthened security for blind signatures. In: Advances in Cryptology – EUROCRYPT '98 (ed. by K. Nyberg; Lecture Notes in Comput. Sci. 1403), 391–405. Springer, Berlin 1998.

[P00] —, The composite discrete logarithm and secure authentication. In: Public Key Cryptography (PKC 2000, Melbourne) (ed. by H. Imai, Y. Zhang; Lecture Notes in Comput. Sci. 1751), 113–128. Springer, Berlin 2000.

[PS96a] Pointcheval, D., Stern, J., Security proofs for signature schemes. In: Advances in Cryptology – EUROCRYPT '96 (ed. by U. Maurer; Lecture Notes in Comput. Sci. 1070), 387–398. Springer, Berlin 1996.

[PS96b] —, —, Provably secure blind signature schemes. In: Advances in Cryptology – ASIACRYPT '96 (ed. by K. Kim, T. Matsumoto; Lecture Notes in Comput. Sci. 1163), 387–393. Springer, Berlin 1996.

[PS00] —, —, Security arguments for digital signatures and blind signatures. J. Cryptology 13 (2000), 361–396.

[RS92] Rackoff, C., Simon, D.R., Non-interactive zero-knowledge proof of knowledge and chosen ciphertext attack. In: Advances in Cryptology – CRYPTO '91 (ed. by J. Feigenbaum; Lecture Notes in Comput. Sci. 576), 433–444. Springer, Berlin 1992.

[Sc91] Schnorr, C.P., Efficient signature generation for smart cards. J. Cryptology 4 (1991), 161–174.

[Sc01] —, Small generic hardcore subsets for the discrete logarithm: short secret DL-keys. Inform. Process. Lett. 79 (2001), 93–98.

[SJ99] Schnorr, C.P., Jakobsson, M., Security of discrete log cryptosystems in the random oracle + generic model. TR report University Frankfurt and Bell Laboratories 1999. `http:www.mi.informatik.uni-frankfurt.de`.

[SJ00] —, —, Security of signed ElGamal encryption. In: Advances in Cryptology – ASIACRYPT 2000 (ed. by T. Okamoto; Lecture Notes in Comput. Sci. 1976), 73–89. Springer, Berlin 2000.

[Sch80] Schwartz, J., Fast probabilistic algorithms for verification of polynomial identities. J. Assoc. Comput. Mach. 27 (1980), 701–717.

[Sh97] Shoup, V., Lower bounds for discrete logarithms and related problems. In: Advances in Cryptology – EUROCRYPT '97 (ed. by W. Fumy; Lecture Notes in Comput. Sci. 1233), 256–266. Springer, Berlin 1997.

[Sh00] —, Using hash functions as a hedge against chosen ciphertext attack. In: Advances in Cryptology – EUROCRYPT 2000 (ed. by B. Preneel; Lecture Notes in Comput. Sci. 1807), 275–288. Springer, Berlin 2000.

[Sh00b] —, OAEP reconsidered. Cryptology ePrint Archive 2000/060.

[SG98] Shoup, V., Gennaro, R., Securing threshold cryptosystems against chosen ciphertext attacks. In: Advances in Cryptology – EUROCRYPT '98 (ed. by K. Nyberg; Lecture Notes in Comput. Sci. 1403), 1–16. Springer, Berlin 1998.

[TY98] Tsiounis, Y., Yung, M., On the security of ElGamal based encryption. In: Public Key Cryptography (PKC '98) (ed. by H. Imai, Y. Zhang; Lecture Notes in Comput. Sci. 1431), 117–134. Springer, Berlin 1998.

[ZS92] Zheng, Y., Seberry, J., Practical approaches to attaining security against adaptively chosen ciphertext attacks. In: Advances in Cryptology – CRYPTO '92 (ed. by E.F. Brickell; Lecture Notes in Comput. Sci. 740), 292–304. Springer, Berlin 1993.

Square-root algorithms for the discrete logarithm problem (a survey)

Edlyn Teske

Abstract. The best algorithms to compute discrete logarithms in arbitrary groups (of prime order) are the baby-step giant-step method, the rho method and the kangaroo method. The first two have (expected) running time $O(\sqrt{n})$ group operations (n denoting the group order), thereby matching Shoup's lower bounds. While the baby-step giant-step method is deterministic but with large memory requirements, the rho and the kangaroo method are probabilistic but can be implemented very space efficiently, and they can be parallelized with linear speed-up. In this paper, we present the state of the art in these methods.

1991 Mathematics Subject Classification: 11Y16, 94A60.

1. Introduction

The discrete logarithm problem (DLP) is one of the most important mathematical problems in temporary cryptography. Important public-key cryptosystems, such as the Diffie–Hellman key exchange protocol, the ElGamal encryption and signature schemes, the U.S. Government's Digital Signature Algorithm (DSA), and its elliptic curve analogue (ECDSA), rely on the difficulty of the DLP [MvOV97]. These and other systems are being deployed on the Internet (to enable secure electronic mail, home banking, and Internet browsers), in the financial services industry (in electronic cash applications, credit card transactions, and instant teller banking), in wireless communications (such as pagers and cellular telephones) and in telecommunications (fax encryptors, modems, secure telephones, cable TV and pay-per-view). The hardness of the DLP is not proven, but researchers are confident that discrete-logarithm-based cryptographic systems provide sufficient security if the parameters are chosen appropriately. For an excellent survey on the discrete logarithm problem in cryptographically significant groups such as groups of points on elliptic curves over finite fields and multiplicative groups of finite fields, see [Odl00].

For the moment, we define the DLP as follows: Let g be a generator of a finite cyclic group denoted by $\langle g\rangle$, and let $h \in \langle g\rangle$. Find the least positive integer x such

Public-Key Cryptography and Computational Number Theory
K. Alster, J. Urbanowicz and H. C. Williams (Eds.)

that $g^x = h$. Then we write $x = \log_g h$. Let $\operatorname{ord} g$ denote the order of g, that is $\operatorname{ord} g = \log_g 1$. Then using exhaustive search, it takes at most $\operatorname{ord} g$ multiplications and equality checks in $\langle g \rangle$ to solve the DLP.

In this survey, we consider *generic square-root algorithms* for the DLP. Here, a generic algorithm is an algorithm that does not exploit any special properties of the objects to which it is applied. More precisely, a generic algorithm for a group G uses only that

- Each group element has a unique encoding as a binary string.
- We have a "black box" oracle for the group operations: multiplication, inversion:
  1. For all $g, h \in G$ we can compute the product $g * h$.
  2. For each $g \in G$ we can compute the inverse g^{-1}.

Notice that from every group element g we can determine the identity element $1 = g * g^{-1}$. Further, the unique encoding of the group elements implies the possibility of an equality check: given $g, h \in G$ we can decide whether $g = h$. The running time complexity of a generic algorithm is measured by counting the number of group multiplications, inversions and equality checks. Similarly, space requirements of generic algorithms are measured in terms of group elements to store. Now, let n be the number of group operations one is expected to invest to solve a certain computational problem using exhaustive search. Then we call an algorithm a square-root algorithm if it needs to perform (expected) $O(\sqrt{n})$ group operations to compute the solution.

All generic square-root algorithms for the DLP known to date are based on only a few methods: the baby-step giant-step method due to Shanks [Sha71], and the rho method and the kangaroo method, due to Pollard [Pol78]. The baby-step giant-step method is a deterministic method that uses a time-memory tradeoff and takes const. $\cdot \sqrt{\operatorname{ord} g}$ group operations and has to store const. $\cdot \sqrt{\operatorname{ord} g}$ group elements. The rho method is a probabilistic method based on the birthday paradoxon. It can be implemented very space efficient, with only a constant number of group elements to store and, under certain heuristic assumptions, has expected running time const. $\cdot \sqrt{\operatorname{ord} g}$ group operations. The kangaroo method is also a probabilistic method designed to compute the discrete logarithm when it is known to lie in an interval $[a, b]$. It also can be implemented very space efficient and has expected running time const. $\cdot \sqrt{b - a}$ group operations, again under certain heuristic assumptions.

Observe that the index calculus is not a generic method; it exploits specifics of the representation of its elements. Index calculus algorithms can be applied to solve the DLP in certain finite abelian groups, including $(\mathbb{Z}/p\mathbb{Z})^*$, ideal class groups of quadratic orders and divisor class groups of hyperelliptic function fields of large genus. They have subexponential running time. What is important for cryptography is that there are no index calculus methods known for elliptic curve groups (with a randomly chosen elliptic curve) and subgroups of finite fields. More precisely, the only known algorithms for these groups are the aforementioned square-root algorithms.

Shoup [Sho97] has shown that if p is the largest prime dividing the group order, a generic algorithm to solve the DLP with a probability bounded away from zero has to perform at least const. $\cdot \sqrt{p}$ group operations. Here it is assumed that $\operatorname{ord} g$ and its factorization into primes is known. In this case, the Pohlig–Hellman method [PH78] applies which is a generic method to reduce the DLP in a group of order n to DLPs in groups of order p where p runs through all the prime factors of n.

So, while Shoup has established an $\Omega(\sqrt{p})$ lower bound for the DLP, the existence of the above generic square-root algorithms implies that we cannot find any group where the DLP is harder than $O(\sqrt{p})$. Thus, given our current knowledge, the DLP in elliptic curve groups and subgroups of multiplicative groups of finite fields is (up to constant factors) as hard as a DLP can be.

In the following, we describe the generic square-root algorithms known to date. We summarize the recent advances that lead to optimized running times, thereby reducing the original running times by constant factors. We show extensions to DLPs where various degrees of knowledge about the discrete logarithm are exploited. In the case of the probabilistic algorithms, we also address the heuristic assumptions underlying the running time analysis, and recent advances in reducing these assumptions. In the next section we re-define the DLP according to what additional knowledge is available. Then, in Section 3 we deal with the baby-step giant-step method, in Section 4 we discuss the rho method, and in Section 5 we cover the kangaroo method.

Acknowledgments. The author wishes to thank Douglas Stinson for a helpful discussion about the low Hamming weight DLP, and Ramarathnam Venkatesan for helpful comments on Cayley graphs and discrete logarithms.

2. Degrees of knowledge

A standard assumption in cryptographic applications is that we know $\operatorname{ord} g$, which implies that $x \in [1, \operatorname{ord} g]$, and that x is chosen uniformly at random from that interval; we call the corresponding DLP the *standard-DLP*. But there are settings where this is not the case. For example, when $h = 1$ then finding $x = \log_g h$ means finding (a multiple of) the order of g. In this case we certainly do not know $\operatorname{ord} g$, but instead we might have some other knowledge about x, such as factors of x, or an approximation. We call the DLP where we do not know $\operatorname{ord} g$ the *unknown-order-DLP*. There are also applications where we have additional knowledge about $x = \log_g h$, such as a certain amount of high or low order bits, or a probability distribution of x. Such knowledge can be exploited to reduce the running time for solving the DLP, which needs to be taken into account in implementations of cryptographic systems based on the DLP.

In the following, we define various DLPs according to what is known about the discrete logarithm prior to the computation. Only in the case of the first DLP below we are not even sure about the mere existence of the discrete logarithms. In all the other definitions, we assume that $x = \log_h g$ exists.

2.1. The decision-DLP

Definition 2.1. The *decision-DLP* is the following: Given group elements g and h. Decide whether there exists an x such that $g^x = h$. If this is the case, find x.

Note that a necessary condition for $h \in \langle g \rangle$ is that $h^{\operatorname{ord} g} = 1$. This can be used as a quick test in the case $\operatorname{ord} g$ is known.

The decision-DLP can be solved in running time $O(\sqrt{\operatorname{ord} g}\,)$ with space requirements $O(\sqrt{\operatorname{ord} g}\,)$.

2.2. An interval is known

Definition 2.2. Let a and b be integers, $0 \le a < b$, such that $x \in [a, b]$. The *interval-$[a, b]$-DLP* is: given a, b, g and $h = g^x$, find x.

If $a = 1$ and $b = \operatorname{ord} g$, the interval-$[a, b]$-DLP is just the standard-DLP. A typical cryptographic application is that $a = 0$ and $b \ll \operatorname{ord} g$, so that a certain number of high-order bits of x are known to be zero. The reason why one would choose the secret key x of such a special form is that then the discrete exponentiation g^x is faster than for randomly chosen x. In the case that $h = 1$, an interval for $x = \operatorname{ord} g$ may be obtained from specific properties of the group. For example, for a cyclic elliptic curve groups over $\mathbb{F}_q$ generated by some point g we know that $|\operatorname{ord} g - (q + 1)| \le 2\sqrt{q}$. Similar bounds hold for the number of points on the Jacobian of a hyperelliptic curve.

The interval-$[a, b]$-DLP can be solved in running time $O(\sqrt{b - a}\,)$ with space requirements $O(\sqrt{b - a}\,)$, or in expected running time $O(\sqrt{b - a}\,)$ with constant space requirements.

2.3. The Hamming weight is known

If $\mu = \lceil \log_2(\operatorname{ord} g) \rceil$, then the binary representation of $x = \log_g h$ requires at most μ bits, and we can write $x = \sum_{i=0}^{\mu-1} x_i 2^i$, where $x_i \in \{0, 1\}$ for $0 \le i \le \mu - 1$. The number of 1's in this representation is called the *Hamming weight* of x, and is denoted by $\operatorname{wt}(x)$.

Definition 2.3. Let $t < \mu$ and such that $\operatorname{wt}(x) = t$. Then the *Hamming-weight-t-DLP* is: given μ, t, g and $h = g^x$, find x.

In a cryptographic application, working with an exponent x of (sufficiently) low Hamming weight has the advantage that the exponentiation g^x is faster than with a randomly chosen x.

The Hamming-weight-t-DLP can be solved in running time $O(\mu\binom{\mu/2}{t/2})$ with space requirements $O(\binom{\mu/2}{t/2})$, or in expected running time $O(\sqrt{t}\,\binom{\mu/2}{t/2})$ with space requirements $O(\binom{\mu/2}{t/2})$.

2.4. The solution lies in some known residue class

Let q be an integer, $2 \le q < \operatorname{ord} g$.

Definition 2.4. Let a, q and x be such that $x \equiv a \mod q$. The *mod-q-DLP* is the following: Given a, q, g and $h = g^x$, find x.

For example, q can be a power of two, which means that a certain number of the lowest bits of x are known.

Observe that if $\operatorname{ord} g$ (or an approximate upper bound of $\operatorname{ord} g$) is known, the mod-q-DLP can be easily reduced to an interval-$[a,b]$-DLP. This works as follows: Without loss of generality, let $q \le n := \operatorname{ord} g$. For $x \equiv c \mod q$, let $l \ge 0$ such that $x = c + lq$. Then $l \le \lfloor (n-c)/q \rfloor$, and $(g^q)^l = h * g^{-c}$. Thus, with $N = \lfloor (n-c)/q \rfloor$, $g' = g^q$ and $h' = h * g^{-c}$ the DLP $g^x = h$ is equivalent to the problem of finding l in the equation $(g')^l = h'$, where l is known to lie in the interval $[0, N]$.

Accordingly, the mod-q-DLP can be solved in (expected) running time $O(\sqrt{\operatorname{ord} g / q}\,)$.

2.5. The probability distribution is known

Let $p_0, p_1, \ldots$ be non-negative real numbers such that $\sum_{i=0}^{\infty} p_i = 1$.

Definition 2.5. Assume $h = g^x$ where x is chosen at random with respect to the probability distribution given by (p_i), where $p_i = \operatorname{Prob}(x = i)$. The *distribution-$(p_i)$-DLP* is: Given (p_i), g and h, find x.

If $p_i = 1/\operatorname{ord} g$ for $1 \le i \le \operatorname{ord} g$ and $p_i = 0$ otherwise, then the distribution-(p_i)-DLP becomes the standard-DLP.

A related scenario is when we just know the expected value of the discrete logarithm x. This gives rise to the following definition:

Definition 2.6. Let $E(x)$ be the expected value of x. The *$E(x)$-DLP* is: given $E(x)$, g and $h = g^x$, find x.

The $E(x)$-DLP can be solved in expected running time $O(\sqrt{E(x)}\,)$.

3. Baby-step giant-step methods

The baby-step giant-step method has been introduced by D. Shanks [Sha71] to compute ideal class numbers of quadratic number fields. Shanks was looking for certain multiples of element orders that were known to lie in certain residue classes and were expected to lie in certain intervals.

Applied to the standard DLP, the baby-step giant-step method can be described as follows: Let g and h be such that $h = g^x$ for some unknown x and let n be an upper bound on x, for example $n = \operatorname{ord} g$. Let $q = \lceil \sqrt{n}\, \rceil$. Then we can write

$$x = dq + s, \qquad \text{where } 0 \le d \le q \text{ and } 0 \le s < q.$$

We need to find d and s, which are unique. For this, for $r = 0, 1, \ldots, q-1$ one computes the group elements $h * g^{-r}$. When doing this, for each r one checks whether $h * g^{-r} = 1$. If this is the case, we set $x = r$ and stop. Otherwise, this element is stored in a table R, together with the corresponding number r. We end up with $R = \{(h * g^{-r}, r) : 0 \leq r \leq q-1\}$. These are the *baby steps*. Then for $c = 1, 2, \ldots, q$ one computes g^{cq}. These are the *giant steps*. For each c, one checks whether $(g^{cq}, r) \in R$ for some $r \in \{0, 1, \ldots, q-1\}$. As soon as this happens, $\log_g h$ is found, since then $g^{cq+r} = h$. Note that $cq + r$ is the smallest positive number for which this is possible. Thus, set $d = c$ and $s = r$.

If $x \geq q$, this algorithm requires one inversion and $q + \lfloor x/q \rfloor + O(\log q)$ group multiplications. One has q elements to store. It requires $\lfloor x/q \rfloor$ table look-ups in a table of size q. Since the group elements can be encoded as unique binary strings, hashing on the group elements is possible so that the table-look ups can be executed efficiently.

The unknown-order DLP can be solved in a similar way. Here we work with an assumed bound n on x and compute a baby-step set as above. Then we compute giant steps until the solution is found. This works even if $n < x$. The running times are the same as above, but note that when $n < x$, the complexity might be worse than $O(\sqrt{\operatorname{ord} g})$.

The baby-step giant-step method can also be used to solve the decision-DLP. If $\operatorname{ord} g$ is known, we proceed just as in case of the standard-DLP, and stop when $x = \log_h g$ is found, or when q giant steps have been executed, whatever happens first. If $\operatorname{ord} g$ is not known, we use an assumed bound n and work with a baby-step set $R = \{(g^{-r}, r) : 0 \leq r \leq q-1\}$. Before adding a new element g^{-r} to R, we check whether $g^{-r} = h^{-1}$ or $g^{-r} = 1$. If one of these is the case, we have found the discrete logarithm, or have found that it does not exists. Otherwise, we proceed. Now, for each giant step element g^{cq}, we check whether $(h^{-1} * g^{cq}, r) \in R$ form some r and, if this is not the case, whether $(g^{cq}, r) \in R$. Eventually, one of these happens. The running time for this is two inversions and $q + 2\lfloor x/q \rfloor + O(\log q)$ multiplications. Here, $x = \min\{\log_g h, \operatorname{ord} g\}$, where $\log_g h = \infty$ if $h \notin \langle g \rangle$.

The interval-$[a, b]$-DLP can be solved by putting $n = b - a$, $q = \lceil n \rceil$ and letting (g^{-a}, a) be the first element in the baby-step set R. Finding $x = \log_g h$ then takes one inversion and $q + \lfloor (x-a)/q \rfloor + O(\log(aq))$ multiplications.

3.1. Dynamic increase of the giant-step width

By a modification of the above method it is possible to obtain complexity results relative to the discrete logarithm rather than relative to an upper bound on it. We discuss two such approaches, which work with increasing the giant-step width in the course of the algorithm. The first [BJT97] works with doubling the step width at certain intervals. We describe a simplified version, which is based on the following statement:

Lemma 3.1 ([BJT97, Lemma 2.1]). *For every positive integer x there exist uniquely determined integers k, d and s with $k \geq 0$, $\lfloor 2^{k-1} \rfloor v^2 \leq d < 2^{k+1}$ and $1 \leq s \leq 2^{k+1}$ such that $x = 2^{k+1}d + s$.*

Notice that an integer x can be written in the form $2^{k+1}d + s$ with d as in Lemma 3.1 if and only if x lies in the interval $I_k = \{\lfloor 2^{k-1} \rfloor 2^{k+1}, \dots, 4^{k+1}\}$.

We next explain how to solve the standard-DLP. The idea to compute $x = \log_g h$ is to search those intervals for $k = 0, 1, 2, \dots$ until x is found. For a fixed k, to check whether $x \in I_k$ we test whether $g^{y+d} = h$ with $y = 2^{k+1}d$, and s and d as in Lemma 3.1. For this purpose, we compute the baby-step set $R_k = \{(h * g^{-r}, r) : 1 \leq r \leq 2^{k+1}\}$. Then for $c = \lfloor 2^{k-1} \rfloor, \lfloor 2^{k-1} \rfloor + 1, \dots, 2^{k+1}$ we compute $g^{2^{k+1}c}$. The elements g^y with $y = 2^{k+1}c$ are the giant steps; their step width is 2^{k+1}. For each such giant step we check whether there exists $(g^y, r) \in R_k$ for some r. If so, $\log_g h = y + r$. When we have searched the whole interval I_k without finding x, we increase k by 1, extend the baby-step set accordingly and perform the next set of giant steps.

This algorithm can be implemented such that it executes one inversion and at most $4\lceil\sqrt{x}\,\rceil + O(\log x)$ multiplications. One has at most $2\lceil\sqrt{x}\,\rceil$ elements to store, and it requires at most $2\lceil\sqrt{x}\,\rceil$ table look-ups.

The second approach uses giant steps whose step width is incremented after *each* baby step [Ter00]. It is based on the following way of representing integers:

Lemma 3.2. *For every non-negative integer x there are uniquely determined integers j and t with $0 \leq t < j$ such that $x = T_{j+1} - t$, where T_j is the j-th triangle number, i.e., $T_1 = 0$, $T_{j+1} = T_j + j$ for $j \geq 1$.*

We show how to solve the standard-DLP. To compute $x = \log_g h$, we need to find the correct value of j in the representation above. This is done alternating baby steps and giant steps as follows: In the beginning, we compute the baby step $g^0 = 1$ and the giant step g^0; the baby step is stored as pair $(g^0, 0)$. Now for $j \geq 1$, the j-th iteration consists of computing the baby step $h * g^j = h * g^{j-1} * g$ (which is stored as pair (g^j, j) in the baby step set R) and the giant step $g^{T_{j+1}} = g^{T_j} * g^j$. If $j \geq 2$, we check whether $(g^{T_{j+1}}, t) \in R$ for some t, $0 \leq t < j$. If this is the case, we put $x = T_{j+1} - t$ and stop. This algorithm terminates when $T_j < x \leq T_{j+1}$. Since $T_{j+1} = j(j+1)/2$ for $j \geq 0$, this happens if $j = \lfloor\sqrt{2x + 1/4}\,\rceil$.[1)]

Since this method requires as many baby steps as giant steps it requires $2\lfloor\sqrt{2x + 1/4}\,\rceil$ group multiplications and $\lfloor\sqrt{2x + 1/4}\,\rceil$ table look-ups in a baby-step set of maximum size $\lfloor\sqrt{2x + 1/4}\,\rceil$. Although these values are by a factor of $\sqrt{2}$ smaller than the upper bounds given for the previous algorithms, in practice both methods perform equally well (see [Ter00]).

Observe that in both variants [BJT97, Ter00], the algorithm for the standard-DLP is the same as for the unknown-order-DLP. The modifications for the decision-DLP and the interval-$[a, b]$-DLP are similar as for the original method, and given above.

1) $\lfloor y \rceil$ is the unique integer y such that $-\frac{1}{2} < y - \lfloor y \rceil \leq \frac{1}{2}$.

3.2. A systematic approach

The original version of baby-step giant-step method with $n = \operatorname{ord} g$ is optimal in the sense that in the worst case, that is, $x \approx n$, it takes the least number of group operations. The modifications [BJT97] and [Ter00] have the advantages that their running times are lower if $x \ll n$. The points of turnover are $x = n/14$ in the case of [BJT97] and $x = n/6$ in the case of [Ter00]. Blackburn and Teske [BT00] took a systematic approach and asked: given the probability distribution for the solution x on $\mathbb{N}_0$, what is the baby-step giant-step strategy that computes x using the smallest expected number of operations? The answer that optimizes the baby-step giant-step method for the distribution-(p_i)-DLP is as follows:

If the distribution has an increasing hazard rate[2)] (IHR), which is the case for the uniform, the normal and the geometric distributions, the optimal strategy is to compute $q = \sqrt{E(x)}$ baby steps at the beginning, and then proceed exclusively with giant steps; here, $E(x)$ denotes the expected value of x. In particular, if x is uniformly distributed over $[1, \operatorname{ord} g]$, then the choice that minimizes the average running time is $q = \sqrt{\operatorname{ord} g/2}$. Thus, remarkably, for IHR distributions the distribution-(p_i)-DLP cannot be solved any faster than the $E(x)$-DLP. In other words, the additional knowledge of the exact probability distribution cannot be exploited to achieve lower expected running times. Only if the distribution has a decreasing hazard rate, which is the case for the discrete Pareto or the Weibull distribution, the optimal strategy is to gradually increase the giant-step width. This optimal strategy is completely determined by the p_i, and there is an explicit formula for it [BT00]. In particular, Terr's method [Ter00] is close to optimal if the distribution of the discrete logarithm is a discrete Pareto distribution.

3.3. Low Hamming weight

The Hamming-weight-t-DLP can be solved by a probabilistic algorithm as follows. Let $X = \mathbb{Z}_\mu$. For a set $Y \subset 2^X$ let $\operatorname{val}(Y) = \sum_{i \in Y} 2^i \in \{0, 1, \ldots, 2^\mu - 1\}$. Assume for simplicity that both μ and t are even. Now we pick a random subset B of X with $\mu/2$ elements and compute the "baby"-step set $R_B = \{(g^{\operatorname{val}(Y)}, \operatorname{val}(Y)) : Y \subset B, |Y| = t/2\}$. Then for all subsets Z of $X \setminus B$ such that $|Z| = t/2$, we check whether $(h * (g^{\operatorname{val}(Z)})^{-1}, r) \in R_B$ for some r. As soon as this happens for some Z, we put $x = r + \operatorname{val}(Z)$ and stop. If this is the case for no such Z, we pick another random subset $B \subset X$, $|B| = \mu/2$, and repeat the computation.

We can efficiently compute R_B using a minimal change ordering on $\{Y \subset B, |Y| = t/2\}$ where any two successive sets differ by only two elements (cf. [KS99]). Then, working with a precomputed list $\{g^{2^i} : i = 0, \ldots, \mu - 1\}$, we can obtain the next term $g^{\operatorname{val}(Y)}$ from the previous one just by multiplying by g^{2^j} and dividing by g^{2^k} for some j and k. Thus, for each B, the computation of R_B takes at most

2) For a probability distribution (p_i), the *hazard rate* (a.k.a. *failure rate*) at k measures the likelihood of x lying in a small interval beyond k, given that $x \geq k$. It is defined as $p_k/(\sum_{i=k}^{\infty} p_i)$ and is the reciprocal of the Mills' ratio.

$2\binom{\mu/2}{t/2} + O(t)$ group operations. A similar method can be applied to compute the elements $h * (g^{\mathrm{val}(Z)})^{-1}$ for all $Z \subset X \setminus B$ with $|Z| = t/2$. Hence, for each B we need to perform at most $4\binom{\mu/2}{t/2} + O(t)$ group operations and at most $\binom{\mu/2}{t/2}$ table look-ups in a table of size $\binom{\mu/2}{t/2}$. Stinson [Sti] shows that this algorithm, which is due to Coppersmith, is expected to succeed after $O(\sqrt{t})$ sets B have been considered. Thus, the expected running time is $O\big(\sqrt{t}\,\binom{\mu/2}{t/2}\big)$ group operations.

A deterministic algorithm can be obtained if instead of randomly picking $B \subset X$ we use (μ, t)-splitting systems. This algorithm is up to a factor of $\mu/\sqrt{t}$ slower. See [Sti] for details.

3.4. Parallelization

If the baby-step giant-step method is applied parallel on m processors, a speed-up by a factor of $\sqrt{m}$ can be easily achieved, simply by dividing the interval $[1, n]$ into m approximately equally long subintervals and applying the method independently on each of these subintervals. Unless the m processors have fully shared memory with unlimited and immediate access to it a $\sqrt{m}$-fold speed-up is the best one can achieve. In other words, the baby-step giant-step method cannot be efficiently parallelized.

4. Pollard's rho method

The rho method has been developed by J. Pollard [Pol78] to compute discrete logarithms in $(\mathbb{Z}/p\mathbb{Z})^*$.

The key idea is that for any finite set W and mapping $F : W \to W$, the sequence $(w_k)_{k \in \mathbb{N}_0}$ formed by the rule $w_0 \in W$, $w_{k+1} = F(w_k)$ $(k \in \mathbb{N}_0)$ is ultimately periodic. Hence, there exist integers $\lambda \geq 1$ and $\mu \geq 0$ such that $w_0, \ldots, w_{\mu+\lambda-1}$ are pairwise distinct and $w_k = w_{k+\lambda}$, $k \geq \mu$. Under the assumption that $w_0 \in W$ is randomly chosen (with respect to the uniform distribution) and F is a random mapping[3)], the expected values for μ and λ are both close to $\sqrt{\pi |W|/8} = 0.626\ldots\sqrt{|W|}$ ([Har60]). The name *rho method* stems from the picture one obtains when drawing the terms of (w_k), starting at the bottom and ending in a cycle, which shows the Greek letter rho.

We describe the rho method in $(\mathbb{Z}/p\mathbb{Z})^*$. Let p be a prime, let h be any integer, and let g be a primitive root modulo p. To compute $x = \log_g h \pmod{p}$, Pollard [Pol78] used the iterating function $F : (\mathbb{Z}/p\mathbb{Z})^* \to (\mathbb{Z}/p\mathbb{Z})^*$ given by

$$F(y) = \begin{cases} gy & \text{if } 0 < y \leq p/3, \\ y^2 & \text{if } p/3 < y \leq 2p/3, \\ hy & \text{if } 2p/3 < y < p. \end{cases} \tag{4.1}$$

3) A *random mapping* is a mapping that is chosen uniformly at random from the set of all $|W|^{|W|}$ mappings $F : W \to W$.

and defined a sequence (y_k) according to the rule $y_0 = 1$, $y_{k+1} \equiv F(y_k) \pmod p$. There are sequences (α_k) and (β_k) such that $y_k = g^{\alpha_k} * h^{\beta_k}$ $(k \in \mathbb{N}_0)$, and these sequences follow the rules

$$\begin{aligned} \alpha_0 &= 0, \quad \alpha_{k+1} \equiv \alpha_k + 1,\ 2\alpha_k,\ \text{or}\ \alpha_k \bmod (p-1), \qquad k \in \mathbb{N}_0, \\ \beta_0 &= 0, \quad \beta_{k+1} \equiv \beta_k,\ 2\beta_k,\ \text{or}\ \beta_k + 1 \bmod (p-1), \qquad k \in \mathbb{N}_0, \end{aligned}$$

according to the three cases above. Notice that the reduction modulo $p-1$ is necessary because otherwise the doubling of α_k and β_k would cause an explosion of the exponents.

A match, that is, two terms y_k, y_l with $y_k = y_l$ and $k \neq l$, leads to an equation $\alpha_k + x\beta_k \equiv \alpha_l + x\beta_l \pmod{p-1}$, from which we can compute x using the Extended Euclidean algorithm if $\gcd(p-1, \beta_k - \beta_l) = 1$.

To find a match, there is, for example, an efficient algorithm due to Brent [Bre80]. It uses an auxiliary variable, say w, which at each stage of the algorithm holds y_j, where j is the largest power of two strictly less than the current index of the sequence k. At the beginning we have $w = h_0$. For each newly computed term of the sequence, the algorithm checks for equality with w. As soon as the current index k equals a power of two, we assign $w = y_k$. This is repeated until a match $(y_k,\ w)$ is found.

Brent's algorithm

Input: Initial value y_0, iterating function $F : W \to W$.

Output: j and k such that $y_j = y_k$.

Algorithm:

```
y = h_0; r = 1; k = 0; done = false
repeat
    w = y, j = k
    r = 2r
    repeat
        k = k + 1, y = F(y)
        done = (w = y)
    until (done) or (k ≥ r)
until (done)
return j, k
```

Under the assumption that F is a random mapping, the first match is expected to occur after $\sqrt{\pi n/2} \approx 1.25\sqrt{n}$ iterations $(n = \operatorname{ord} g)$, and then with Brent's algorithm the first match is found after an expected number of $\approx 1.97\sqrt{n}$ iterations. Modifications of Brent's algorithm that require slightly more storage but less iterations can be found in [SL84] and [Tes98a].

Pollard's original application can easily be generalized to any other group. All we need is a rule how to partition the group into 3 disjoint sets of equally large size, which can be done based on the unique encoding of the group elements as binary strings. Experiments show, however, that iterating functions of the form (4.1) do not behave like random functions. For example, in $(\mathbb{Z}/p\mathbb{Z})^*$ with the rule of partitioning given above, matches occur only after an average number of

$1.37\sqrt{p-1}$ iterations. In other groups, or for other ways of partitioning $(\mathbb{Z}/p\mathbb{Z})^*$, they occur even later, after an (experimental) average of $\approx 1.56\sqrt{n}$ iterations, where $n = \operatorname{ord} g$ (see [Tes00]).

4.1. Better random walks

Better walks for the rho method have been introduced that yield the same performance as we expect from a random mapping, which in the case of an arbitrary group means a speed-up by about 20% compared to a function of the type (4.1). Let g, h be group elements, $G = \langle g, h \rangle$. Let r be a small positive integer: $r \in [3, 100]$. We need a partitioning of $G = T_1 \cup \cdots \cup T_r$ with roughly equally large sets T_s. For this, we take a hash function $v : G \to \{1, \ldots, r\}$ and define $T_s := \{y \in G : v(y) = s\}$, $s = 1, \ldots, r$. A popular example (see [Tes98b]) is the 20-adding walk, which works with $r = 20$ sets T_s and is generated by an iterating function of the form $F(y) = y * M_{v(y)}$, where the multipliers $M_1, \ldots, M_{20}$ are computed according to the rule $M_s = g^{m_s} * h^{n_s}$ with m_s and n_s randomly chosen from $\{1, \ldots, |G|\}$. We compute y_0 by taking a random integer in $\{1, \ldots, |G|\}$ and putting $y_0 = g^\alpha$, and then iteratively compute the sequence (y_k) following the rule $y_{k+1} = F(y_k)$. Then each term y_k can be represented in the form $y_k = g^{\alpha_k} * h^{\beta_k}$, where $\beta_0 = 0$ and

$$\alpha_{k+1} = \alpha_k + m_{v(Y_k)} \quad \text{and} \quad \beta_{k+1} = \beta_k + n_{v(Y_k)} \,. \tag{4.2}$$

These addings (whence the name of the walks) are computed modulo $|G|$. However, since the increase of the terms in (4.2) is only linear in the number of iterations, it is not even necessary to perform the reduction modulo $|G|$ while computing the terms of the sequences (α_k) and (β_k). This means that r-adding walks can be computed even if the group order is not known. Indeed, 8-adding walks have been first used by Sattler and Schnorr [SS85] to compute element orders.

In experiments with cyclic elliptic curve (sub)groups, matches in 20-adding walks occur after an average of $1.26\sqrt{n}$ iterations. This is approximately the performance of a random walk.

4.2. Degrees of knowledge

Let us now look at what knowledge is needed to apply the rho method, and to what extent additional knowledge can be exploited. In principle, the rho method only suits to solve the standard-DLP.

First of all, even if the sequence (y_k) and the corresponding sequences of exponents can be computed without knowledge of the order $n = \operatorname{ord} g$, we need to know n to find the discrete logarithm once a match has occurred. This is because a match only gives an equation of the form $bx \equiv a \mod n$, which is very unlikely to hold over the integers, and thus can only be solved for x if n is known. (One certainly can, however, apply the rho method twice, once to compute $\operatorname{ord} g$ and then to compute $x = \log_g h$.)

Also, the rho method cannot be used to solve the decision-DLP. Assume we do not know whether $h \in \langle g \rangle$. Without loss of generality we can assume that $\operatorname{ord} h$ divides $\operatorname{ord} g$. We apply the rho method until we find a match $g^{\alpha_l} * h^{\beta_l} = g^{\alpha_k} * h^{\beta_k}$. Let $a = \alpha_l - \alpha_k \pmod{|G|}$ and $b = \beta_k - \beta_l$. Then $g^a = h^b$. Let $n = \operatorname{ord} g$ and let $c = \gcd(b, n)$. If $c = 1$, we know that $h \in \langle g \rangle$. In any case, we use the extended Euclidean algorithm to compute two integers s and t such that $c = sb + tn$ and obtain that $g^{sa} = h^c$. If $c \neq 1$, it might be that $h \notin \langle g \rangle$, or we just might have had bad luck. There are two ways to proceed. The first is to repeat the rho method several times until we either have found that $h \in \langle g \rangle$, or until we can conclude with a high enough probability that $h \notin \langle g \rangle$. The second way is to check for the c solutions of the equation $sa \equiv cz \pmod{n}$ whether $g^z = h$. This method works very well in practice. Notice that $h \in \langle g \rangle$ if and only if $g^z = h$ for one of those c values z. However, in the worst case where $\operatorname{ord} g = \operatorname{ord} h = p$ (p prime) and $\langle g \rangle \cap \langle h \rangle = \emptyset$, it may take p such checks!

Knowledge of an interval cannot be exploited to obtain a significant speed-up of the rho method (but there is the kangaroo method that takes care of this setting). It is not known how the Hamming-weight-t-DLP, the distribution-(p_i)-DLP or the $E(x)$-DLP could be solved any faster than the standard-DLP using the rho method. The knowledge of $x \bmod q$ can be used to decrease the running time by a factor of $\sqrt{\gcd(q, \operatorname{ord} g)}$, by working in the group generated by g^q (which gives no speed-up if $\operatorname{ord} g$ is prime). Also note that it is not possible to modify the method to obtain running time estimates in terms of the actual logarithm rather than in terms of $\operatorname{ord} g$.

A major advantage of the rho method over the baby-step giant-step method is that it is space efficient. It only needs a constant number of group elements to store, independently of the size of the group in which one works. Moreover, the rho method can be efficiently parallelized, where storage requirements on a central server are somewhat higher but can effectively be monitored.

4.3. Parallelization

Van Oorschot and Wiener [vOW99] have shown that the rho method can be parallelized with linear speed-up.

Assume we are given m processors. On each processor we use the same iterating function, initialize a sequence (y_k) by computing a random power of g and then proceed as in the serial rho method. However, we look for a match not among the terms of that single sequence (which would give only a speed-up of a factor of $\sqrt{m}$ [Bre90]), but among all terms computed on all m processors. This is done using the *distinguished-point method* which works as follows: One defines a set $\mathcal{D} \in \langle g \rangle$ that consists of all group elements that satisfy a certain distinguishing property. The elements in $\mathcal{D}$ are called distinguished point. A popular way of defining $\mathcal{D}$ is to fix an integer f and to define that $z \in \mathcal{D}$ if and only if the f least significant bits in the representation of z as a binary string are zero. Notice that this definition allows a fast check for the distinguishing property to hold, and the size of $\mathcal{D}$ can be easily monitored as well. Now, while computing a sequence

(y_k) together with the corresponding exponent sequences (α_k) and (β_k), for each y_k the processor checks whether y_k is a distinguished point. If this is the case, it submits the triple (y_k, α_k, β_k) to a central server, and then goes on. The central server, which collects all distinguished points found by all m processors, searches for a match among those points. That can be done totally independent of the m processors.

Van Oorschot and Wiener have analyzed this method and showed that, under the assumption that the iterating function behaves like a random mapping, the expected running time on each processor is $\sqrt{\pi\,(\operatorname{ord} g)/2}\,/m + 1/\Theta$, where Θ is the proportion of distinguished points (i.e., $\Theta = 2^{-f}$ in the above example for $\mathcal{D}$). The central server has to store an expected number of $\Theta\sqrt{\pi\,(\operatorname{ord} g)/2}$ distinguished points.

In November 1997, the Mississauga company Certicom issued a challenge: a list of elliptic curve DLPs in elliptic curve groups of various sizes [Cer]. The DLPs in the smallest groups up to a prime group order of 108 bits have been solved by parallelized attacks distributed over the Internet. The extensive experimental work related to this challenge (see, for example, [ESST98]) confirms the theoretically predicted linear speed-up for the parallelized rho method.

4.3.1. Inverse-point strategy. In certain groups, a speed-up up to a factor of $\sqrt{2}$ can be achieved by applying the *inverse-point strategy.* In this method, each group element is paired with its inverse: The iterating function is defined such that if the group element a is mapped to the group element b, its inverse a^{-1} is mapped to b^{-1}. Further, the distinguished point set is defined such that if b is a distinguished point, then b^{-1} is also a distinguished point. Also, we require from the representation of the group elements that it can be detected if the inverse of a previously submitted point is submitted, and then such a collision also suits to reveal the discrete logarithm.

The inverse point strategy has been successfully applied in groups of elliptic curves over finite fields [ESST98] (see also [WZ98]).

A related technique has been applied in the special case of elliptic curves over $\mathbb{F}_{2^{nk}}$ defined over $\mathbb{F}_{2^n}$ [GLV00, WZ98]. Here, a speed-up up to a factor of $\sqrt{2k}$ is achieved by assembling the group elements in equivalence classes via the Frobenius map. Non-trivial automorphisms have also been exploited to speed up the rho method for discrete logarithm computation on certain hyperelliptic curves [DGM99].

4.4. On the analysis of the running time

The usual heuristic assumption in the analysis of the rho method is that the iterating function is a random mapping, and then the birthday paradoxon can be applied. However, in practice, one is restricted to a small set of mappings $F : G \to G$, namely those which can be evaluated efficiently and allow us to keep track of the exponent sequences (α_k) and (β_k).

The iterating functions that correspond to r-adding walks as defined in Section 4.1 seem to be a good choice despite the fact that they are not random mappings. Indeed, let $M_1, \ldots, M_r$ be randomly chosen elements in $\langle g \rangle$ and let $n = \operatorname{ord} g$ be prime. Then from Hildebrand's work [Hil94] on random walks on the integers modulo n, Teske [Tes00] concludes that if in the iteration $y_{k+1} = y_k * M_{\nu(y_k)}$ the $\nu(y_k)$ are randomly chosen from $\{1, \ldots, r\}$, r-adding walks are getting close to uniformly distributed on $\langle g \rangle$ after $O(n^{2/(r-1)})$ steps. Consequently, for $r \geq 6$ the number of steps in the rho method is expected to be $O(\sqrt{n})$, under the above randomness assumption on the $\nu(y_k)$. In practice, for $r \geq 16$ and with ν a hash function we find the same performance as expected from a random mapping, that is, cycles occur after an expected number of $1.25\sqrt{n}$ iterations ([Tes00]).

Horwitz and Venkatesan [HV] view r-adding walks as random walks on r-regular Cayley graphs over $\langle g \rangle$ that are generated by r random group elements $M_1, \ldots, M_r$. Then solving the DLP with the rho method can be related to finding non-trivial cycles in random Cayley graphs, and the rapid mixing property of such Cayley graphs with $O(\log n)$ generators, which are expander graphs, can be exploited to analyze the running time of the rho method. Indeed, for n prime, $r = \text{const.} \cdot \log n$, and under the assumption that a t-wise independent hash function is used to navigate on the Cayley graph ($t = \text{const.} \cdot \log^2 n$), the running time is shown to be $O((\log^c n)\sqrt{n}\,)$ operations (c a small constant). Horwitz and Venkatesan also give methods to implement t-wise independent hash functions.

5. Pollard's kangaroo method

The kangaroo method has been invented by Pollard [Pol78] as a space efficient method to solve the interval-$[a, b]$-DLP. In its original version, it works with a tame kangaroo T with starting point $t_0 = g^b$ and a wild kangaroo W with starting point $w_0 = h$. In terms of the exponents of g, T starts at the upper end of the interval $[a, b[$, while W starts at an unknown spot x. Let $S = \{g^{s_1}, \ldots, g^{s_r}\}$ ($s_i > 0$ and of size $O(\sqrt{b-a}\,)$, $r = O(\log(b-a))$) be a set of jumps, and let $\nu : G \to \{1, \ldots, r\}$ be a hash function. We think of the exponents s_i as travel distances. The tame kangaroo travels through the group following the path

$$t_{j+1} = t_j * g^{s_{\nu(t_j)}}, \qquad j \in \mathbb{N}_0.$$

While computing the path (t_j), we keep track of the travelled distances $\delta_j(T)$, where $\delta_0(T) = 0$ and

$$\delta_{j+1}(T) = \delta_j(T) + s_{\nu(t_j)}\,, \qquad j \in \mathbb{N}_0.$$

Hence, $t_j = g^{b+\delta_j(T)}$, for each $j \in \mathbb{N}_0$. After a certain number of jumps, the tame kangaroo stops and installs a trap at its final spot, say t_M. Then the wild kangaroo is set off, following the path

$$w_{j+1} = w_j * g^{s_{\nu(w_j)}}, \qquad j \in \mathbb{N}_0,$$

with the distances $\delta_j(W)$ given by $\delta_0(W) = 0$ and

$$\delta_{j+1}(W) = \delta_j(W) + s_{\nu(w_j)}, \qquad j \in \mathbb{N}_0.$$

Then $w_j = h * g^{\delta_j(W)}$ ($j \in \mathbb{N}_0$). Hence, in terms of the exponents of g, at each stage we know the exact position of T, but we do not know the position of W since its starting spot x is unknown; this is why W is wild. After each of W's jumps it is checked whether W has fallen into the trap. If this is the case, say at w_N, the equation $t_M = w_N$ immediately yields a solution of $g^x = h$, namely $x = \delta_M(T) - \delta_N(W)$. Otherwise, after a certain number of jumps the wild kangaroo is halted, and a new wild kangaroo with starting point $w_0 = h * g^z$ and initial distance $\delta_0(W) = z$ (z small) is set off. If W falls into T's trap, this means that the paths of the two kangaroos must have met earlier during the travel, from which point on they have been identical. In this case, if we draw the two paths on a piece of paper starting at the bottom left and the bottom right, respectively, we obtain the Greek letter lambda, which explains why the method is also called the lambda method. Van Oorschot and Wiener [vOW99] have analyzed this method and found out that the running time is expected to be minimal if the mean value of the s_i is approximately $\sqrt{b-a}/2$, and if T makes approximately $0.7\sqrt{b-a}$ jumps before placing the trap. After about $2.7\sqrt{b-a}$ jumps of the wild kangaroo, W either must have fallen into the trap or it must have safely passed the trap and then should be halted. The former happens with probability ≈ 0.75 after an expected number of $1.7\sqrt{b-a}$ jumps, while the latter is expected to happen about 0.33 times. Thus, the expected total running time amounts to $(0.7 + 0.33 \cdot 2.7 + 1.7)\sqrt{b-a} \approx 3.3\sqrt{b-a}$ group operations. The algorithm has to store only the set of jumps and the current positions of the two kangaroos. If one allows more storage, one can use the distinguished point method to obtain an algorithm that takes expected $2\sqrt{b-a}$ operations. That algorithm is a special case ($m = 2$) of the parallelized version described next.

5.1. Parallelization

The distinguished point method that is used to parallelize the rho method can also be used for the kangaroo method [vOW99]. Assume there are m processors, m even. (The serial case corresponds to $m = 2$, with two processors simulated on one machine.) On each processor, we have one kangaroo, tame or wild. Everywhere we use the same set of jumps S whose jump distances s_i are chosen such that their mean value, say β, satisfies

$$\beta = m\sqrt{b-a}/4.$$

Let $\sigma \approx \frac{1}{m/2}\beta$ be an integer that indicates the spacing between members of the same herd. That is, $m/2$ tame and $m/2$ wild kangaroos $T_1, \ldots, T_{m/2}$ and $W_1, \ldots, W_{m/2}$ are set off at the group elements

$$t_0(T_i) = g^{(a+b)/2+(i-1)\sigma} \quad \text{and} \quad w_0(W_i) = h * g^{(i-1)\sigma}, \quad i = 1, \ldots, m/2,$$

one on each processor. (Observe that the tame kangaroos are set off near the middle of the interval rather than its end—this leads to lowest average running times.) The initial travel distances are set to $\delta_{0,\text{tame}}(T_i) = \delta_{0,\text{wild}}(W_i) = (i-1)\sigma$ and each kangaroo is provided with a tag indicating whether it is tame or wild. Then each kangaroo jumps on its processor, and after each jump it is checked whether the new kangaroo spot is a distinguished point. If this is the case, that distinguished point together with the corresponding travel distance and the tame/wild tag is sent to a central server. The central server checks whether that point has already been previously submitted (in order to detect a collision between a tame and a wild kangaroo), and stores it if this is not the case. Under certain randomness assumptions (see Section 5.2), such a collision is expected to occur after $2\sqrt{b-a}/m$ jumps of each kangaroo. Thus, with Θ denoting the proportion of distinguished points in $\langle g\rangle$, the expected running time on each processor amounts to $2\sqrt{b-a}/m + 1/\Theta$ group operations.

It is possible in the above situation that kangaroos of the same herd collide. Such a collision is useless, because it does not reveal anything about the discrete logarithm. An analysis of this phenomenon [Tes01] shows that at most 2 useless collisions are expected to occur, independent of the number of processors.

Pollard [Pol00] has developed a version of parallelization where useless collisions *cannot* occur. Here one works with u tame and v wild kangaroos, where u and v are coprime and such that $u \approx v \approx m/2$ and $u+v \leq m$. The r jump distances in the set of jumps are multiples of uv, say $s_i = q_i uv$ for positive integers $q_1, \ldots, q_r$. Here, the mean value of the q_i should be close to $\beta = \sqrt{(b-a)/(uv)}/2$ for optimal results. The set-off points of the tame and wild kangaroos are $g^{(a+b)/2+iv}$ ($i = 0, \ldots, u-1$) and $h * g^{ju}$ ($j = 0, \ldots, v-1$), respectively. This implies that any two tame kangaroos (and also any two wild kangaroos) travel with travel distances that are in pairwise distinct residue classes modulo uv. As above, the actual travel is determined by a hash function $\nu : G \to \{1, \ldots, r\}$. Since the equation $\frac{a+b}{2} + iv \equiv x + ju \pmod{uv}$ has a unique solution in i and j, there is exactly one pair of tame and wild kangaroos that travel in the same residue class modulo uv, and this is the only pair that can collide. The expected running time in this case is $\sqrt{(b-a)/(uv)} + 1/\Theta$ jumps on each processor. If u and v are close to $m/2$, this yields approximately the same expected running time as for the van Oorschot-Wiener parallelization.

In the above versions of the kangaroo method, proper choices for the jump distances s_i, respectively q_i in the Pollard parallelization, are (a) consecutive powers of two starting from $2^0 = 1$, or (b) integers randomly chosen from $[1, 2\beta]$ subject to the condition that they are pairwise distinct and have no non-trivial common divisor.

Experimental work with the (parallelized) kangaroo method shows that the theoretically predicted running times can be met. See [Tes01] for a corresponding report.

5.2. On the running time analysis

All expected running times given above have been derived under the assumption that the discrete logarithm is uniformly distributed over the interval $[a, b]$. For an arbitrary distribution, the expected running time can always be bounded by $3\sqrt{b-a}\,/m + 1/\Theta$ group operations.

Under the assumption that the hash function ν behaves like a random function $G \to \{1, \ldots, r\}$, Pollard [Pol00] shows that a collision between a wild and a tame kangaroo is expected to occur after about $2\sqrt{b-a}\,B_S$ iterations (in the serial version). Here, B_S is a constant that depends on the set of jump distances s_i and can be determined by solving a certain large set of equations. For the case that the s_i are the consecutive powers of two starting from 1, Pollard shows that B_S is close to one if $|S| \geq 6$. We cannot compute B_S for the case that the jump distances are randomly chosen integers from $[1, 2\beta]$. Thus, in this case, the running times are subject to the additional assumption that for each jump of a fixed wild and tame kangaroo in the same region (on the line through $[a, b]$), a collision occurs with probability $1/\beta$. This assumption is obviously fulfilled if each jump were with a distance randomly chosen from $[1, 2\beta]$.

5.3. Degrees of knowledge

The kangaroo method is not well suited be to solve the decision-DLP. Of course, a collision between a tame and a wild kangaroo can occur if and only if $h \in \langle g \rangle$. But if no collision has occurred after a certain number of steps, this might be due to extremely bad luck or to $h \notin \langle g \rangle$. For example, if $h \in \langle g \rangle$, the probability that a collision has occurred after $4\sqrt{b-a}$ steps is about 0.9989. (This probability has been derived under the aforementioned assumption that a collision occurs with probability $1/\beta$ in each pair of jumps, and that the discrete logarithm is uniformly distributed over $[a, b]$. See [Tes01] for details.)

We can solve the unknown-order-DLP with the kangaroo method: As soon as a collision between a tame and wild kangaroo has happened, a solution x for $g^x = h$ can be derived from the respective travel distances without knowing $\operatorname{ord} g$. If $b - a \leq \operatorname{ord} g$, with a probability at least $1 - 10^{-6}$ it even yields the least positive such value x.

The kangaroo method has been designed as a square-root algorithm to solve the interval-$[a, b]$-DLP. It is faster than the rho method if $b - a < \pi/8 \cdot \operatorname{ord} g$. Thus, if the discrete logarithm is only known to lie in an interval of length between $\pi/8 \cdot \operatorname{ord} g$ and $\operatorname{ord} g$, this additional knowledge cannot be exploited for a speed-up

Due to the reduction from the mod-q-DLP to the interval-$[a, b]$-DLP, the kangaroo method can be used to solve the mod-q-DLP in expected running time $2\sqrt{\operatorname{ord} g/q} + 1/\Theta$ group operations if $\operatorname{ord} g$ is known. For the same reasons as before, this only gives a speed-up if $q > 8/\pi$.

In the case of the distribution-(p_i)-DLP, we can speed-up the kangaroo method by modifying the set-off points for the tame kangaroos accordingly, which reduces the average running time to cover the initial distance between the tame and the wild herds. Here, lowest expected running times are achieved if the tame kangaroos are set off at the expected value for the discrete logarithm. If we are given a distribution of the discrete logarithms over the entire interval $[1, \operatorname{ord} g]$, instead of simply using the rho method we can use the kangaroo method with an interval $[a, b]$ that comprises a high percentage of the solutions. This can yield a speed-up if the distribution is not too wide-spread.

Finally, there is no algorithm known to date to solve the Hamming-weight-t-DLP with some kangaroo method. Since the same is the case for the rho method, this means that so far no space efficient, parallelizable algorithm for the Hamming-weight-t-DLP has been found.

References

[BJT97] Buchmann, J., Jacobson, Jr., M.J., Teske, E., On some computational problems in finite abelian groups. Math. Comp. 66 (1997), 1663–1687.

[Bre80] Brent, R.P., An improved Monte Carlo factorization algorithm. BIT 20 (1980), 176–184.

[Bre90] —, Parallel algorithms for integer factorization. In: Number Theory and Cryptography (ed. by J. H. Loxton; London Math. Soc. Lecture Note Ser. 154), 26–37. Cambridge University Press, 1990.

[BT00] Blackburn, S.R., Teske, E., Baby-step giant-step algorithms for non-uniform distributions. In: Algorithmic Number Theory – ANTS IV (ed. by W. Bosma; Lecture Notes in Comput. Sci. 1838), 153–168. Springer, Berlin 2000.

[Cer] The Certicom ECC Challenge.
On `http://www.certicom.com/research/ecc_challenge.html`.

[DGM99] Duursma, I., Gaudry, P., Morain, F., Speeding up the discrete log computation on curves with automorphisms. In: Advances in Cryptology – ASIACRYPT '99 (ed. by K.Y. Lam, E. Okamoto, C. Xing; Lecture Notes in Comput. Sci. 1716), 103–121. Springer, Berlin 1999.

[ESST98] Escott, A.E., Sager, J.C., Selkirk, A.P.L., Tsapakidis, D., Attacking elliptic curve cryptosystems using the parallel Pollard rho method. CryptoBytes (The technical newsletter of RSA laboratories) 4 (1998), no. 2, 15–19. Available at `http://www.rsa.com/rsalabs/pubs/cryptobytes`.

[GLV00] Gallant, R., Lambert, R., Vanstone, S., Improving the parallelized Pollard lambda search on binary anomalous curves. Math. Comp. 69 (2000), 1699–1705.

[Har60] Harris, B., Probability distributions related to random mappings. Ann. Math. Stat. 31 (1960), 1045–1062.

[Hil94] Hildebrand, M., Random walks supported on the random points of $\mathbb{Z}/n\mathbb{Z}$. Probab. Theory Related Fields 100 (1994), 191–203.

[HV] Horwitz, J., Venkatesan, R., Random Cayley graphs and the discrete log. Preprint, 2000.

[KS99] Kreher, D.L., Stinson, D.R., Combinatorial Algorithms. CRC Press, Boca Raton 1999.

[MvOV97] Menezes, A., van Oorschot, P., Vanstone, S.A., Handbook of Applied Cryptography. CRC Press, Boca Raton 1997.

[Odl00] Odlyzko, A.M., Discrete logarithms: the past and the future. Des. Codes Cryptogr. 19 (2000), 129–145.

[vOW99] van Oorschot, P.C., Wiener, M.J., Parallel collision search with cryptanalytic applications. J. Cryptology 12 (1999), 1–28.

[PH78] Pohlig, S.C., Hellman, M.E., An improved algorithm for computing logarithms over $GF(p)$ and its cryptographic significance. IEEE Trans. Information Theory 24 (1978), 106–110.

[Pol78] Pollard, J.M., Monte Carlo methods for index computation (mod p). Math. Comp. 32 (1978), 918–924.

[Pol00] —, Kangaroos, Monopoly and discrete logarithms. J. Cryptology 13 (2000), 437–447.

[SS85] Sattler, J., Schnorr, C.P., Generating random walks in groups. Ann. Univ. Sci. Budapest. Sect. Comput. 6 (1985), 65–79.

[SL84] Schnorr, C.P., Lenstra, Jr., H.W., A Monte Carlo factoring algorithm with linear storage. Math. Comp. 43 (1984), 289–311.

[Sha71] Shanks, D., Class number, a theory of factorization and genera. In: 1969 Number Theory Institute (Proc. Symp. Pure Math. 20), 415–440. Amer. Math. Soc., Providence 1971.

[Sho97] Shoup, V., Lower bounds for discrete logarithms and related problems. In: Advances in Cryptology – EUROCRYPT '97 (ed. by W. Fumy; Lecture Notes in Comput. Sci. 1233), 256–266. Springer, Berlin 1997.

[Sti] Stinson, D., Some baby-step giant-step algorithms for the low Hamming weight discrete logarithm problem. Math. Comp., posted on 7 March 2001, PII S 0025-5718(01)01310-2 (to appear in print).

[Ter00] Terr, D.C., A modification of Shanks' baby-step giant-step algorithm. Math. Comp. 69 (2000), 767–773.

[Tes98a] Teske, E., A space efficient algorithm for group structure computation. Math. Comp. 67 (1998), 1637–1663.

[Tes98b] —, Speeding up Pollard's rho method for computing discrete logarithms. In: Algorithmic Number Theory – ANTS III (ed. by J.P. Buhler; Lecture Notes in Comput. Sci. 1423), 541–554. Springer, Berlin 1998.

[Tes00] —, On random walks for Pollard's rho method. Math. Comp. 70 (2001), 809–825.

[Tes01] —, Computing discrete logarithms with the parallelized kangaroo method. Research Report CORR 01-01, Department of Combinatorics and Optimization, University of Waterloo, Waterloo, Ontario, Canada, 2001. 18 pages.

[WZ98] Wiener, M., Zuccherato, R., Faster attacks on elliptic curve cryptosystems. In: Selected Areas in Cryptography (SAC '98) (ed. by S. Tavares, H. Meijer; Lecture Notes in Comput. Sci. 1556), 190–200. Springer, Berlin 1999.

Height functions on elliptic curves

Horst G. Zimmer

1. Introduction

Let K be a global field with $\mathcal{M}_K$, a set of pairwise inequivalent additive absolute values v satisfying the sum formula (resp. product formula)

$$\sum_{v \in \mathcal{M}_k} \lambda_v v(z) = 0 \quad (\text{resp.} \prod_{v \in \mathcal{M}_k} |z|_v^{\lambda_v} = 1)$$

with multiplicities $\lambda_v \in \mathbb{R}_{>0}$ (see [5], [7]). Here, we have set $v(z) := -\log |z|_v$. Further, we may take

$$\lambda_v = \begin{cases} n_v = [K_v : \mathbb{Q}_p] \text{ if } K \text{ is an algebraic number field} \\ \qquad \text{with local degree } n_v \\ f_v = \text{residue degree of } K \text{ if } K/k \text{ is an algebraic} \\ \qquad \text{function field with field of constants } k. \end{cases}$$

In this context, K_v is the completion of K with respect to v, $K_v = \mathbb{R}$ or $\mathbb{C}$ if v is an archimedean absolute value and $\mathbb{Q}_\infty := \mathbb{R}$, so that

$$n_v = \begin{cases} [\mathbb{R} : \mathbb{R}] = 1 \\ [\mathbb{C} : \mathbb{R}] = 2 \end{cases}$$

if v is archimedean. In the number field case, the restriction of v to $\mathbb{Q}$ is $v|_\mathbb{Q} = v_p$, where p is a prime of $\mathbb{Q}$ or ∞. Note that, in any case, $n_v = e_v f_v$, where e_v is the ramification order and f_v is the residue degree of K (cf. [5]).

Let

$$E: \quad Y^2 + a_1XY + a_3Y = X^3 + a_2X^2 + a_4X + a_6 \qquad (a_i \in K)$$

be an elliptic curve in general Weierstrass form over a global field K. Tate's quantities are

$$\begin{aligned} b_2 &= a_1^2 + 4a_2, \\ b_4 &= 2a_4 + a_1a_3, \\ b_6 &= a_3^2 + 4a_6, \\ b_8 &= a_1^2a_6 + 4a_2a_6 - a_1a_3a_4 + a_2a_3^2 - a_4^2 \end{aligned}$$

Public-Key Cryptography and Computational Number Theory
K. Alster, J. Urbanowicz and H. C. Williams (Eds.)

so that

$$\begin{aligned} 4b_8 &= b_2b_6 - b_4^2. \\ c_4 &= b_2^2 - 24b_4, \\ c_6 &= -b_2^3 + 36b_2b_4 - 216b_6. \end{aligned}$$

The discriminant of E/K is

$$\Delta = -b_2^2b_8 - 8b_4^3 - 27b_6^2 + 9b_2b_4b_6$$

so that $12^3\Delta = c_4^3 - c_6^2$, and the absolute invariant of E/K is

$$j = \frac{c_4^3}{\Delta}.$$

Suppose that the characteristic of $K \neq 2$. Then the transformation

$$X \mapsto X, \quad Y \mapsto \frac{1}{2}(Y - a_1X - a_3)$$

leads to a defining equation

$$Y^2 = 4X^3 + b_2X^2 + 2b_4X + b_6$$

for E/K, and if the characteristic of $K \neq 2, 3$, then the transformation

$$X \mapsto \frac{1}{36}(X - 3b_2), \quad Y \mapsto \frac{1}{108}Y$$

yields a defining equation

$$Y^2 = X^3 - 27c_4X - 54c_6$$

for E/K (cf. [11]).

According to the Mordell–Weil Theorem,

$$E(K) \cong E_{\text{tors}}(K) \oplus \mathbb{Z}^r$$

is a finitely generated additive abelian group if K is an algebraic number field or an algebraic function field and the elliptic curve is non-constant over K (i.e., not defined over k) in the latter case (see [7], [15]). If $P = (x, y) \in E(K)$, then by the universal multiplication formulae, the m-multiple ($m \in \mathbb{N}$) of P is

$$mP = \left(\frac{\Phi_m(X)}{\Psi_m^2(X)}, \frac{\Omega_m(X)}{\Psi_m^3(X)}\right) \in E(K)$$

with universal polynomials Φ_m, Ψ_m, $\Omega_m \in \mathbb{Z}[X]$ in one variable X over $\mathbb{Q}$ where, more precisely, the other variable Y over $\mathbb{Q}$ appears in Ψ_m if m is even (resp. in Ω_m if m is odd) provided that the characteristic of K is different from 2. For those polynomials, we have the following recursion formulae:

$$\begin{aligned} \Psi_1 &= 1, \ \Phi_1 = X, \ \Omega_1 = Y, \\ \Psi_2 &= 2Y + a_1X + a_3, \ \Phi_2 = X^4 - b_4X^2 - 2b_6X - b_8, \\ \Psi_3 &= 3X^4 + b_2X^3 + 3b_4X^2 + 3b_6X + b_8, \\ \Psi_4 &= \Psi_2\big(2X^6 + b_2X^5 + 5b_4X^4 + 10b_6X^3 + 10b_8X^2 \\ &\qquad + (b_2b_8 - b_4b_6)X + (b_4b_8 - b_6^2)\big). \end{aligned}$$

For $m \geq 2$,

$$\begin{aligned}
\Phi_m &= x\Psi_m^2 - \Psi_{m-1}\Psi_{m+1},\\
\Psi_{2m+1} &= \Psi_m^3\Psi_{m+2} - \Psi_{m-1}\Psi_{m+1}^3,\\
\Psi_{2m} &= 2\Psi_m\Omega_m + \Psi_m^2(a_1\Phi_m + a_3\Psi_m^2),\\
\Psi_2\Psi_{2m} &= \Psi_m(\Psi_{m-1}^2\Psi_{m+2} - \Psi_{m-2}\Psi_{m+1}^2),\\
2\Psi_2\Omega_m &= \Psi_{m-1}^2\Psi_{m+2} - \Psi_{m-2}\Psi_{m+1}^2 - \Psi_2\Psi_m(a_1\Phi_m + a_3\Psi_m^2),
\end{aligned}$$

where

$$\Psi_2^2(X) = 4X^3 + b_2X^2 + 2b_4X + b_6$$

(cf. the equation for E/K if $\mathrm{char}(K) \neq 2$).

2. The various height functions

Logarithmic heights are additive. For $P = (x_0 : x_1 : \ldots : x_n) \in \mathbb{P}_K^n$, the projective space of dimension n over K, we have locally at v

$$H_v(P) = \max\{|x_0|_v, |x_1|_v, \ldots |x_n|_v\}$$

and globally (a posteriori)

$$H(P) = \prod_{v\in\mathcal{M}_K} H_v(P)^{\lambda_v}.$$

Taking logarithms yields locally

$$h_v(P) = \log H_v(P) = -\min\{-\log|x_0|_v, -\log|x_1|_v, \ldots, -\log|x_n|_v\}$$

and then globally

$$h(P) = \log H(P) = \sum_{v\in\mathcal{M}_K} \lambda_v h_v(P).$$

We have

$$v(x_i) = -\log|x_i|_v \quad (i = 0, 1, \ldots, n)$$

so that

$$h_v(P) = -\min\{v(x_0), v(x_1), \ldots, v(x_n)\}.$$

Put, at $v \in \mathcal{M}_K$,

$$\mu_v = \min\Big\{v(b_2), \frac{1}{2}v(b_4), \frac{1}{3}v(b_6), \frac{1}{4}v(b_8)\Big\}.$$

The μ_v's have to do with the *additive height* of the elliptic curve E/K. Observe that for $x \in K$, one has $n = 1$ and $P = (1 : x) \in \mathbb{P}_K^1$. Define now the *local Weil height* of $P = (x, y) \in E(K)$ by

$$h_v(P) = -\frac{1}{2}\min\{0, v(x)\}$$

and the *modified local Weil height* by

$$d_v(P) = -\frac{1}{2}\min\{\mu_v, v(x)\}$$

(with a normalizing factor $\frac{1}{2}$) setting $d_v(P) := -\frac{1}{2}\mu_v$ for $P = \mathcal{O} = (\infty, \infty)$. The *global Weil height* of $P \in E(K)$ is then

$$h(P) = \sum_{v \in \mathcal{M}_K} \lambda_v h_v(P)$$

and the *modified global Weil height*

$$d(P) = \sum_{v \in \mathcal{M}_K} \lambda_v d_v(P).$$

Both heights h and d are good for calculations, but bad for the theory.

Now we can define the *global Néron–Tate height* in two ways:

$$\widehat{h}(P) = \lim_{m \to \infty} \frac{h(2^m P)}{2^{2m}} = \lim_{m \to \infty} \frac{d(2^m P)}{2^{2m}}.$$

The height $\widehat{h}$ is conversely bad for calculations, but good for the theory.

The reason becomes clear from the following observation.

Locally, the Néron–Tate height is (see [12], [15], [16])

$$\widehat{h}_v(P) = \lim_{m \to \infty} \left\{ \frac{d_v(2^m P) - \frac{1}{2}v(\Psi_{2^m}^2(x))}{2^{2m}} \right\} + \frac{1}{12}v(\Delta).$$

Of course, $\widehat{h}(P) = \sum_{v \in \mathcal{M}_K} \lambda_v \widehat{h}_v(P)$. Furthermore, as can be proven (see [12]),

$$\frac{1}{6}(6\mu_v - v(\Delta)) + \frac{4}{3}\alpha_v \leq \widehat{h}_v(P) - \left\{d_v(P) + \frac{1}{12}v(\Delta)\right\} \leq -\frac{1}{2}\alpha_v$$

with $\alpha_v = -\log(2)$. This observation shows why one should choose d_v (resp. d) or $\widehat{h}_v$ (resp. $\widehat{h}$) depending on the purpose one wants to pursue.

The function $\widehat{h}$ has the following properties. $\widehat{h}$ is a quadratic form on $E(K)$, that is:

(Ia) for P, $Q \in E(K)$, $\widehat{h}(P + Q) + \widehat{h}(P - Q) = 2\widehat{h}(P) + 2\widehat{h}(Q)$,

(Ib) for $P \in E(K)$, $\widehat{h}(mP) = m^2\widehat{h}(P)$,

(II) $\widehat{h}$ is invariant under isomorphisms of E/K,

(III) $\widehat{h}(P) = 0 \Leftrightarrow P \in E_{\mathrm{tors}}(K)$ if K is a number field.

Property (Ib) is implied by (Ia), property (III) by (Ib).

3. Estimates between the Néron–Tate height and the two Weil heights

We shall need the following formulae. Let $P = (x_P, y_P)$, $Q = (x_Q, y_Q) \in E(K)$.

(1) For $P = Q$,

$$x_{2P} = \frac{\Phi_2(x_P)}{\Psi_2^2(x_P)} = \frac{x_P^4 - b_4 x_P^2 - 2b_6 x_P - b_8}{4x_P^3 + b_2 x_P^2 + 2b_4 x_P + b_6}.$$

(2) For $P \neq \pm Q$,

$$x_{P-Q} x_{P+Q} = \frac{x_P^2 x_Q^2 - b_4 x_P x_Q - b_6(x_P + x_Q) - b_8}{(x_P - x_Q)^2}.$$

(3) For $P \neq \pm Q$,

$$x_{P+Q} + x_{P-Q} = \frac{2x_P^2 x_Q + 2x_P x_Q^2 + b_2 x_P x_Q + b_4(x_P + x_Q) + b_6}{(x_P - x_Q)^2}.$$

(4) For $P \neq \pm Q$,

$$x_{P+Q} - x_{P-Q} = -\frac{\Psi_2(x_P, y_P)\Psi_2(x_Q, y_Q)}{(x_P - x_Q)^2},$$

(5) $$\begin{aligned}\Delta = {} & [-48X^2 - 8b_2 X + (b_2^2 - 32b_4)]\Phi_2(X) \\ & + [12X^3 - b_2 X^2 - 10b_4 X + (b_2 b_4 - 27b_6)]\Psi_2^2(X),\end{aligned}$$

$(\Rightarrow (\Phi_2, \Psi_2^2) = 1)$

(6) $$\begin{aligned}\Delta X^7 = {} & [\Delta X^3 + 4(b_2 b_4 b_8 - b_6 b_8 - b_2 b_6^2)X^2 \\ & + (12b_4^2 b_8 - b_2 b_6 b_8 - 11b_4 b_6^2)X + 6b_6(b_4 b_8 - b_6^2)]\Phi_2(X) \\ & + [-(b_2 b_4 b_8 - b_6 b_8 - b_2 b_6^2)X^3 + (5b_4^2 b_8 - b_2 b_6 b_8 - 4b_4 b_6^2)X^2 \\ & + (13b_4 b_6 b_8 - b_2 b_8^2 - 12b_6^3)X + 6b_8(b_4 b_8 - b_6^2)]\Psi_2^2(X).\end{aligned}$$

On the basis of these equations, we get

Theorem 1. *Locally,*

(a) $$\begin{aligned}\tfrac{1}{2}\big(6\mu_v - v(\Delta)\big) + 5\alpha_v \le {} & d_v(P+Q) + d_v(P-Q) \\ & - 2d_v(P) - 2d_v(Q) - v(x_P - x_Q) \le -2\alpha_v \quad \textit{for } P \neq \pm Q,\end{aligned}$$

(b) $$\begin{aligned}& \tfrac{1}{2}\big(6\mu_v - v(\Delta)\big) + 4\alpha_v \\ & \le d_v(2P) - 4d_v(P) - v(\Psi_2(x_P, y_P)) \le -\frac{3}{2}\alpha_v \qquad \textit{for } P = Q,\end{aligned}$$

where $\alpha_v = -\log(2)$ *for* v *archimedean and* 0 *for* v *non-archimedean.*

Proof (cf. [12]). Put, for $P, Q \in E(\mathbb{Q})$, locally at $v \in \mathcal{M}_K$

$$\begin{aligned} d_v(P,Q) &= d_v(P+Q) + d_v(P-Q) - 2d_v(P) - 2d_v(Q) - v(x_P - x_Q) \quad \text{for } P \neq \pm Q, \\ d_v(P,P) &= d_v(2P) - 4d_v(P) - v(\Psi_2(x_P, y_P)) \quad \text{for } P = Q \end{aligned}$$

and globally

$$\begin{aligned} d(P,Q) &= d(P+Q) + d(P-Q) - 2d(P) - 2d(Q) \quad \text{for } P \neq \pm Q, \\ d(P,P) &= d(2P) - 4d(P) \quad \text{for } P = Q, \end{aligned}$$

so that by the sum formula

$$\begin{aligned} d(P,Q) &= \sum_{v \in \mathcal{M}_K} \lambda_v d_v(P,Q) \quad \text{for } P \neq \pm Q, \\ d(P,P) &= \sum_{v \in \mathcal{M}_K} \lambda_v d_v(P,P) \quad \text{for } P = Q. \end{aligned}$$

We first prove assertion (a).

(i) Estimate above:

Let $P, Q \in E(K)$, $P, Q, P \pm Q \neq \mathcal{O}$, and take $v \in \mathcal{M}_K$. We distinguish four cases.

Case 1. $v(x_{P+Q}) < \mu_v$, $v(x_{P-Q}) < \mu_v$.

Then $d_v(P \pm Q) = -\frac{1}{2} v(x_{P \pm Q})$. Formula (2) yields

$$\begin{aligned} &v(x_{P+Q}) + v(x_{P-Q}) \\ &\quad \geq \min\{2v(x_P) + 2v(x_Q), 2\mu_v + v(x_P) + v(x_Q), \\ &\qquad\qquad 3\mu_v + v(x_P), 3\mu_v + v(x_Q), 4\mu_v\} + 3\alpha_v - 2v(x_P - x_Q) \\ &\quad \geq \min\{2v(x_P), \mu_v + v(x_P), 2\mu_v\} \\ &\qquad\qquad + \min\{2v(x_Q), \mu_v + v(x_Q), 2\mu_v\} + 3\alpha_v - 2v(x_P - x_Q) \\ &\quad = 2\min\{\mu_v, v(x_P)\} + 2\min\{\mu_v, v(x_Q)\} + 3\alpha_v - 2v(x_P - x_Q). \end{aligned}$$

This shows that

$$d_v(P+Q) + d_v(P-Q) \leq 2d_v(P) + 2d_v(Q) - \frac{3}{2}\alpha_v + v(x_P - x_Q)$$

and thus

$$d_v(P,Q) \leq -\frac{3}{2}\alpha_v \leq -2\alpha_v.$$

Case 2. $v(x_{P+Q}) < \mu_v$, $v(x_{P-Q}) \geq \mu_v$.

Then $d_v(P+Q) = -\frac{1}{2} v(x_{P+Q})$, $\quad d_v(P-Q) = -\frac{1}{2}\mu_v$. From formula (3) we derive

$$\begin{aligned} x_{P+Q} = (x_P - x_Q)^{-2}(&-x_{P-Q}x_P^2 + 2x_{P-Q}x_Px_Q - x_{P-Q}x_Q^2 \\ &+ 2x_P^2x_Q + 2x_Px_Q^2 + b_2x_Px_Q + b_4(x_P + x_Q) + b_6). \end{aligned}$$

Hence

$$\begin{aligned} v(x_{P+Q}) + \mu_v \geq \min\{&\mu_v + 2v(x_P), \mu_v + v(x_P) + v(x_Q), \mu_v + 2v(x_Q), \\ &2v(x_P) + v(x_Q), v(x_P) + 2v(x_Q), \mu_v + v(x_P) + v(x_Q), \\ &2\mu_v + v(x_P), 2\mu_v + v(x_Q), 3\mu_v\} \\ &+ 4\alpha_v + \mu_v - 2v(x_P - x_Q) \\ \geq\ & 2\min\{\mu_v, v(x_P)\} + 2\min\{\mu_v, v(x_P)\} + 4\alpha_v - 2v(x_P - x_Q). \end{aligned}$$

In this case, we obtain therefore

$$d_v(P+Q) + d_v(P-Q) \leq 2d_v(P) + 2d_v(Q) - 2\alpha_v + v(x_P - x_Q)$$

and thus

$$d_v(P,Q) \leq -2\alpha_v.$$

Case 3. $v(x_{P+Q}) \geq \mu_v$, $v(x_{P-Q}) < \mu_v$.

For symmetry reasons, the same arguments as in case 2 lead to the same estimate as in case 2.

Case 4. $v(x_{P+Q}) \geq \mu_v$, $v(x_{P-Q}) \geq \mu_v$.

Of course, $d_v(P \pm Q) = -\frac{1}{2}\mu_v$. We have trivially

$$\begin{aligned} v(x_P - x_Q) &\geq \min\{\mu_v, v(x_P)\} + \min\{\mu_v, v(x_Q)\} - \mu_v + \alpha_v \\ &= \min\{\mu_v, v(x_P)\} + \min\{\mu_v, v(x_P)\} \\ &\quad - \frac{1}{2}\min\{\mu_v, v(x_{P+Q})\} - \frac{1}{2}\min\{\mu_v, v(x_{P-Q})\} + \alpha_v, \end{aligned}$$

since for any point $R \in E(K)$,

$$-\min\{\mu_v, v(x_R)\} \geq -\mu_v.$$

This entails

$$v(x_P - x_Q) \geq -2d_v(P) - 2d_v(Q) + d_v(P+Q) + d_v(P-Q) + \alpha_v$$

and thus

$$d_v(P,Q) \leq -\alpha_v \leq -2\alpha_v.$$

(ii) Estimate below:

From the estimate above, we know that

$$d_v(P+Q, P-Q) \leq -2\alpha_v.$$

Therefore, according to the definition of $d_v(P+Q, P-Q)$,

$$\begin{aligned} (7)\qquad & d_v(2P) + d_v(2Q) \\ &\leq 2d_v(P+Q) + 2d_v(P-Q) + v(x_{P+Q} - x_{p-Q}) - 2\alpha_v \\ &= 2d_v(P+Q) + 2d_v(P-Q) + v(\Psi_2(x_P, y_P)) \\ &\quad + v(\Psi_2(x_Q, y_Q)) - 2v(x_P - x_Q) - 2\alpha_v, \end{aligned}$$

the latter by formula (4).

For $d_v(2P)$ (and $d_v(2Q)$), in view of (7), it suffices to derive the following estimates below. We consider two cases.

Case 1. $v(x_P) < \mu_v$.

Then from (6) and the multiplication formula (1) for x_{2P},

$$\begin{aligned} v(\Delta) + 7v(x_P) \geq \min\bigl\{&6\mu_v + 3v(x_P) + 7\alpha_v + v(\Phi_2(x_P)), \\ &7\mu_v + 3v(x_P) + 6\alpha_v + v(\Psi_2^2(x_P))\bigr\} + \alpha_v \\ \geq\ & 6\mu_v + 3v(x_P) + v(\Psi_2^2(x_P)) + 8\alpha_v + \min\{\mu_v, v(x_{2P})\}. \end{aligned}$$

Hence

$$d_v(2P) - 4d_v(P) - v(\Psi_2(x_P, y_P)) \geq \frac{1}{2}\bigl(6\mu_v - v(\Delta)\bigr) + 4\alpha_v$$

and thus

$$d_v(P, P) \geq \frac{1}{2}\bigl(6\mu_v - v(\Delta)\bigr) + 4\alpha_v.$$

Case 2. $v(x_P) \geq \mu_v$.

Then, from (5) and the multiplication formula (1) for x_{2P},

$$\begin{aligned} v(\Delta) &\geq \min\{2\mu_v + 7\alpha_v + v(\Phi_2(x_P)), 3\mu_v + 6\alpha_v + v(\Psi_2^2(x_P))\} + \alpha_v \\ &\geq 2\mu_v + 8\alpha_v + v(\Psi_2^2(x_P)) + \min\{\mu_v, v(x_{2P})\}. \end{aligned}$$

This implies again

$$d_v(2P) - 4d_v(P) - v(\Psi_2(x_P, y_P)) \geq \frac{1}{2}\bigl(6\mu_v - v(\Delta)\bigr) + 4\alpha_v$$

and thus

$$d_v(P, P) \geq \frac{1}{2}\bigl(6\mu_v - v(\Delta)\bigr) + 4\alpha_v.$$

(Of course, the inequalities derived in these two cases hold as well for Q in place of P.)

We insert these estimates below for the functions $d_v(2P)$ and $d_v(2Q)$ in (7) and get

$$\begin{aligned} 4d_v(P) + 4d_v(Q) + v(\Psi_2(x_P, y_P)) + v(\Psi_2(x_Q, y_Q)) + \bigl(6\mu_v - v(\Delta)\bigr) + 8\alpha_v \\ \leq 2d_v(P+Q) + 2d_v(P-Q) + v(\Psi_2(x_P, y_P)) + v(\Psi_2(x_Q, y_Q)) \\ - 2v(x_P - x_Q) - 2\alpha_v. \end{aligned}$$

From this inequality it follows that

$$\frac{1}{2}\bigl(6\mu_v - v(\Delta)\bigr) + 5\alpha_v \leq d_v(P, Q).$$

Altogether, this proves assertion (a).

To prove the easier assertion (b), we first note that

$$d_v(2P) - 4d_v(P) - v(\Psi_2(x_P, y_P)) \geq \frac{1}{2}\big(6\mu_v - v(\Delta)\big) + 4\alpha_v$$

has already been shown thus establishing the estimate below. For the estimate above, we again distinguish two cases.

Case 1. $v(x_{2P}) < \mu_v$.

From formula (1), we derive that

$$v(x_{2P}) \geq 4\min\{\mu_v, v(x_P)\} + 3\alpha_v - v(\Psi_2^2(x_P))$$

and hence

$$d_v(2P) \leq 4d_v(P) + v(\Psi_2(x_P, y_P)) - \frac{3}{2}\alpha_v.$$

Case 2. $v(x_{2P}) \geq \mu_v$.

Since, by the multiplication formulae,

$$\Psi_2^2(x_P) = 4x_P^3 + b_2 x_P^2 + 2b_4 x_P + b_6,$$

we have

$$\begin{aligned} v(\Psi_2^2(x_P)) &\geq 3\min\{\mu_v, v(x_P)\} + 3\alpha_v \\ &\geq 4\min\{\mu_v, v(x_P)\} - \mu_v + 3\alpha_v. \end{aligned}$$

Hence,

$$d_v(2P) - 4d_v(P) - v(\Psi_2(x_P, y_P)) \leq -\frac{3}{2}\alpha_v$$

and thus

$$d_v(P, P) \leq -\frac{3}{2}\alpha_v.$$

In both cases, assertion (b) is established. □

Put

$$\mu = -\sum_{v \in \mathcal{M}_K} \lambda_v \mu_v, \quad \alpha = -\sum_{v \in \mathcal{M}_K} \lambda_v \alpha_v \in \mathbb{R}.$$

Then, observing the relations $v(b_{2\nu}) \geq \nu\mu_v$, $(\nu = 1, 2, 3, 4)$ and $v(0) = \infty$ for $v \in \mathcal{M}_K$, we see that

$$\mu \geq 0, \quad \alpha \geq 0.$$

More precisely, we have

$$\alpha = \left\{ \begin{array}{ll} [K : \mathbb{Q}] \log(2) & \text{in the number field case} \\ 0 & \text{in the function field case} \end{array} \right\}.$$

Theorem 2. *Globally,*

(a) $-3\mu - 5\alpha \le d(P+Q) + d(P-Q) - 2d(P) - 2d(Q) \le 2\alpha$ *for* $P \ne Q$,

(b) $-3\mu - 4\alpha \le d(2P) - 4d(P) \le \frac{3}{2}\alpha$ *for* $P = Q$.

Proof. This is clear from Theorem 1 and by the sum formula. □

The latter identity (Theorem 2(b)) immediately gives

Theorem 3. *Globally,*

$$-\mu - \frac{4}{3}\alpha \le \widehat{h}(P) - d(P) \le \frac{1}{2}\alpha \quad \textit{for } P \in E(K).$$

Notice that

$$\frac{d(2^m P)}{2^{2m}} - d(P) = \sum_{\nu=1}^{m} \frac{d(2^{\nu-1}P, 2^{\nu-1}P)}{2^{2\nu}},$$

where $\sum_{\nu=1}^{\infty} \frac{1}{2^{2\nu}} = \frac{1}{3}$.

We now define

$$\mathcal{M}_K^0 := \text{non-archimedean places of } K,$$
$$\mathcal{M}_K^\infty := \text{archimedean places of } K$$

so that

$$\mathcal{M}_K = \mathcal{M}_K^0 \cup \mathcal{M}_K^\infty.$$

Further, we put

$$\mu^0 := \sum_{v \in \mathcal{M}_K^0} \lambda_v \mu_v, \quad \mu^\infty = -\sum_{v \in \mathcal{M}_K^\infty} \lambda_v \mu_v$$

so that

$$\mu = -\mu^0 + \mu^\infty.$$

(Trivially, we put $\alpha^0 = 0$, $\alpha^\infty = \alpha$.)

Then, if the coefficients a_i of E/K and hence the b_i and the c_i are integral elements of K, we have globally the following assertion (see [10]).

Proposition 1. *Suppose that the coefficients a_i of the curve E/K are integers in K. Then,*

$$-\frac{1}{2}\mu^0 \le d(P) - h(P) \le \frac{1}{2}\mu^\infty \quad \textit{for } P \in E(K).$$

These inequalities, together with Theorem 3, lead to a coarse result:

Theorem 4. *Globally,*

$$-\frac{1}{2}(\mu + \mu^\infty) - \frac{4}{3}\alpha \le \widehat{h}(P) - h(P) \le \frac{1}{2}(\mu^\infty + \alpha) \quad \textit{for } P \in E(K).$$

If E has a special equation, one gets better estimates. Especially if E is one of the special CM–curves over $\mathbb{Q}$

$$(8) \qquad E: \quad y^2 = x^3 + m, \qquad m \in \mathbb{N}, \quad \text{(Mordell's equation)}$$

or

$$(9) \qquad E: \quad y^2 = x(x^2 - m), \qquad m \in \mathbb{N}.$$

Then we have

$$\mu = 0,$$

so that in these cases, Theorem 3 reads

$$-\frac{4}{3}\alpha \le \widehat{h}(P) - d(P) \le \frac{1}{2}\alpha \qquad \text{for } P \in E(K).$$

This result is independent of the parameter

$$m \in K.$$

Grigorov and Rizov [4] showed for the curve (9) over the field $K := \mathbb{Q}$, that

$$|\widehat{h}(P) - h(P)| \le \frac{1}{2}\alpha \qquad \text{for } P \in \Gamma,$$

Γ being a certain subgroup of $E(\mathbb{Q})$. But remarkably the estimate is independent of the parameter $m \in \mathbb{Z}_{>2}$ and hence of the curve (9).

That is why we want to plead for the use of the modified Weil height d instead of the ordinary Weil height h, even though the modified Weil height contains some sort of the height of the elliptic curve E/K and hence in a way depends on E/K.

4. Motivation

Why does one want such estimates? They have important applications! We mention here only a few of them.

- Determination of the Mordell–Weil group $E(K)$. Mordell–Weil's theorem says:

$$E(K) \cong E_{\text{tors}}(K) \oplus \mathbb{Z}^r$$

if E is non-constant over K in the function field case, where $E_{\text{tors}}(K)$ is a *finite group* and $r \in \mathbb{N}_0$ is the *rank* of E/K. If K is a number field (also if K/k is a function field), we have the Conjecture of Birch and Swinnerton-Dyer:

$$\lim_{s\to 1} \frac{L(s, E/K)}{(s-1)^r} = \frac{\omega R |Ш(E/K)| \prod_{v \in \mathcal{M}_K} c_v}{|E_{\text{tors}}(K)|^2} \frac{2^{r_2}}{\sqrt{D_K}}.$$

In particular, $L(s, E/K)$ is conjectured to possess a zero of order r at $s = 1$. Here, R is the *regulator* of E/K given by the Néron–Tate heights of a basis (of infinite order) $P_1, \ldots, P_r$ of $E(K)$, ω is the *real period*, $Ш(E/K)$ denotes the

Tate–Shafarevich group of E/K, c_v are the *Tamagawa numbers* of E/K, and D_K denotes the *discriminant* of K and r_2 the number of complex embeddings of K.

The program of determining the Mordell–Weil group $E(K)$, if K is a number field, was carried out by Susanne Schmitt [9]. She used the Birch and Swinnerton-Dyer conjecture assuming it to be true.

- Calculation of all integral and, more generally, all S-integral points on an elliptic curve E/K if $K = \mathbb{Q}$, especially if E is given in short Weierstrass normal form (cf. the end of this paper):

$$E: \quad y^2 = x^3 + ax + b \qquad (a, b \in \mathbb{Q}).$$

By a theorem of Siegel, generalized by Mahler, both these sets are finite, but the proofs are not constructive. Later on A. Baker [1] and J. Coates [2] proved effective versions of this result but the bounds for integral and S-integral points should be efficient, and examples should be given.

The latter program was carried out, among others, by A. Pethő, J. Gebel, E. Herrmann and the author (see [3] and [8]).

However, here only the (trivial) right hand inequality of Theorem 3 is utilized.

- The constructive version of Falting's theorem (in very special cases) for $K = \mathbb{Q}$.

Kulesz [6], employing the method of Dem'janenko–Manin (as opposed to the method of Chabauty), proved the following:

Suppose that the elliptic curve (of genus $g = 1$)

$$E_{m^2}: \quad y^2 = x(x^2 - m^2), \qquad m \in \mathbb{N},$$

has rank $r = 1$ over $\mathbb{Q}$. Consider the parametrized curves

$$C_m: \quad my^2 = 6(x^2 - 4)(x^2 + 2)(x^2 + 8)$$

of genus $g = 2$ over $\mathbb{Q}$ and

$$D_m: \quad my^2 = -x(x^2 - 1)(x + 2)(2x + 1)(x^2 + x + 1)$$

of genus $g = 3$ over $\mathbb{Q}$. Then, the rational points are:

$$\begin{aligned} C_m(\mathbb{Q}) &= \{(2,0), (-2,0)\} \text{ if } m \neq 6, \\ C_m(\mathbb{Q}) &= \{(\pm 2, 0), (\pm 4, \pm 72), \pm\infty\} \text{ if } m = 6, \\ D_m(\mathbb{Q}) &= \{(0,0), (\pm 1, 0), (-2, 0), (-\tfrac{1}{2}, 0)\}. \end{aligned}$$

Here, he referred to the height estimates of Grigorov and Rizov [4] mentioned earlier.

5. A special Weierstrass form

For the special Weierstrass form

$$E': \quad Y^2 = X(X^2 + cX + d) \qquad (c, d \in K)$$

over K, a global field of characteristic $\neq 2$, the point $P_0 = (0,0) \in E(K)$ is a torsion point of order 2. This is one reason for the importance of these curves. We have then

$$\begin{aligned} b_2 &= 4c, \\ b_4 &= 2d, \\ b_6 &= 0, \\ b_8 &= -d^2, \\ c_4 &= 16(c^2-3d), \\ c_6 &= -32(2c^3-9cd), \\ \Delta' &= -d^2(c^2-4d) = \frac{1}{16}\Delta \neq 0, \\ j &= 16^2\frac{(c^2-3d)^3}{-\Delta'}. \end{aligned}$$

In this situation, the formulae needed read:

(1′) For $P = Q$, $\quad x_{2P} = \dfrac{\Phi_2(x_P)}{\Psi_2^2(x_P)} = \dfrac{x_P^4 - 2dx_P^2 + d^2}{4x_P^3 + 4cx_P^2 + 4dx_P}$.

(2′) For $P \neq \pm Q$, $\quad x_{P+Q}x_{P-Q} = \dfrac{x_P^2x_Q^2 - 2dx_Px_Q + d^2}{(x_P - x_Q)^2}$.

(3′) For $P \neq \pm Q$,

$$x_{P+Q} + x_{P-Q} = \frac{2x_P^2x_Q + 2x_Px_Q^2 + 4cx_Px_Q + 2d(x_P + x_Q)}{(x_P - x_Q)^2}.$$

(4′) For $P \neq \pm Q$, $\quad x_{P+Q} - x_{P-Q} = -\dfrac{\Psi_2(x_P, y_P)\psi_2(x_Q, y_Q)}{(x_P - x_Q)^2}$

with

$$\Psi_2^2(X) = 4X^3 + 4cX^2 + 4dX.$$

(5′) $\Delta' = (-3X^2 - 2cX + (c^2-4d))\Phi_2(X) + 3X^3 - cX^2 - 5dX + 2cd) \cdot \dfrac{\Psi_2^2(X)}{4}$.

(6′) $\Delta' X^7 = (\Delta' X^3 - 2cd^3X^2 - 3d^4X)\Phi_2(X)$

$$+ (2cd^3x_P^3 - 5d^4x_P^2 - cd^4X + 3d^5) \cdot \frac{\Psi_2^2(X)}{4}$$

with $\dfrac{\Psi_2^2(X)}{4} = X^3 + cX^2 + dX = X(X^2 + cX + d) = Y^2$.

Locally, we put

$$\mu_v' := \min\Big\{v(c), \frac{1}{2}v(d)\Big\}.$$

Globally, we put then

$$\mu' = -\sum_{v\in\mathcal{M}_K} \lambda_v \mu'_v \geq 0.$$

We remark that μ'_v deviates from μ_v only by rational multiple of the value $v(2)$.

Locally, for $P = (x, y) \in E'(K)$, we set further

$$d'_v(P) = -\frac{1}{2}\min\{\mu'_v, v(x)\}$$

and define $d'_v(P, Q)$, $d'_v(P, P)$ as in the general situation.

Globally, we set then

$$d'(P) = \sum_{v\in\mathcal{M}_K} \lambda_v d'_v(P)$$

and again define $d'(P, Q)$, $d'(P, P)$ as in the general situation.

Clearly,

$$\widehat{h}(P) = \lim_{m\to\infty} \frac{d'(2^m P)}{2^{2m}}$$

(see Theorem 3′ below).

Theorem 1′. *Locally,*

(a) $$\frac{1}{2}(6\mu'_v - v(\Delta')) + \frac{7}{2}\alpha_v - v_*(2) \leq d'_v(P+Q) + d'_v(P-Q) - 2d'_v(P) - 2d'_v(Q) - v(x_P - x_Q) \leq -2\alpha_v \quad \textit{for } P \neq \pm Q \textit{ in } E'(K),$$

(b) $$\frac{1}{2}(6\mu'_v - v(\Delta')) + \frac{5}{2}\alpha_v - v_*(2) \leq d'_v(2P) - 4d'_v(P) - v(\Psi_2(x_p, y_p)) \leq -2\alpha_v \qquad \textit{for } P = Q \textit{ in } E'(K),$$

where

$$v_*(2) = \left\{\begin{array}{ll} v(2) & \textit{if } v \textit{ is non-archimedean} \\ 0 & \textit{if } v \textit{ is archimedean} \end{array}\right\}.$$

Proof. We start with proving (a).

(i) Estimate above.

Case 1. $v(x_{P\pm Q}) < \mu'_v$.

We get from (2′)

$$\begin{aligned} v(x_{P+Q}) + v(x_{P-Q}) &\geq \min\{2v(x_P) + 2v(x_Q), 2\mu'_v + v(x_P) + v(x_Q), 4\mu'_v\} \\ &\quad + 2\alpha_v - 2v(x_P - x_Q) \\ &\geq 2\min\{\mu'_v, v(x_P)\} + 2\min\{\mu'_v, v(x_P)\} \\ &\quad + 2\alpha_v - 2v(x_P - x_Q). \end{aligned}$$

It follows that

$$d'_v(P+Q) + d'_v(P-Q) \le 2d'_v(P) + 2d'_v(Q) - \alpha_v + v(x_P - x_Q).$$

Case 2. $v(x_{P+Q}) < \mu'_v$, $v(x_{P-Q}) \ge \mu'_v$.
We get from (3′)

$$x_{P+Q} = (x_P - x_Q)^{-2}(-x_{P-Q}x_P^2 + 2x_{P-Q}x_Px_Q - x_{P-Q}x_Q^2 \\ + 2x_P^2x_Q + 2x_Px_Q^2 + 4cx_Px_Q + 2d(x_P + x_Q)).$$

Hence,

$$v(x_{P+Q}) + \mu'_v \ge \min\{\mu'_v + 2v(x_P), \mu'_v + v(x_P) + v(x_Q), \\ \mu'_v + 2v(x_Q), 2v(x_P) + v(x_Q), v(x_P) + 2v(x_Q), \\ \mu' + v(x_P) + v(x_Q), 2\mu'_v + v(x_P), 2\mu'_v + v(x_Q)\} \\ + 4\alpha_v - 2v(x_P - x_Q) + \mu'_v.$$

Therefore, as in the general situation,

$$d'_v(P+Q) + d'_v(P-Q) \le 2d'_v(P) + 2d'_v(Q) - 2\alpha_v + v(x_P - x_Q).$$

Case 3, where $v(x_{P+Q}) \ge \mu'_v$, $v(x_{P-Q}) < \mu'_v$, is treated like case 2 and leads to the same result.

Case 4. $v(x_{P\pm Q}) \ge \mu'_v$.
We again have, as in the general situation,

$$v(x_P - x_Q) \ge -2d'_v(P) - 2d'_v(Q) + d'_v(P+Q) + d'_v(P-Q) + \alpha_v.$$

(ii) Estimate below.
From (i) we know that

$$d'_v(P+Q, P-Q) \le -2\alpha_v,$$

that is,

$$d'_v(2P) + d'_v(2Q) \le 2d'_v(P+Q) + 2d'_v(P-Q) + v(x_{P+Q} - x_{P-Q}) - 2\alpha_v.$$

The right hand side of this inequality equals by (4′)

$$(7') \qquad 2d'_v(P+Q) + 2d'_v(P-Q) + v(\Psi_2(x_P, y_P)) + v(\Psi_2(x_Q, y_Q)) \\ - 2v(x_P - x_Q) - 2\alpha_v.$$

Again, in view of (7′), for $d'(2P)$ and also for $d'_v(2Q)$ we derive the subsequent estimates below. We again consider two cases.

Case 1. $v(x_P) < \mu'_v$.
From (6′) and the multiplication formula (1′) for x_{2P},

$$v(\Delta') + 7v(x_P) \ge \min\{6\mu'_v + 3v(x_P) + 3\alpha_v + v(\Psi_2^2(x_P)) + v(x_{2P}), \\ 7\mu'_v + 3v(x_P) + 4\alpha_v + v(\Psi_2^2(x_P)/4)\} + \alpha_v \\ \ge 6\mu'_v + 3v(x_P) + v(\Psi_2^2(x_P)) + 5\alpha_v - 2v(2) + \min\{\mu'_v, v(x_{2P})\}.$$

Hence,

$$d'_v(2P) - 4d'_v(P) - v(\Psi_2(x_P, y_P)) \geq \frac{1}{2}\big(6\mu'_v - v(\Delta')\big) + \frac{5}{2}\alpha_v - v_*(2),$$

where

$$v_*(2) = \left\{ \begin{array}{ll} v(2) & \text{if } v \text{ is non-archimedean} \\ 0 & \text{if } v \text{ is archimedean} \end{array} \right\}.$$

Case 2. $v(x_P) \geq \mu'_v$.

From $(5')$ and the multiplication formula $(1')$ for x_{2P},

$$\begin{aligned} v(\Delta') &\geq \min\{2\mu'_v + 4\alpha_v + v(\Phi_2(x_P)), 3\mu' + 4\alpha_v + v(\Psi_2^2(x_P)) - 2v(2)\} + \alpha_v \\ &\geq 2\mu' + 5\alpha_v + \min\{\mu'_v, v(x_{2P})\} + v(\Psi_2^2(x_P)) - 2v_*(2). \end{aligned}$$

This entails

$$d'_v(2P) - 4d'_v(P) - v(\Psi_2(x_P, y_P)) \geq \frac{1}{2}\big(6\mu'_v - v(\Delta')\big) + \frac{5}{2}\alpha_v - v_*(2).$$

Inserting these estimates (which hold for arbitrary points $P \in E(K)$ and thus also for $Q \in E(K)$) in $(7')$ yields

$$\begin{aligned} 4d'_v(P) + 4d'_v(Q) &+ v(\Psi_2(x_P, y_P)) + v(\Psi_2(x_Q, y_Q)) \\ &\qquad + \big(6\mu'_v - v(\Delta')\big) + 5\alpha_v - 2v_*(2) \\ &\leq 2d'_v(P+Q) + 2d'_v(P-Q) + v(\Psi_2(x_P, y_P)) + v(\Psi_2(x_Q, y_Q)) \\ &\qquad - 2v(x_P - x_Q) - 2\alpha_v. \end{aligned}$$

This establishes assertion (a).

To prove the easier assertion (b), we note that

$$d'_v(2P) - 4d'_v(P) - v(\Psi_2(x_P, y_P)) \geq \frac{1}{2}\big(6\mu'_v - v(\Delta')\big) + \frac{5}{2}\alpha_v - v_*(2)$$

and thus the estimate below has already been shown to be true.

For (b), it remains to prove the claimed estimate above. We distinguish two cases.

Case 1. $v(x_{2P}) < \mu'_v$.

From $(1')$, we conclude that

$$v(x_{2P}) \geq 4\min\{\mu'_v, v(x_P)\} + 2\alpha_v - v(\Psi_2^2(x_P))$$

and hence that

$$d(2P) \leq 4d'_v(P) + v(\Psi_2(x_P, y_P)) - \alpha_v,$$

where $-\alpha_v \leq -2\alpha_v$.

Case 2. $v(x_{2P}) \geq \mu'_v$.

Since, by the multiplication formula (see $(1')$),

$$\Psi_2^2(x_P) = 4(x_P^3 + cx_P^2 + dx_P),$$

we obtain

$$\begin{aligned} v(\Psi_2^2(x_P)) &\geq 3\min\{\mu_v', v(x_P)\} + 4\alpha_v \\ &\geq 4\min\{\mu_v', v(x_P)\} - \mu_v' + 4\alpha_v, \end{aligned}$$

thus,

$$d_v'(2P) - 4d_v'(P) - v(\Psi_2(x_P, y_P)) \leq -2\alpha_v.$$

Taking into account the sum formula for the absolute values v of K and the fact that, by this formula,

$$-\sum_{v\in\mathcal{M}_K} \lambda_v v_*(2) = -\sum_{v\in\mathcal{M}_K^0} \lambda_v v(2) = \sum_{v\in\mathcal{M}_K^\infty} \lambda_v v(2) = \sum_{v\in\mathcal{M}_K} \lambda_v \alpha_v = -\alpha,$$

we arrive at

Theorem 2′. *Globally,*

(a) $3\mu' - \frac{9}{2}\alpha \leq d'(P+Q) + d'(P-Q) - 2d'(P) - 2d'(Q) \leq 2\alpha$ *for* $P \neq \pm Q$ *in* $E'(K)$,

(b) $3\mu' - \frac{7}{2}\alpha \leq d'(2P) - 4d'(P) \leq 2\alpha$ *for* $P = Q$ *in* $E'(K)$.

Assertion (b) of Theorem 2′ gives rise to

Theorem 3′. *Globally,*

$$-\mu' - \frac{7}{6}\alpha \leq \widehat{h}(P) - d'(P) \leq \frac{2}{3}\alpha \quad \textit{for } P \in E'(K).$$

These estimates are stronger than those in Theorem 3, but the μ_v''s, compared to the μ_v's, are differently defined.

More precisely, the difference between μ and μ' is

$$-2\alpha \leq \mu - \mu' \leq 2\alpha$$

and accordingly, the difference between the two modified Weil heights for $P \in E'(K)$

$$-\alpha \leq d(P) - d'(P) \leq \alpha.$$

For theoretical, not for computational reasons, we state

Theorem 4′. *Globally, if the coefficients c and d of E'/K are integers in K,*

$$-\frac{1}{2}\left(\mu' + \mu'^{\infty}\right) - \frac{7}{6}\alpha \leq \widehat{h}(P) - h(P) \leq \frac{1}{2}\left(\mu'^{\infty} + \frac{8}{3}\alpha\right)$$

for $P \in E'(K)$.

In analogy to Proposition 1, we get here

Proposition 1′. *Suppose that the coefficients c, d of the curve E'/K are integers in K. Then,*

$$-\frac{1}{2}\mu'^0 \leq d'(P) - h(P) \leq \frac{1}{2}\mu'^\infty \quad \textit{for } P \in E'(K).$$

Again, we have put

$$\mu'^0 = \sum_{v \in \mathcal{M}_K^0} \lambda_v {\mu'}_v, \quad \mu'^\infty = -\sum_{v \in \mathcal{M}_K^\infty} \lambda_v {\mu'}_v,$$

so that

$$\mu' = -\mu'^0 + \mu'^\infty.$$

In case that

$$c = 0,$$

we have

$$\mu' = 0.$$

This yields the corresponding bounds in Theorems 2′–4′.

Another special elliptic curve E'' over a global field K of characteristic $\neq 2, 3$ is, of course, given by the short Weierstrass equation

$$E'': \quad Y^2 = X^3 + aX + b \quad (a, b \in K).$$

We only remark the following. If we take the short Weierstrass form

$$E'': \quad Y^2 = X^3 - 27c_4X - 54c_6,$$

define locally

$$\begin{aligned}
\mu_v''' &= \min\{\frac{1}{2}v(3^3c_4), \frac{1}{3}v(2 \cdot 3^3c_6)\} \\
&= \min\{\frac{1}{2}v(2^4 \cdot 3^4a), \frac{1}{3}(2^6 \cdot 3^6b)\} \\
&= \mu_v'' + 2v(2) + 2v(3)
\end{aligned}$$

and put globally (as usual)

$$\mu''' = -\sum_{v \in \mathcal{M}_K} \lambda_v \mu_v''',$$

we end up with the identity

$$\mu''' = \mu''.$$

The short Weierstrass form is treated in [16]. The μ_v'' are there

$$\mu_v'' := \min\Big\{\frac{1}{2}v(a), \frac{1}{3}v(b)\Big\},$$

and for a point $P = (x, y) \in E''(K)$, we define locally at $v \in \mathcal{M}_K$

$$d''_v(P) = -\frac{1}{2}\min\{\mu''_v, v(x)\}$$

and globally

$$d''(P) = \sum_{v \in \mathcal{M}_K} \lambda_v d''_v(P).$$

The main result of [16] (see the Theorem there) is

Theorem 3″. *For a point $P \in E''(K)$, globally*

$$-\mu'' - \frac{5}{3}\alpha \leq \widehat{h}(P) - d''(P) \leq \frac{2}{3}\alpha.$$

Here, we have put

$$\mu'' = -\sum_{v \in \mathcal{M}_K} \lambda_v \mu''_v \quad (\geq 0)$$

and remind of the fact that

$$\alpha = \left\{ \begin{array}{cl} [K:\mathbb{Q}] \cdot \log(2) & \text{in the number field case,} \\ 0 & \text{in the function field case.} \end{array} \right\}$$

But we do not go into this matter in more detail.

References

[1] Baker, A., The diophantine equation $y^2 = ax^3 + bx^2 + cx + d$. J. London Math. Soc. 43 (1968), 1–9.

[2] Coates, J., An effective p-adic analogue of a theorem of Thue. III. The diophantine equation $y^2 = x^3 + k$. Acta Arith. 16 (1970), 425–435.

[3] Gebel, H., Pethő, A., Zimmer, H.G., Computing integral points on elliptic curves. Acta Arith. 68 (1994), 171–192.

[4] Grigorov, G., Rizov, J., Heights on elliptic curves and the diophantine equation $x^4 + y^4 = cz^4$. Manuscript, Sofia University 1998.

[5] Hasse, H., Number Theory. Springer, Berlin-Heidelberg-New York 1980.

[6] Kulesz, L., Application de la méthode de Dem'janenko-Manin à certaines familles de courbes de genre 2 et 3. J. Number Theory 76 (1999), 130–146.

[7] Lang, S., Fundamentals of Diophantine Geometry. Springer, New York–Berlin–Heidelberg–Tokyo 1983.

[8] Pethő, A., Zimmer, H.G., Gebel, J., Herrmann, E., Computing all S-integral points on elliptic curves. Math. Proc. Cambridge Philos. Soc. 127 (1999), 383–402.

[9] Schmitt, S., Bestimmung der Mordell-Weil Gruppe elliptischer Kurven über algebraischen Zahlkörpern. Ph.D. Thesis, Saarbrücken 1999.

[10] Schmitt, S., Zimmer, H.G., Elliptic Curves – a Computational Approach. Walter de Gruyter, to appear 2002.

[11] Silverman, J.H., The Arithmetic of Elliptic Curves. Springer, New York–Berlin–Heidelberg–Tokyo 1986.

[12] Tschöpe, H.M., Berechnung der lokalen und globalen Néron-Tate-Höhe von Punkten auf elliptischen Kurven. Diploma Thesis, Saarbrücken 1985.

[13] Zimmer, H.G.., On the difference of the Weil height and the Néron-Tate height. Math. Z. 147 (1976), 35–51.

[14] —, Lokale und globale Höhenfunktionen auf elliptischen Kurven. Abh. Braunschweig. Wiss. Ges. 33 (1982), 253–260.

[15] —, A limit formula for the canonical height of an elliptic curve and its application to height computations. In: Number Theory (Banff, 1988) (ed. by R.A. Mollin), 641–659. Walter de Gruyter, Berlin 1990.

[16] Zimmer, H.G., Schmitt, S., Height estimates for elliptic curves in short Weierstrass form over global fields and a comparison. Arch. Math. 77 (2001), 22–31.

List of participants

Kazimierz Alster, kalster@impan.gov.pl
Institute of Mathematics, Polish Academy of Sciences, Warszawa, Poland

Grzegorz Banaszak, banaszak@amu.edu.pl
Faculty of Mathematics and Computer Science, Adam Mickiewicz University, Poznań, Poland

Marek Bartosiewicz, mbartosi@elka.pw.edu.pl
Faculty of Electronics and Information Technology, Warsaw University of Technology, Warszawa, Poland

Anna Biedrzycka
National Bank of Poland, Warszawa, Poland

Rafał Bieniek, rbieniek@elka.pw.edu.pl
Faculty of Electronics and Information Technology, Warsaw University of Technology, Warszawa, Poland

Joanna Bissinger, mont@amu.edu.pl
Faculty of Mathematics and Computer Science, Adam Mickiewicz University, Poznań, Poland

Marcin Błaszczak, mblaszc1@elka.pw.edu.pl
Faculty of Electronics and Information Technology, Warsaw University of Technology, Warszawa, Poland

Monika Bogusławska, monbogus@alpha.net.pl
Warszawa, Poland

Andrzej M. Borzyszkowski, A.Borzyszkowski@ipipan.gda.pl
Institute of Computer Science, Polish Academy of Sciences, Gdańsk, Poland

Anetta Brekiewicz, abrek@ksinet.univ.gda.pl
Institute of Mathematics, Faculty of Mathematics and Physics, University of Gdańsk, Poland

Jerzy Browkin, bro@mimuw.edu.pl
Institute of Mathematics, Faculty of Mathematics, Informatics and Mechanics, Warsaw University, Warszawa, Poland

Juliusz Brzeziński, jub@math.chalmers.se
Department of Mathematics, Chalmers University of Technology and Göteborg University, Göteborg, Sweden

Włodzimierz Chocianowicz, filsys@unet.pl
Software and Data Security Department, Technical University of Szczecin, Szczecin, Poland

Nicolas T. Courtois, courtois@minrank.org
Bull CP8 R&D, Paris VI University, Toulon University, France

Karl Dilcher, dilcher@mscs.dal.ca
Department of Mathematics and Statistics, Dalhousie University, Halifax, N.S., Canada

Hans Dobbertin, hdobbert@uni-klu.ac.at
Department of Mathematics, University of Klagenfurt, Austria

Andreas Enge, enge@math.uni-augsburg.de
Institute of Mathematics, Faculty of Mathematics and Natural Sciences, University of Augsburg, Germany

Gerhard Frey, frey@exp-math.uni-essen.de
Institute of Experimental Mathematics, Faculty of Mathematics and Informatics, University of Essen, Germany

Tadeusz L. Fryska, tfryska@main.amu.edu.pl
Faculty of Mathematics and Computer Science, Adam Mickiewicz University, Poznań, Poland

Steven Galbraith, galbra@exp-math.uni-essen.de
Institute of Experimental Mathematics, Faculty of Mathematics and Informatics, University of Essen, Germany

Karol Gołąb
Institute of Informatics, Faculty of Mathematics, Informatics and Mechanics, Warsaw University, Warszawa, Poland

Grzegorz Grudziński, gsg@mimuw.edu.pl
Institute of Informatics, Faculty of Mathematics, Informatics and Mechanics, Warsaw University, Warszawa, Poland

Maciej Grześkowiak, maciejg@amu.edu.pl
Faculty of Mathematics and Computer Science, Adam Mickiewicz University, Poznań, Poland

Wojciech Guzicki, guzicki@mimuw.edu.pl
Institute of Mathematics, Faculty of Mathematics, Informatics and Mechanics, Warsaw University, Warszawa, Poland

Marzena Guzowska, maga@argon.kul.lublin.pl
Department of Mathematics, Catholic University of Lublin, Poland

Peter Hegarty, hegarty@math.chalmers.se
Department of Mathematics, Chalmers University of Technology, Göteborg, Sweden

Michael John Jacobson, mjjacobs@cs.umanitoba.ca
Department of Computer Science, University of Manitoba, Winnipeg, Manitoba, Canada

Stanisław Jarecki, stasio@theory.lcs.mit.edu
Laboratory of Computer Science, Massachusetts Institute of Technology, Cambridge, MA, U.S.A.

Tomasz Jurdziński, tju@ii.uni.wroc.pl
Institute of Computer Science, Faculty of Mathematics and Computer Science, University of Wrocław, Poland

Jerzy Kaczorowski, kjerzy@math.amu.edu.pl
Faculty of Mathematics and Computer Science, Adam Mickiewicz University, Poznań, Poland

Michał Karoński, karonski@amu.edu.pl
Faculty of Mathematics and Computer Science, Adam Mickiewicz University, Poznań, Poland

Waldemar Korczyński
Department of Management, Akademia Świętokrzyska, Kielce, Poland

Jolanta Korniluk, korniluk@ksinet.univ.gda.pl
Institute of Mathematics, Faculty of Mathematics and Physics, University of Gdańsk, Poland

Ryszard Kossowski, ryszard@ia.pw.edu.pl
Faculty of Electronic and Information Technology, Warsaw University of Technology, Warszawa, Poland

Przemysław Kubiak, hokum@amu.edu.pl
Faculty of Mathematics and Computer Science, Adam Mickiewicz University, Poznań, Poland

Daniel Kucner, daniel@ii.uni.wroc.pl
Institute of Computer Science, Faculty of Mathematics and Computer Science, University of Wrocław, Poland

Mieczysław Kula, kula@ux2.math.us.edu.pl
Institute of Mathematics, Faculty of Mathematics, Physics and Chemistry, Silesian University, Katowice, Poland

Mirosław Kutyłowski, mirekk@ii.uni.wroc.pl
Faculty of Mathematics and Computer Science, Adam Mickiewicz University, Poznań and Institute of Computer Science, Faculty of Mathematics and Computer Science, University of Wrocław, Poland

Paweł Latkowski, platkow1@elka.edu.pw.pl
Faculty of Electronics and Information Technology, Warsaw University of Technology, Warszawa, Poland

Arjen K. Lenstra, arjen.lenstra@citicorp.com
Citibank, N.A., Mendham, NJ, U.S.A.

Krzysztof Loryś, lorys@tcs.uni.wroc.pl
Institute of Computer Science, Faculty of Mathematics and Computer Science, University of Wrocław, Poland

Marcin Machnio, mmachnio@elka.edu.pw.pl
Faculty of Electronics and Information Technology, Warsaw University of Technology, Warszawa, Poland

Robert Maron, robmar@robmar.net
Institute of Informatics, Faculty of Mathematics, Informatics and Mechanics, Warsaw University, Warszawa, Poland

Ján Mináč, minac@uwo.ca
Department of Mathematics, Middlesex College, The University of Western Ontario, London, Ontario, Canada

Pieter Moree, moree@wins.uva.nl
Faculty of Mathematics, Computer Science, Physics and Astronomy, University of Amsterdam, The Netherlands

Siguna Müller, smueller@uni-klu.ac.at
Department of Mathematics, University of Klagenfurt, Austria

Rafał Nessel, R.Nessel@impan.gov.pl
Institute of Mathematics, Polish Academy of Sciences, Warszawa, Poland

Harald Niederreiter, hnied@gauus.dismat.oeaw.ac.at
Institute of Discrete Mathematics, Austrian Academy of Sciences, Vienna, Austria

Damian Niwiński, niwinski@mimuw.edu.pl
Institute of Informatics, Faculty of Mathematics, Informatics and Mechanics, Warsaw University, Warszawa, Poland

Adam Obtułowicz, adamo@impan.gov.pl
Institute of Mathematics, Polish Academy of Sciences, Warszawa, Poland

Andrew Odlyzko, amo@research.att.com
Shannon Laboratory, AT&T Labs — Research, Florham Park, NJ, U.S.A.

Arkadiusz Orłowski,
Institute of Physics, Polish Academy of Sciences, Warszawa, Poland

Jacques Patarin, J.Patarin@frlv.bull.fr
Bull CP8, Louveciennes, France

Jerzy Pejaś, jerzy.pejas@wi.ps.pl
Software and Data Security Department, Technical University of Szczecin, Szczecin, Poland

Attila Pethő , pethoe@turan.math.klte.hu
Institute of Mathematics and Informatics, University of Debrecen, Hungary

Józef Pieprzyk, josef@uow.edu.au
School of Information Technology and Computer Science, University of Wollongong, Australia

Andrzej Rotkiewicz, rotkiewi@impan.gov.pl
Institute of Mathematics, Polish Academy of Sciences, Warszawa, Poland

Tibor Daniel Różański, bitro@tcs.uni.wroc.pl
Institute of Computer Science, Faculty of Mathematics and Computer Science, University of Wrocław, Poland

Danuta Rutkowska, danara@amu.edu.pl
Faculty of Mathematics and Computer Science, Adam Mickiewicz University, Poznań, Poland

Claus P. Schnorr, schnorr@cs.uni-frankfurt.de
Department of Mathematics, University of Frankfurt, Frankfurt am Main, Germany

Nigel P. Smart, nigel@cs.bristol.ac.uk
Computer Science Department, University of Bristol, U.K.

Artur Andrzej Soroczuk, a.soroczuk@mp.wsp.czest.pl
Institute of Mathematics and Informatics, Faculty of Mathematics and Natural Sciences, Pedagogical University, Częstochowa, Poland

Stanisław Spież, spiez@impan.gov.pl
Institute of Mathematics, Polish Academy of Sciences, Warszawa, Poland

Marian Srebrny, marians@ipipan.waw.pl
Institute of Computer Science, Polish Academy of Sciences, Warszawa, Poland

Michał Szafrański
Institute of Informatics, Faculty of Mathematics, Informatics and Mechanics, Warsaw University, Warszawa, Poland

Krzysztof Szczypiorski, krzysiek@tele.pw.edu.pl
Institute of Telecommunication, Faculty of Electronic and Information Technology, Warsaw University of Technology, Warszawa, Poland

Andrzej Szepietowski, matszp@ksinet.univ.gda.pl
Institute of Mathematics, Faculty of Mathematics and Physics, University of Gdańsk, Poland

Paweł Tatarczak
Institute of Informatics, Faculty of Mathematics, Informatics and Mechanics, Warsaw University, Warszawa, Poland

Wojciech Tomanik, tomanik@acm.org
Institute of Computer Science, Faculty of Mathematics and Computer Science, University of Wrocław, Poland

Robert Tracz, rtracz@elka.edu.pl
Faculty of Electronics and Information Technology, Warsaw University of Technology, Warszawa, Poland

Tomasz Tyksiński, gandalf@amu.edu.pl
Faculty of Mathematics and Computer Science, Adam Mickiewicz University, Poznań, Poland

Janusz A. Urbanowicz, alex@bofh.torun.pl
Faculty of Physics and Astronomy, Nicholas Copernicus University, Toruń, Poland

Jerzy Urbanowicz, urbanowi@impan.gov.pl
Institute of Mathematics, Polish Academy of Sciences, Warszawa, Poland

Milan Vojvoda, vojvoda@pobox.sk
Department of Mathematics, Faculty of Electrical Engineering and Information Technology, Slovak University of Technology, Bratislava, Slovakia

Gary Walsh, gwalsh@mathstat.uottawa.ca
Department of Mathematics and Statistics, University of Ottawa, Canada

Paul van Wamelen, wamelen@marais.math.lsu.edu
Department of Mathematics, Louisiana State University, Baton Rouge, LA, U.S.A.

Aleksander Wittlin, wittlin@ifpan.edu.pl
Institute of Physics, Polish Academy of Sciences, Warszawa, Poland

Artur Zgoda, azgoda@ii.uni.wroc.pl
Institute of Computer Science, Faculty of Mathematics and Computer Science, University of Wrocław, Poland

Horst G. Zimmer, zimmer@math.uni-sb.de
Faculty of Mathematics, Saarland University, Saarbrücken, Germany

Daniel Żukowski, dzukowsk@elka.edu.pl
Faculty of Electronics and Information Technology, Warsaw University of Technology, Warszawa, Poland

Grażyna Zwoźniak, grazyna@ii.uni.wroc.pl
Institute of Computer Science, Faculty of Mathematics and Computer Science, University of Wrocław, Poland

List of contributors

Johannes Buchmann, Technische Universität Darmstadt, Fachbereich Informatik, Alexanderstr. 10, 64283 Darmstadt, Germany
buchmann@cdc.informatik.tu-darmstadt.de 1

Jean-Marc Couveignes, Groupe de Recherche en Informatique et Mathématiques du Mirail, Université de Toulouse II, Le Mirail, France
couveig@univ-tlse2.fr 17

Karl Dilcher, Department of Mathematics and Statistics, Dalhousie University, Halifax, Nova Scotia, B3H 3J5, Canada
dilcher@mathstat.dal.ca 29

Andreas Enge, Lehrstuhl für Diskrete Mathematik, Optimierung und Operations Research, Universität Augsburg, 86135 Augsburg, Germany
enge@math.uni-augsburg.de
http://www.math.uni-augsburg.de/~enge
and Laboratoire d'Informatique, École Polytechnique, 91128 Palaiseau CEDEX, France
enge@lix.polytechnique.fr
http://www.lix.polytechnique.fr/Labo/Andreas.Enge 49

Steven D. Galbraith, Department of Computer Science, University of Bristol, Merchant Venturers Building, Woodland Rd, Bristol BS8 1UB, UK
steven@cs.bris.ac.uk 59

Otokar Grošek, Department of Mathematics, FEI, Slovak Technical University, 812 19 Bratislava, Slovakia
grosek@kmat.elf.stuba.sk 71

Safuat Hamdy, Technische Universität Darmstadt, Fachbereich Informatik, Alexanderstr. 10, 64283 Darmstadt, Germany
hamdy@cdc.informatik.tu-darmstadt.de 1

Jeffrey Hoffstein, NTRU Cryptosystems, Inc., 5 Burlington Woods, Burlington, MA 01803, U.S.A.
jhoffstein@ntru.com 77

Michael John Jacobson, Jr., Department of Computer Science, University of Manitoba, Winnipeg, Manitoba R3T 2N2, Canada
jacobs@cs.umanitoba.ca 89

Przemysław Kubiak, Cryptography Laboratory, Department of Mathematics and Computer Science, Adam Mickiewicz University, Poznań, Poland
pkubiak@amu.edu.pl 113

Daniel Kucner, Institute of Computer Science, University of Wrocław, Poland
daniel@ii.uni.wroc.pl 137

Mirosław Kutyłowski, Cryptography Laboratory, Department of Mathematics and Computer Science, Adam Mickiewicz University, Poznań, Poland
and Mathematical Institute, Wrocław University of Technology, Wrocław, Poland
mirekk@amu.edu.pl 137

Arjen K. Lenstra, Citibank, N.A., 1 North Gate Road, Mendham, NJ 07945-3104, U.S.A., and Technische Universiteit Eindhoven, The Netherlands
arjen.lenstra@citicorp.com 151

Spyros S. Magliveras, Department of Mathematical Sciences, Florida Atlantic University, Boca Raton, FL 33431, U.S.A.
spyros@pandora.math.fau.edu 71

Siguna Müller, University of Klagenfurt, Department of Mathematics, A-9020 Klagenfurt, Austria
siguna.mueller@uni-klu.ac.at 181

Volker Müller, Universitas Kristen Duta Wacana, Jl. Dr. Wahidin 5-19, Yogyakarta, 55224, Indonesia
vmueller@ukdw.ac.id 197

Harald Niederreiter, Department of Mathematics, National University of Singapore, 2 Science Drive 2, Singapore 117543, Republic of Singapore
nied@math.nus.edu.sg 209

Andrew Odlyzko, Shannon Laboratory, AT&T Labs – Research, Florham Park, NJ, U.S.A.
amo@research.att.com ix

Jacques Patarin, Bull CP8, 68 route de Versailles - BP 45, 78434 Louveciennes Cedex, France
J.Patarin@frlv.bull.fr 221

Attila Pethő, University of Debrecen, Department of Computer Science, H–4010 Debrecen, P.O.B. 12, Hungary
pethoe@math.klte.hu 239

Herman te Riele, Center for Mathematics and Computer Science (CWI), Kruislaan 413, 1098 SJ Amsterdam, The Netherlands
herman.te.riele@cwi.nl 249

Renate Scheidler, Department of Mathematical Sciences, University of Delaware, Newark, Delaware 19716, U.S.A.
scheidle@math.udel.edu 89

Claus Peter Schnorr, Fachbereiche Mathematik/Informatik, Universität Frankfurt, PSF 111932, 60054 Frankfurt am Main, Germany
schnorr@cs.uni-frankfurt.de 257

Joseph Silverman, NTRU Cryptosystems, Inc., 5 Burlington Woods, Burlington, MA 01803, U.S.A.
jsilverman@ntru.com 77

Edlyn Teske, Dept. of Combinatorics and Optimization, University of Waterloo, Waterloo, ON N2L 3G1, Canada
eteske@cacr.math.uwaterloo.ca 283

Eric R. Verheul, PricewaterhouseCoopers, GRMS Crypto Group, Goudsbloemstraat 14, 5644 KE Eindhoven, The Netherlands
eric.verheul@[nl.pwcglobal.com,pobox.com] 151

Wandi Wei, Department of Computer Science, Western Illinois University, Macomb, IL 61455, U.S.A.
wandi_wei@hotmail.com 71

Hugh C. Williams, Department of Computer Science, University of Manitoba, Winnipeg, Manitoba R3T 2N2, Canada
williams@cs.umanitoba.ca 89

Horst G. Zimmer, FR 6.1 – Mathematik, Universität des Saarlandes, Postfach 15 11 50, 66041 Saarbrücken, Germany
zimmer@math.uni-sb.de 303